EFFECTIVE GROUNDWATER MODEL CALIBRATION

THE WILEY BICENTENNIAL—KNOWLEDGE FOR GENERATIONS

Each generation has its unique needs and aspirations. When Charles Wiley first opened his small printing shop in lower Manhattan in 1807, it was a generation of boundless potential searching for an identity. And we were there, helping to define a new American literary tradition. Over half a century later, in the midst of the Second Industrial Revolution, it was a generation focused on building the future. Once again, we were there, supplying the critical scientific, technical, and engineering knowledge that helped frame the world. Throughout the 20th Century, and into the new millennium, nations began to reach out beyond their own borders and a new international community was born. Wiley was there, expanding its operations around the world to enable a global exchange of ideas, opinions, and know-how.

For 200 years, Wiley has been an integral part of each generation's journey, enabling the flow of information and understanding necessary to meet their needs and fulfill their aspirations. Today, bold new technologies are changing the way we live and learn. Wiley will be there, providing you the must-have knowledge you need to imagine new worlds, new possibilities, and new opportunities.

Generations come and go, but you can always count on Wiley to provide you the knowledge you need, when and where you need it!

WILLIAM J. PESCE
PRESIDENT AND CHIEF EXECUTIVE OFFICER

PETER BOOTH WILEY
CHAIRMAN OF THE BOARD

EFFECTIVE GROUNDWATER MODEL CALIBRATION

With Analysis of Data, Sensitivities, Predictions, and Uncertainty

MARY C. HILL
CLAIRE R. TIEDEMAN

WILEY-INTERSCIENCE
A JOHN WILEY & SONS, INC., PUBLICATION

Published 2007 by John Wiley & Sons, Inc., Hoboken, New Jersey.
Published simultaneously in Canada.

No part of this publication may be reproduced, stored in a retrieval system, or transmitted in any form or by any means, electronic, mechanical, photocopying, recording, scanning, or otherwise, except as permitted under Section 107 or 108 of the 1976 United States Copyright Act, without either the prior written permission of the Publisher, or authorization through payment of the appropriate per-copy fee to the Copyright Clearance Center, Inc., 222 Rosewood Drive, Danvers, MA 01923, 978-750-8400, fax 978-750-4470, or on the web at www.copyright.com. Requests to the Publisher for permission should be addressed to the Permissions Department, John Wiley & Sons, Inc., 111 River Street, Hoboken, NJ 07030, 201-748-6011, fax 201-748-6008, or online at http://www.wiley.com/go/permission.

Limit of Liability/Disclaimer of Warranty: While the publisher and author have used their best efforts in preparing this book, they make no representations or warranties with respect to the accuracy or completeness of the contents of this book and specifically disclaim any implied warranties of merchantability or fitness for a particular purpose. No warranty may be created or extended by sales representatives or written sales materials. The advice and strategies contained herein may not be suitable for your situation. You should consult with a professional where appropriate. Neither the publisher nor author shall be liable for any loss of profit or any other commercial damages, including but not limited to special, incidental, consequential, or other damages.

For general information on our other products and services or for technical support, please contact our Customer Care Department within the United States at 877-762-2974, outside the United States at 317-572-3993 or fax 317-572-4002.

Wiley also publishes its books in a variety of electronic formats. Some content that appears in print may not be available in electronic formats. For more information about Wiley products, visit our web site at www.wiley.com.

Library of Congress Cataloging-in-Publication Data:

Hill, Mary C. (Mary Catherine)
   Effective groundwater model calibration: with analysis of data, sensitivities, predictions, and uncertainty/Mary C. Hill, Claire R. Tiedeman.
       p. cm.
   Includes index.
   ISBN-13:   978-0-471-77636-9 (cloth)
   ISBN-10:   0-471-77636-X (cloth)
   1.   Groundwater--Mathematical models.   2. Hydrologic models.   I. Tiedeman, Claire R. II. Title.
   GB1001.72.M35H55 2006
   551.4901'5118--dc22
                                    2005036657

Printed in the United States of America

10 9 8 7 6 5 4 3 2 1

We dedicate this book to the groundwater modelers and software developers of the U.S. Geological Survey. These men and women devote their careers to providing sound scientific analyses for policy makers and to enabling others in the government and the private sector to do the same. We are honored to be their colleagues.

We also dedicate this book to the United States taxpayers, to whom we are ultimately accountable. They have supported our educations, salaries, field work and students. We hope our efforts have improved the understanding and management of their groundwater resources.

With love, we also dedicate this book to our husbands and families.

CONTENTS

Preface   xvii

1 Introduction   1

    1.1 Book and Associated Contributions: Methods, Guidelines, Exercises, Answers, Software, and PowerPoint Files, 1

    1.2 Model Calibration with Inverse Modeling, 3

        1.2.1 Parameterization, 5

        1.2.2 Objective Function, 6

        1.2.3 Utility of Inverse Modeling and Associated Methods, 6

        1.2.4 Using the Model to Quantitatively Connect Parameters, Observations, and Predictions, 7

    1.3 Relation of this Book to Other Ideas and Previous Works, 8

        1.3.1 Predictive Versus Calibrated Models, 8

        1.3.2 Previous Work, 8

    1.4 A Few Definitions, 12

        1.4.1 Linear and Nonlinear, 12

        1.4.2 Precision, Accuracy, Reliability, and Uncertainty, 13

    1.5 Advantageous Expertise and Suggested Readings, 14

    1.6 Overview of Chapters 2 Through 15, 16

2 Computer Software and Groundwater Management Problem Used in the Exercises — 18

2.1 Computer Programs MODFLOW-2000, UCODE_2005, and PEST, 18

2.2 Groundwater Management Problem Used for the Exercises, 21
    2.2.1 Purpose and Strategy, 23
    2.2.2 Flow System Characteristics, 23

2.3 Exercises, 24
    Exercise 2.1: Simulate Steady-State Heads and Perform Preparatory Steps, 25

3 Comparing Observed and Simulated Values Using Objective Functions — 26

3.1 Weighted Least-Squares Objective Function, 26
    3.1.1 With a Diagonal Weight Matrix, 27
    3.1.2 With a Full Weight Matrix, 28

3.2 Alternative Objective Functions, 28
    3.2.1 Maximum-Likelihood Objective Function, 29
    3.2.2 L_1 Norm Objective Function, 29
    3.2.3 Multiobjective Function, 29

3.3 Requirements for Accurate Simulated Results, 30
    3.3.1 Accurate Model, 30
    3.3.2 Unbiased Observations and Prior Information, 30
    3.3.3 Weighting Reflects Errors, 31

3.4 Additional Issues
    3.4.1 Prior Information, 32
    3.4.2 Weighting, 34
    3.4.3 Residuals and Weighted Residuals, 35

3.5 Least-Squares Objective-Function Surfaces, 35

3.6 Exercises, 36
    Exercise 3.1: Steady-State Parameter Definition, 36
    Exercise 3.2: Observations for the Steady-State Problem, 38
    Exercise 3.3: Evaluate Model Fit Using Starting Parameter Values, 40

4 Determining the Information that Observations Provide on Parameter Values using Fit-Independent Statistics — 41

4.1 Using Observations, 42
    4.1.1 Model Construction and Parameter Definition, 42
    4.1.2 Parameter Values, 43

4.2 When to Determine the Information that Observations Provide About Parameter Values, 44
4.3 Fit-Independent Statistics for Sensitivity Analysis, 46
    4.3.1 Sensitivities, 47
    4.3.2 Scaling, 48
    4.3.3 Dimensionless Scaled Sensitivities (dss), 48
    4.3.4 Composite Scaled Sensitivities (css), 50
    4.3.5 Parameter Correlation Coefficients (pcc), 51
    4.3.6 Leverage Statistics, 54
    4.3.7 One-Percent Scaled Sensitivities, 54
4.4 Advantages and Limitations of Fit-Independent Statistics for Sensitivity Analysis, 56
    4.4.1 Scaled Sensitivities, 56
    4.4.2 Parameter Correlation Coefficients, 58
    4.4.3 Leverage Statistics, 59
4.5 Exercises, 60
    Exercise 4.1: Sensitivity Analysis for the Steady-State Model with Starting Parameter Values, 60

5 Estimating Parameter Values   67

5.1 The Modified Gauss–Newton Gradient Method, 68
    5.1.1 Normal Equations, 68
    5.1.2 An Example, 74
    5.1.3 Convergence Criteria, 76
5.2 Alternative Optimization Methods, 77
5.3 Multiobjective Optimization, 78
5.4 Log-Transformed Parameters, 78
5.5 Use of Limits on Estimated Parameter Values, 80
5.6 Exercises, 80
    Exercise 5.1: Modified Gauss–Newton Method and Application to a Two-Parameter Problem, 80
    Exercise 5.2: Estimate the Parameters of the Steady-State Model, 87

6 Evaluating Model Fit   93

6.1 Magnitude of Residuals and Weighted Residuals, 93
6.2 Identify Systematic Misfit, 94
6.3 Measures of Overall Model Fit, 94
    6.3.1 Objective-Function Value, 95

- 6.3.2 Calculated Error Variance and Standard Error, 95
- 6.3.3 AIC, AIC_c, and BIC Statistics, 98
- 6.4 Analyzing Model Fit Graphically and Related Statistics, 99
  - 6.4.1 Using Graphical Analysis of Weighted Residuals to Detect Model Error, 100
  - 6.4.2 Weighted Residuals Versus Weighted or Unweighted Simulated Values and Minimum, Maximum, and Average Weighted Residuals, 100
  - 6.4.3 Weighted or Unweighted Observations Versus Simulated Values and Correlation Coefficient R, 105
  - 6.4.4 Graphs and Maps Using Independent Variables and the Runs Statistic, 106
  - 6.4.5 Normal Probability Graphs and Correlation Coefficient R_N^2, 108
  - 6.4.6 Acceptable Deviations from Random, Normally Distributed Weighted Residuals, 111
- 6.5 Exercises, 113
  - Exercise 6.1: Statistical Measures of Overall Fit, 113
  - Exercise 6.2: Evaluate Graph Model fit and Related Statistics, 115

7 Evaluating Estimated Parameter Values and Parameter Uncertainty 124

- 7.1 Reevaluating Composite Scaled Sensitivities, 124
- 7.2 Using Statistics from the Parameter Variance–Covariance Matrix, 125
  - 7.2.1 Five Versions of the Variance–Covariance Matrix, 125
  - 7.2.2 Parameter Variances, Covariances, Standard Deviations, Coefficients of Variation, and Correlation Coefficients, 126
  - 7.2.3 Relation Between Sample and Regression Statistics, 127
  - 7.2.4 Statistics for Log-Transformed Parameters, 130
  - 7.2.5 When to Use the Five Versions of the Parameter Variance–Covariance Matrix, 130
  - 7.2.6 Some Alternate Methods: Eigenvectors, Eigenvalues, and Singular Value Decomposition, 132
- 7.3 Identifying Observations Important to Estimated Parameter Values, 132
  - 7.3.1 Leverage Statistics, 134
  - 7.3.2 Influence Statistics, 134
- 7.4 Uniqueness and Optimality of the Estimated Parameter Values, 137
- 7.5 Quantifying Parameter Value Uncertainty, 137

7.5.1 Inferential Statistics, 137
7.5.2 Monte Carlo Methods, 140
7.6 Checking Parameter Estimates Against Reasonable Values, 140
7.7 Testing Linearity, 142
7.8 Exercises, 145
  Exercise 7.1: Parameter Statistics, 145
  Exercise 7.2: Consider All the Different Correlation Coefficients Presented, 155
  Exercise 7.3: Test for Linearity, 155

8 Evaluating Model Predictions, Data Needs, and Prediction Uncertainty   158

8.1 Simulating Predictions and Prediction Sensitivities and Standard Deviations, 158
8.2 Using Predictions to Guide Collection of Data that Directly Characterize System Properties, 159
  8.2.1 Prediction Scaled Sensitivities (pss), 160
  8.2.2 Prediction Scaled Sensitivities Used in Conjunction with Composite Scaled Sensitivities, 162
  8.2.3 Parameter Correlation Coefficients without and with Predictions, 162
  8.2.4 Composite and Prediction Scaled Sensitivities Used with Parameter Correlation Coefficients, 165
  8.2.5 Parameter–Prediction (ppr) Statistic, 166
8.3 Using Predictions to Guide Collection of Observation Data, 170
  8.3.1 Use of Prediction, Composite, and Dimensionless Scaled Sensitivities and Parameter Correlation Coefficients, 170
  8.3.2 Observation–Prediction (opr) Statistic, 171
  8.3.3 Insights About the opr Statistic from Other Fit-Independent Statistics, 173
  8.3.4 Implications for Monitoring Network Design, 174
8.4 Quantifying Prediction Uncertainty Using Inferential Statistics, 174
  8.4.1 Definitions, 175
  8.4.2 Linear Confidence and Prediction Intervals on Predictions, 176
  8.4.3 Nonlinear Confidence and Prediction Intervals, 177
  8.4.4 Using the Theis Example to Understand Linear and Nonlinear Confidence Intervals, 181
  8.4.5 Differences and Their Standard Deviations, Confidence Intervals, and Prediction Intervals, 182

8.4.6 Using Confidence Intervals to Serve the Purposes of Traditional Sensitivity Analysis, 184
8.5 Quantifying Prediction Uncertainty Using Monte Carlo Analysis, 185
   8.5.1 Elements of a Monte Carlo Analysis, 185
   8.5.2 Relation Between Monte Carlo Analysis and Linear and Nonlinear Confidence Intervals, 187
   8.5.3 Using the Theis Example to Understand Monte Carlo Methods, 188
8.6 Quantifying Prediction Uncertainty Using Alternative Models, 189
8.7 Testing Model Nonlinearity with Respect to the Predictions, 189
8.8 Exercises, 193
   Exercise 8.1: Predict Advective Transport and Perform Sensitivity Analysis, 195
   Exercise 8.2: Prediction Uncertainty Measured Using Inferential Statistics, 207

9 Calibrating Transient and Transport Models and Recalibrating Existing Models   213

9.1 Strategies for Calibrating Transient Models, 213
   9.1.1 Initial Conditions, 213
   9.1.2 Transient Observations, 214
   9.1.3 Additional Model Inputs, 216
9.2 Strategies for Calibrating Transport Models, 217
   9.2.1 Selecting Processes to Include, 217
   9.2.2 Defining Source Geometry and Concentrations, 218
   9.2.3 Scale Issues, 219
   9.2.4 Numerical Issues: Model Accuracy and Execution Time, 220
   9.2.5 Transport Observations, 223
   9.2.6 Additional Model Inputs, 225
   9.2.7 Examples of Obtaining a Tractable, Useful Model, 226
9.3 Strategies for Recalibrating Existing Models, 227
9.4 Exercises (optional), 228
   Exercises 9.1 and 9.2: Simulate Transient Hydraulic Heads and Perform Preparatory Steps, 229
   Exercise 9.3: Transient Parameter Definition, 230

Exercise 9.4: Observations for the Transient Problem, 231
Exercise 9.5: Evaluate Transient Model Fit Using Starting Parameter Values, 235
Exercise 9.6: Sensitivity Analysis for the Initial Model, 235
Exercise 9.7: Estimate Parameters for the Transient System by Nonlinear Regression, 243
Exercise 9.8: Evaluate Measures of Model Fit, 244
Exercise 9.9: Perform Graphical Analyses of Model Fit and Evaluate Related Statistics, 246
Exercise 9.10: Evaluate Estimated Parameters, 250
Exercise 9.11: Test for Linearity, 253
Exercise 9.12: Predictions, 254

10 Guidelines for Effective Modeling 260

10.1 Purpose of the Guidelines, 263
10.2 Relation to Previous Work, 264
10.3 Suggestions for Effective Implementation, 264

11 Guidelines 1 Through 8—Model Development 268

Guideline 1: Apply the Principle of Parsimony, 268
    G1.1 Problem, 269
    G1.2 Constructive Approaches, 270
Guideline 2: Use a Broad Range of System Information to Constrain the Problem, 272
    G2.1 Data Assimilation, 273
    G2.2 Using System Information, 273
    G2.3 Data Management, 274
    G2.4 Application: Characterizing a Fractured Dolomite Aquifer, 277
Guideline 3: Maintain a Well-Posed, Comprehensive Regression Problem, 277
    G3.1 Examples, 278
    G3.2 Effects of Nonlinearity on the css and pcc, 281
Guideline 4: Include Many Kinds of Data as Observations in the Regression, 284
    G4.1 Interpolated "Observations", 284
    G4.2 Clustered Observations, 285
    G4.3 Observations that Are Inconsistent with Model Construction, 286

G4.4 Applications: Using Different Types of Observations to Calibrate Groundwater Flow and Transport Models, 287

Guideline 5: Use Prior Information Carefully, 288

G5.1 Use of Prior Information Compared with Observations, 288

G5.2 Highly Parameterized Models, 290

G5.3 Applications: Geophysical Data, 291

Guideline 6: Assign Weights that Reflect Errors, 291

G6.1 Determine Weights, 294

G6.2 Issues of Weighting in Nonlinear Regression, 298

Guideline 7: Encourage Convergence by Making the Model More Accurate and Evaluating the Observations, 306

Guideline 8: Consider Alternative Models, 308

G8.1 Develop Alternative Models, 309

G8.2 Discriminate Between Models, 310

G8.3 Simulate Predictions with Alternative Models, 312

G8.4 Application, 313

12 Guidelines 9 and 10—Model Testing   315

Guideline 9: Evaluate Model Fit, 316

G9.1 Determine Model Fit, 316

G9.2 Examine Fit for Existing Observations Important to the Purpose of the Model, 320

G9.3 Diagnose the Cause of Poor Model Fit, 320

Guideline 10: Evaluate Optimized Parameter Values, 323

G10.1 Quantify Parameter-Value Uncertainty, 323

G10.2 Use Parameter Estimates to Detect Model Error, 323

G10.3 Diagnose the Cause of Unreasonable Optimal Parameter Estimates, 326

G10.4 Identify Observations Important to the Parameter Estimates, 327

G10.5 Reduce or Increase the Number of Parameters, 328

13 Guidelines 11 and 12—Potential New Data   329

Guideline 11: Identify New Data to Improve Simulated Processes, Features, and Properties, 330

Guideline 12: Identify New Data to Improve Predictions, 334

G12.1 Potential New Data to Improve Features and Properties Governing System Dynamics, 334

G12.2 Potential New Data to Support Observations, 335

14 Guidelines 13 and 14—Prediction Uncertainty 337

Guideline 13: Evaluate Prediction Uncertainty and Accuracy Using Deterministic Methods, 337

    G13.1 Use Regression to Determine Whether Predicted Values Are Contradicted by the Calibrated Model, 337

    G13.2 Use Omitted Data and Postaudits, 338

Guideline 14: Quantify Prediction Uncertainty Using Statistical Methods, 339

    G14.1 Inferential Statistics, 341

    G14.2 Monte Carlo Methods, 341

15 Using and Testing the Methods and Guidelines 345

15.1 Execution Time Issues, 345

15.2 Field Applications and Synthetic Test Cases, 347

    15.2.1 The Death Valley Regional Flow System, California and Nevada, USA, 347

    15.2.2 Grindsted Landfill, Denmark, 370

Appendix A: Objective Function Issues 374

A.1 Derivation of the Maximum-Likelihood Objective Function, 375

A.2 Relation of the Maximum-Likelihood and Least-Squares Objective Functions, 376

A.3 Assumptions Required for Diagonal Weighting to be Correct, 376

A.4 References, 381

Appendix B: Calculation Details of the Modified Gauss–Newton Method 383

B.1 Vectors and Matrices for Nonlinear Regression, 383

B.2 Quasi-Newton Updating of the Normal Equations, 384

B.3 Calculating the Damping Parameter, 385

B.4 Solving the Normal Equations, 389

B.5 References, 390

Appendix C: Two Important Properties of Linear Regression and the Effects of Nonlinearity 391

C.1 Identities Needed for the Proofs, 392

    C.1.1 True Linear Model, 392

    C.1.2 True Nonlinear Model, 392

C.1.3 Linearized True Nonlinear Model, 392
C.1.4 Approximate Linear Model, 392
C.1.5 Approximate Nonlinear Model, 393
C.1.6 Linearized Approximate Nonlinear Model, 393
C.1.7 The Importance of X and X, 394
C.1.8 Considering Many Observations, 394
C.1.9 Normal Equations, 395
C.1.10 Random Variables, 395
C.1.11 Expected Value, 395
C.1.12 Variance–Covariance Matrix of a Vector, 395
C.2 Proof of Property 1: Parameters Estimated by Linear Regression are Unbiased, 395
C.3 Proof of Property 2: The Weight Matrix Needs to be Defined in a Particular Way for Eq. (7.1) to Apply and for the Parameter Estimates to have the Smallest Variance, 396
C.4 References, 398

Appendix D: Selected Statistical Tables   **399**

D.1 References, 406

References   **407**

Index   **427**

PREFACE

This book is intended for use in undergraduate and graduate classes, and is also appropriate for use as a reference book and for self-study. Minimal expertise in statistics and mathematics is required for all except a few advanced, optional topics. Knowledge of groundwater principles is needed to understand some parts of the exercises and some of the examples, but students from other fields of science have found classes based on drafts of the book to be very useful.

This book has been more than 12 years in the making. Progressively more mature versions have been used to teach short courses most years since 1991. The short courses have been held at the U.S. Geological Survey National Training Center in Denver, Colorado; the International Ground Water Modeling Center at the Colorado School of Mines in Golden, Colorado; the South Florida Water Management District in West Palm Beach, Florida; the University of Minnesota, in Minneapolis, Minnesota; the Delft University of Technology, The Netherlands; Charles University in Prague, the Czech Republic; University of the Western Cape in Belleville, South Africa; and Utrecht University, The Netherlands. A version also was used to teach a semester course at the University of Colorado in Boulder, Colorado in the fall of 2000. Much of what the book has become results from our many wonderful students. We thank them for their interest, enthusiasm, good humor, and encouragement as we struggled to develop many of the ideas presented in this book.

We also are deeply indebted to the following colleagues for insightful discussions and fruitful collaborations: Richard L. Cooley, Richard M. Yager, Frank A. D'Agnese, Claudia C. Faunt, Arlen W. Harbaugh, Edward R. Banta, Marshall W. Gannett, and D. Matthew Ely of the U.S. Geological Survey, Eileen P. Poeter of the Colorado School of Mines, Evan R. Anderman formerly of Calibra Consultants and McDonald-Morrissey Associates, Inc., Heidi Christiansen Barlebo of the

Geological Survey of Denmark and Greenland, John Doherty of Watermark Numerical Computing and the University of Queensland (Australia), Karel Kovar of MNP (The Netherlands), Steen Christensen of Aarhus University (Denmark), Theo Olsthoorn of Amsterdam Water Supply (The Netherlands), Richard Waddel of HSI-Geotrans, Inc., Frank Smits formerly of Witteveen + Bos, James Rumbaugh of ESI, Inc., Norm Jones of Utah State University, and Jeff Davis of EMS. In addition, thought-provoking questions from users of MODFLOWP, MODFLOW-2000, PEST, UCODE, and UCODE_2005 throughout the years have been invaluable.

The book benefited from the careful reviews provided by Peter Kitanidis of Stanford University, Eileen Poeter of the Colorado School of Mines and the International GroundWater Modeling Center (USA), Steen Christensen of the University of Aarhus (Denmark), Roseanna Neupauer of the University of Virginia (USA) (now at the University of Colorado, USA), Luc Lebbe of Ghent University (Belgium), David Lerner of the University of Sheffield (England), Chunmiao Zheng of the University of Alabama (USA), and Howard Reeves and Marshall Gannett of the U.S. Geological Survey. It also benefitted from the kind, professional editors and copyeditor at Wiley: Jonathan Rose, Rosalyn Farkas, and Christina Della Bartolomea.

All errors and omissions are the sole responsibility of the authors.

MARY C. HILL
CLAIRE R. TIEDEMAN

1

INTRODUCTION

In many fields of science and engineering, mathematical models are used to represent complex processes and results are used for system management and risk analysis. The methods commonly used to develop and apply such models often do not take full advantage of either the data available for model construction and calibration or the developed model. This book presents a set of methods and guidelines that, it is hoped, will improve how data and models are used.

This introductory chapter first describes the contributions of the book, including a description of what is on the associated web site. Sections 1.2 and 1.3 provide some context for the book by reviewing inverse modeling and considering the methods covered by the book relative to other paradigms for integrating data and models. After providing a few definitions, Chapter 1 concludes with a discussion of the expertise readers are expected to possess and some suggested readings and an overview of Chapters 2 through 15.

1.1 BOOK AND ASSOCIATED CONTRIBUTIONS: METHODS, GUIDELINES, EXERCISES, ANSWERS, SOFTWARE, AND POWERPOINT FILES

The methods presented in the book include (1) sensitivity analysis for evaluating the information content of data, (2) data assessment strategies for identifying (a) existing measurements that dominate model development and predictions

---

Effective Groundwater Model Calibration: With Analysis of Data, Sensitivities, Predictions, and Uncertainty. By Mary C. Hill and Claire R. Tiedeman
Published 2007 by John Wiley & Sons, Inc.

and (b) potential measurements likely to improve the reliability of predictions, (3) calibration techniques for developing models that are consistent with the data in some optimal manner, and (4) uncertainty evaluation for quantifying and communicating the potential error in simulated results (e.g., predictions) that often are used to make important societal decisions.

The fourteen guidelines presented in the book focus on practical application of the methods and are organized into four categories: (1) model development guidelines, (2) model testing guidelines, (3) potential new data guidelines, and (4) prediction uncertainty guidelines.

Most of the methods presented and referred to in the guidelines are based on linear or nonlinear regression theory. While this body of knowledge has its limits, it is very useful in many circumstances. The strengths and limitations of the methods presented are discussed throughout the book. In practice, linear and nonlinear regression are best thought of as imperfect, insightful tools. Whether regression methods prove to be beneficial in a given situation depends on how they are used. Here, the term beneficial refers to increasing the chance of achieving one or more useful models given the available data and a reasonable model development effort. The methods, guidelines, and related exercises presented in this book illustrate how to improve the chances of achieving useful models, and how to address problems that commonly are encountered along the way.

Besides the methods and guidelines, the book emphasizes the importance of how results are presented. To this end, the book can be thought of as emphasizing two criteria: valid statistical concepts and effective communication with resource managers. The most advanced, complex mathematics and statistics are worth very little if they cannot be used to address the societal needs related to the modeling objectives.

The methods and guidelines in this book have wide applicability for mathematical models of many types of systems and are presented in a general manner. The expertise of the authors is in the simulation of groundwater systems, and most of the examples are from this field. There are also some surface-water examples and a few references to other fields such as geophysics and biology. The fundamental aspects of systems most advantageously addressed by the methods and guidelines presented in this work are those typical of groundwater systems and shared by many other natural systems. Of relevance are that groundwater systems commonly involve (1) solutions in up to three spatial dimensions and time, (2) system characteristics that can vary dramatically in space and time, (3) knowledge about system variability in addition to the data used directly in regression methods, (4) available data sets that are typically sparse, and (5) nonlinearities that are often significant but not extreme.

Four important additional aspects of the book are the exercises, answers, software, and PowerPoint files available for teaching.

The exercises focus on a groundwater flow system and management problem to which students apply all the methods presented in the book. The system is simple, which allows basic principles to be clearly demonstrated, and is designed to have aspects that are directly relevant to typical systems. The exercises can be conducted

1.2 MODEL CALIBRATION WITH INVERSE MODELING

using the material provided in the book, or as hands-on computer exercises using instructions and files available on the web site http://water.usgs.gov/lookup/get?crresearch/hill_tiedeman_book.

The web site includes instructions for doing the exercises using files directly and/or using public-domain interface and visualization capabilities. It may also include instructions for using selected versions of commercial interfaces. The instructions are designed so that students can maximize the time spent understanding the ideas and the capabilities discussed in the book.

Answers to selected exercises are provided on the web site.

The software used for the exercises is freely available, open source, well documented, and widely used. The groundwater flow system is simulated using the Ground-Water Flow Process of MODFLOW-2000 (Harbaugh et al., 2000; Hill et al., 2000). The sensitivity analysis, calibration, and uncertainty aspects of the exercises can be accomplished using MODFLOW-2000's Observation, Sensitivity, and Parameter-Estimation Processes or UCODE_2005 (Poeter et al., 2005). Most of the sensitivity analysis, calibration, and uncertainty aspects of the exercises also can be conducted using PEST (Doherty, 1994, 2005). Relevant capabilities of MODFLOW-2000 and UCODE_2005 are noted as methods and guidelines are presented; relevant capabilities of PEST are noted in some cases. The public-domain programs for interface and visualization are MFI2K (Harbaugh, 2002), GWChart (Winston, 2000), and ModelViewer (Hsieh and Winston, 2002). The web sites from which these programs can be downloaded are listed with the references and on the book web site listed above.

The methods and guidelines presented in this book are broadly applicable. Throughout the book they are presented in the context of the capabilities of the computer codes mentioned above to provide concrete examples and encourage use.

PowerPoint files designed for teaching of the material in the book are provided on the web site. The authors invite those who use the PowerPoint files to share their additions and changes with others, in the same spirit with which we share these files with you.

The use of trade, firm, or product names in this book is for descriptive purposes only and does not imply endorsement by the U.S. Government.

The rest of this introductory chapter provides a brief overview of how regression methods fit into model calibration (Section 1.2), some perspective of how the ideas presented here relate to other ideas and past work (Section 1.3), some definitions (Section 1.4), a description of expertise that would assist readers and how to obtain that expertise (Section 1.5), and an overview of Chapters 2 through 15 (Section 1.6).

1.2 MODEL CALIBRATION WITH INVERSE MODELING

During calibration, model input such as system geometry and properties, initial and boundary conditions, and stresses are changed so that the model output matches related measured values. Many of the model inputs that are changed can be characterized using what are called "parameters" in this work. The measured values related to model outputs often are called "observations" or "observed values," which are equivalent terms and are used interchangeably in this book.

The basic steps of model calibration are shown in Figure 1.1. In the context of the entire modeling process, effectively using system information and observations to constrain the model is likely to produce a model that more accurately represents the simulated system and produces more accurate predictions, compared to a modeling procedure that uses these types of data less effectively. The ideas, methods, and guidelines presented in this book are aimed at helping to achieve more effective use of data.

The difficulties faced in simulating natural systems are demonstrated by the complex variability shown in Figure 1.2 as discussed by Zhang et al. (2006).

Four issues fundamental to model calibration are discussed in the next four sections. These include parameter definition or parameterization, which is the mechanism used to obtain a tractable and hopefully meaningful representation of

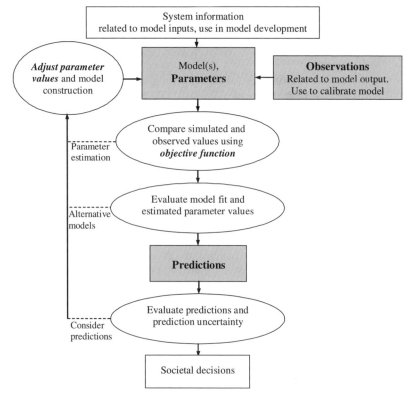

FIGURE 1.1 Flowchart showing the major steps of calibrating a model and using it to make predictions. Bold, italicized terms indicate the steps that are directly affected by nonlinear regression, including the use of an objective function to quantify the comparison between simulated and observed values. Predictions can be used during calibration as described in Chapter 8. (Adapted from Herb Buxton, U.S. Geological Survey, written communication, 1990.)

1.2 MODEL CALIBRATION WITH INVERSE MODELING

FIGURE 1.2 Experimental results from a subsiding tank, showing the kind of complexity characteristic of deltaic deposits in a subsiding basin. (Reproduced with permission from Paola et al. 2001.)

systems such as that shown in Figure 1.2; the objective function mentioned in Figure 1.1; the utility of inverse modeling, which is also called parameter estimation in this book; and using the model to quantitatively connect observations, parameters, and predictions.

1.2.1 Parameterization

The model inputs that need to be estimated are often distributed spatially and/or temporally, so that the number of parameter values could be infinite. The observations, however, generally are limited in number and support the estimation of relatively few parameters. Addressing this discrepancy is one of the greatest challenges faced by modelers in many fields. Typically, so-called parameterization is introduced that allows a limited number of parameter values to define model inputs throughout the spatial domain and time of interest. In this book, the term "parameter" is reserved for the values used to define model inputs. Consider the parameters defined in three groundwater model examples.

Example 1: One parameter represents the hydraulic conductivity of a hydrogeologic unit that occupies a prescribed volume of the model domain and is hydraulically distinctive and relatively uniform.

Example 2: One parameter represents a scalar multiplier of spatially varying recharge rates initially specified by the modeler for a given geographic area on the basis of precipitation, vegetation, elevation, and topography.

Example 3: One parameter represents the hydraulic head at a constant-head boundary that is used to simulate the water level in a lake.

This book focuses primarily on models for which a limited number of parameters are defined. Alternative methods are discussed in Section 1.3.2.

Historically, observed and simulated values, such as hydraulic heads, flows, and concentrations for groundwater systems, often were compared subjectively, so that it was difficult to determine how well one model was calibrated relative to another. In addition, in modeling of groundwater and other types of systems, adjustments of parameter values and other model characteristics were accomplished mostly by trial and error, which is time consuming, subjective, and inconclusive.

Formal methods have been developed that attempt to estimate parameter values given a mathematical model of system processes and a set of relevant observations. These are called inverse methods, and generally they are limited to the estimation of parameters as defined above. Thus, the terms "inverse modeling" and "parameter estimation" commonly are synonymous, as in this book. For some models, the inverse problem is linear, in that the observed quantities are linear functions of the parameters. In many circumstances of practical interest, however, the inverse problem is nonlinear, and its solution is not as straightforward as for linear problems. This book discusses methods for nonlinear inverse problems. One method of solving such problems is nonlinear regression, which is the primary solution method discussed in this book.

The complexity of many real systems and the scarcity of available data sets result in inversions that are often plagued by problems of insensitivity, nonuniqueness, and instability, regardless of how model calibration is achieved. Insensitivity occurs when the observations do not contain enough information to support estimation of the parameters. Nonuniqueness occurs when different combinations of parameter values match the observations equally well. Instability occurs when slight changes in, for example, parameter values or observations radically change simulated results. All these problems are exacerbated when the system is nonlinear. These problems are usually more easily detected when using formal inverse modeling and associated methods than when using trial-and-error methods for calibration. Detecting these problems is important to understanding the value of the resulting model.

1.2.2 Objective Function

In inverse modeling, the comparison of simulated and observed values is accomplished quantitatively using an objective function (Figure 1.1). The simulated and observed values include system-dependent variables (e.g., hydraulic head for the groundwater flow equation or concentration for the groundwater transport equation) and other system characteristics as represented by prior information on parameters. Parameter values that produce the "best fit" are defined as those that produce the smallest value of the objective function.

1.2.3 Utility of Inverse Modeling and Associated Methods

Recent work has clearly demonstrated that inverse modeling and associated sensitivity analysis, data needs assessment, and uncertainty evaluation methods provide

1.2 MODEL CALIBRATION WITH INVERSE MODELING

capabilities that help modelers take greater advantage of their models and data, even for simulated systems that are very complex (i.e., Poeter and Hill, 1997; Faunt et al., 2004). The benefits include

1. Clear determination of parameter values that produce the best possible fit to the available observations.
2. Graphical analyses and diagnostic statistics that quantify the quality of calibration and data shortcomings and needs, including analyses of model fit, model bias, parameter estimates, and model predictions.
3. Inferential statistics that quantify the reliability of parameter estimates and predictions.
4. Other evaluations of uncertainty, including deterministic and Monte Carlo methods.
5. Identification of issues that are easily overlooked when calibration is conducted using trial and error methods alone.

Quantifying the quality of calibration, data shortcomings and needs, and uncertainty of parameter estimates and predictions is important to model defensibility and transparency and to communicating the results of modeling studies to managers, regulators, lawyers, concerned citizens, and to the modelers themselves.

Despite its apparent utility, in many fields, such as groundwater hydrology, the methods described in this book are not routinely used, and calibration using only trial-and-error methods is more common. This, in part, is due to lack of familiarity with the methods and the perception that they require more time than trial-and-error methods. It is also because inverse modeling and related sensitivity analysis methods clearly reveal problems such as insensitivity and nonuniqueness, and thereby reveal inconvenient model weaknesses. Yet if they are revealed, such weaknesses often can be reduced or eliminated. This occurs because knowledge of the weaknesses can be used to determine data collection and model development effort needed to strengthen the model. We hope this text will encourage modelers to use, and resource managers to demand, the more transparent and defensible models that result from using the types of methods and ideas described in this book.

1.2.4 Using the Model to Quantitatively Connect Parameters, Observations, and Predictions

The model quantitatively connects the system information and the observations to the predictions and their uncertainty. The entities Parameters, Observations, and Predictions are in bold type in Figure 1.1 because these entities are directly used by or produced by the model, whereas the system information often is indirectly used to create model input. Many of the methods presented in this book take advantage of the quantitative links the model provides between what is referred to in this book as the triad of the observations, parameters, and predictions.

The depiction of model calibration shown in Figure 1.1 is unusual in that it suggests simulating predictions and prediction uncertainty as model calibration proceeds. When execution times allow, it is often useful to include predictive analyses during model calibration so that the dynamics affecting model predictions can be better understood. Care must be taken, of course, not to use such simulations to bias model predictions.

1.3 RELATION OF THIS BOOK TO OTHER IDEAS AND PREVIOUS WORKS

This section relates the ideas of this book to predictive models and other literature.

1.3.1 Predictive Versus Calibrated Models

When simulating natural systems, the objective is often to produce a model that can predict, accurately enough to be useful, for assessing the consequences of introducing something new in the system. In groundwater systems, this may entail new pumpage or transport of recently introduced or potential contamination.

Ideally, model inputs would be determined accurately and completely enough from directly related field data to produce useful model results. This is advantageous because the resulting model is likely to be able to predict results in a wide range of circumstances, and for this reason such models are called *predictive models* (e.g., see Wilcock and Iverson, 2003; National Research Council, 2002). However, commonly quantities simulated by the model can be more readily measured than model inputs. The best possible determination of model inputs based on directly related field data can produce model outputs that match the measured equivalents poorly. If the fit is poor enough that the utility of model predictions is questionable, then a decision needs to be made about how to proceed. The choices are to use the predictive model, which has been shown to perform poorly in the circumstances for which testing is possible, or to modify the model so that, at the very least, it matches the available measured equivalents of model results. A model modified in this way is called a *calibrated model*.

There is significant and important debate about the utility of predictive and calibrated models, and it is our hope that the debate will lead to better methods of measuring quantities directly related to model inputs. We would rejoice with all others in the natural sciences to be able to always use predictive models. Until then, however, it is our opinion that methods and guidelines that promote the best possible use of models and data in the development of calibrated models are critical. It is also our belief that such methods and guidelines can play a role in informing and focusing the efforts of developing field methods that may ultimately allow predictive models to be used in more circumstances.

1.3.2 Previous Work

For the most part, comments in this introductory chapter are limited to the history, evolution, and status of nonlinear regression and modeling as related to groundwater systems. Comments about how specific methods or ideas relate to previous

1.3 RELATION OF THIS BOOK TO OTHER IDEAS AND PREVIOUS WORKS

publications appear elsewhere in the book. This section contains the broadest discussion of parameterization methods presented in the book.

The topics covered by this book have been addressed by others using a variety of different methods, and have been developed for and applied to many different fields of science and engineering. We do not attempt to provide a full review of all work on these topics. Selected textbooks are as follows. Parker (1994), Sun (1994), Lebbe (1999), and Aster et al. (2005) discuss nonlinear regression in the field of geophysics. More general references for nonlinear regression and associated analyses include Bard (1974), Beck and Arnold (1977), Belsley et al. (1980), Seber and Wild (1989), Dennis and Schnabel (1996), and Tarantola (2005). Saltelli et al. (2000, 2004) provide comprehensive overviews of sensitivity-analysis methods. This book focuses on what Saltelli et al. describe as local sensitivity methods, and includes new sensitivity-analysis methods not included in the previous books.

The pioneers of using regression methods in groundwater modeling were Cooley (1977) and Yeh and Yoon (1981). Some of the material in this book was first published in U.S. Geological Survey reports (Cooley and Naff, 1990; Hill, 1992; Hill, 1994; Hill, 1998). Cooley and Naff (1990) presented a modified Gauss–Newton method of nonlinear regression that with some modification is used in Chapter 5, and residual analysis ideas derived from early editions of Draper and Smith (1998) that are used in Chapter 6. Hill (1992) presents sensitivity-analysis and residual-analysis methods used in Chapters 4 and 6. Cooley and Naff (1990), and Hill (1992), and Hill (1994) present methods of residual analysis and linear uncertainty analysis that are used in Chapters 6 and 8. Hill (1998) enhanced the methods presented in the previous works and presents the first version of the guidelines that are described in Chapters 10 through 14. Various aspects of the guidelines have a long history, and relevant references are cited in later chapters. To the authors' knowledge, these guidelines provide a more comprehensive foundation for the calibration and use of models of complex systems than any similar set of published guidelines. In general, the book expands the previously presented material, presents some new methods, and includes an extensive set of exercises.

Achieving Tractable Problems Regression is a powerful tool for using data to test hypothesized physical relations and to calibrate models in many fields (Seber and Wild, 1989; Draper and Smith, 1998). Despite its introduction into the groundwater literature in the 1970s (reviewed by McLaughlin and Townley, 1996), regression is only starting to be used with any regularity to develop numerical models of complicated groundwater systems. The scarcity of data, nonlinearity of the regression, and complexity of the physical systems cause substantial difficulties. Obtaining tractable models that represent the true system well enough to yield useful results is arguably the most important problem in the field. The only options are (1) improving the data, (2) ignoring the nonlinearity, and/or (3) carefully ignoring some of the system complexity. Scarcity of data is a perpetual problem not likely to be alleviated at most field sites despite recent impressive advances in geophysical data collection and analysis (e.g., Eppstein and Dougherty, 1996; Hyndman and Gorelick, 1996; Lebbe, 1999; Dam and Christensen, 2003). Methods that ignore nonlinearity are presented by, for example, Kitanidis (1997) and Sun (1994, p. 182). The large changes in parameter values that occur in most nonlinear regressions of many problems after the first iteration, however, indicate that linearized methods are unlikely to produce satisfactory results in many circumstances. This leaves option 3, which is discussed in the following paragraphs.

Defining a tractable and useful level of parameterization for groundwater inverse problems has been an intensely sought goal, focused mostly on the representation of hydraulic conductivity or transmissivity. Suggested approaches vary considerably. The most complex parameterizations are cell- or pixel-based methods in which hydraulic conductivity or transmissivity parameters are defined for each model cell, element, or other basic model entity, and prior information or regularization is used to stabilize the solution (e.g., see Tikhonov and Arsenin, 1977; Clifton and Neuman, 1982; Backus, 1988; McLaughlin and Townley, 1996). The simplest parameterizations require homogeneity, such that, at the extreme, one parameter specifies hydraulic conductivity throughout the model.

As more parameters are defined and the information contained in the observations is overwhelmed, prior information on parameters and/or regularization on observations and/or parameters become necessary to attain a tractable problem. In this book, we use definitions of prior information and regularization derived from Backus (1988). When applied to parameters, prior information and regularization produce similar penalty-function terms in the objective function. For prior information, the weighting used approximates the reliability of the prior information based on either classical or Bayesian statistical arguments. Essentially, classical statistical arguments are based on sampling methods; Bayesian statistical arguments are, at least in part, based on belief (Bolstad, 2004). In contrast, for regularization the weighting generally is determined as required to produce a tractable problem, as represented by a unique set of estimated parameter values. The resulting weights generally are much larger than can be justified based on what could possibly be known or theorized about the parameter values and distribution. For both prior information and regularization, the values used in the penalty function need to be unbiased (see the definition in Section 1.4.2).

Between the two extreme parameterizations mentioned previously, there is a wide array of designs ranging from interpolation methods such as pilot points (RamaRoa et al., 1995; Doherty, 2003; Moore and Doherty, 2005, 2006) to zones of constant value designed using geologic information (see Chapter 15 for examples). For example, the Regularization Capability of the computer code PEST (Doherty, 1994, 2005) typically allows many parameters to be estimated. Indeed, the number of parameters may exceed the number of observations. Parameter estimation is made possible by requiring that the parameter values satisfy additional considerations. Most commonly, the parameter distribution is required to be smooth. This and other considerations are discussed by Tikhonov and Arsenin (1977) and Menke (1989). More recent approaches include the superparameters of Tonkin and Doherty (2006) and the representer method of Valstar et al. (2004). The former uses singular value decomposition to identify a few major eigenvectors from sensitivity matrices; only the "superparameters" defined by the eigenvectors are estimated by regression.

1.3 RELATION OF THIS BOOK TO OTHER IDEAS AND PREVIOUS WORKS

Parameterizations with many parameters are advantageous in that they minimize user-imposed simplifications, but they have the following problems: (1) they do not eliminate the scale problem if heterogeneities smaller than the grid or parameter scale are important, as they often are in transport problems, for example; (2) they generally require more and better hydraulic-conductivity or transmissivity data than are available in most circumstances or unsupportable assumptions about smoothness; and (3) they can easily lead to overfitting the observations and a resulting decline in predictive accuracy. Historically, parameterization methods that resulted in many parameters also were unable to accommodate easily knowledge about geologic structure. Gradually, the ability to apply geologic constraints within the context of many defined parameters is being developed and provides exciting possibilities.

Simpler parameterizations (simpler in that there are fewer defined parameters) can be achieved using zonation, interpolation, or eigenvectors of the variance–covariance matrix of grid-scale parameters (e.g., Jacobson, 1985; Sun and Yeh, 1985; Cooley et al., 1986; RamaRao et al., 1995; Eppstein and Dougherty, 1996; Reid, 1996; D'Agnese et al., 1999; Tonkin and Doherty, 2006). Stochastic methods (e.g., Gelhar, 1993; Kitanidis, 1995; Yeh et al., 1995; Carle et al., 1998) also generally fall into this category, although they share some of the characteristics of the grid-based methods. These simpler parameterizations produce a more tractable problem, but it is not clear what level of simplicity diminishes utility.

The principle of parsimony (Box et al., 1994; Parker, 1994) suggests that simple models should be considered, but the perception remains that many complex systems cannot be adequately represented using parsimonious models. For example, Gelhar (1993, p. 341) claims that for groundwater systems "there is no clear evidence that [nonlinear regression] methods [using simple parameterizations] actually work under field conditions." Indeed, Beven and Binley (1992) even suggest that for some problems it may be best to abandon the concept of parameterizations simple enough to produce an optimal set of parameter values.

A concept as useful as parsimony should not be given up lightly, yet there have been few conclusive evaluations of the parameter complexity needed to produce useful results for groundwater models (Hill et al., 1998). In this book we proceed from the point of view that it is best to introduce complexity slowly and carefully, which is taken to mean increase the number of parameters slowly and carefully. One reason for this approach is that models with a few parameters can be used to learn things about a system that are true for all parameterizations but are more difficult to determine when many parameters are defined. As related to the famous quote by George E. P. Box, "All models are wrong, but some are useful," the idea is that parsimony is likely to play an important role in achieving useful models. We suggest that simpler parameterizations are useful for many models and for the initial phases of development of all models.

Direct and Indirect Inverse Modeling In groundwater inverse modeling, methods have been classified as indirect and direct (Neuman, 1973; Yeh, 1986; Sun, 1994). This book considers indirect inverse modeling, which uses available observation data and optimization techniques to estimate model input values.

Direct inverse modeling is dramatically different: available, usually sparse observations are interpolated or extrapolated everywhere in the model domain to create "observations" throughout the system. Using these "observations," the differential equations describing the simulated processes (such as groundwater flow or transport) are used to calculate the model input values (parameters) directly. The direct inverse modeling methods have been in existence longer than the indirect methods but have been shown consistently to be unstable in the presence of common measurement errors (Yeh, 1986). The direct methods do not use sensitivities and rarely calculate them, so these methods cannot be used to compute many of the statistics used for model evaluation that are presented in this book.

1.4 A FEW DEFINITIONS

This section defines what is meant by a linear and a nonlinear model in the context of parameter estimation. It also defines four terms that are often confusing and states how the terms are used in this book.

1.4.1 Linear and Nonlinear

As discussed in Section 1.1, this book focuses on models for which parameter estimation is nonlinear. In this context, nonlinearity results when simulated equivalents to observations are nonlinearly related to parameters. For example, consider groundwater flow.

In a confined groundwater flow system, hydraulic head is a linear function of space and time, which is why superposition can be used (Reilly et al., 1987). In contrast, for the same circumstances, head is a nonlinear function of many parameter values of interest, such as hydraulic conductivity. The simplest form of the groundwater flow equation, Darcy's Law, can be used to demonstrate both linearity with respect to the spatial dimension and nonlinearity with respect to hydraulic conductivity. This was shown by Hill et al. (2000, pp. 16–18) and is presented here in a modified form.

Darcy's Law relates the hydraulic head along the length of a cylinder packed with saturated porous media and flow through the cylinder. Darcy's Law can be expressed as

$$Q = -KA \frac{\partial h}{\partial X} \tag{1.1}$$

where Q = flow produced by imposing different hydraulic heads at opposite ends of a cylinder containing homogenous, saturated, porous media [L^3/T];

K = hydraulic conductivity of the saturated porous media [L/T];

A = cross-sectional area of the cylinder [L^2];

X = distance along an axis parallel to the length of the cylinder and, therefore, parallel to the direction of flow [L];

h = hydraulic head at any distance X along the cylinder [L].

1.4 A FEW DEFINITIONS

Equation (1.1) can be solved for the hydraulic head at any distance, X, to achieve

$$h = h_0 - \frac{Q}{KA} X \tag{1.2}$$

where h_0 is the hydraulic head at $X = 0$.

The derivatives $\partial h/\partial Q$ and $\partial h/\partial K$ are sensitivities in a parameter-estimation problem in which Q and K are estimated. By using partial derivative notation, the derivatives of Eq. (1.2) with respect to X, Q, and K are

$$\frac{\partial h}{\partial X} = -\frac{Q}{KA} \tag{1.3}$$

$$\frac{\partial h}{\partial Q} = -\frac{1}{KA} X \tag{1.4}$$

$$\frac{\partial h}{\partial K} = -\frac{Q}{K^2 A} X \tag{1.5}$$

The hydraulic head is considered to be a linear function of X because $\partial h/\partial X$ is independent of X. Hydraulic head also is a linear function of Q, because $\partial h/\partial Q$ is independent of Q. However, hydraulic head is a nonlinear function of K because $\partial h/\partial K$ is a function of K. As in this simple example, sensitivities with respect to flows, such as Q, are nearly always functions of aquifer properties; sensitivities with respect to aquifer properties, such as K, are nearly always functions of the aquifer properties and the flows. If Q and K are both estimated, both situations make the regression nonlinear.

1.4.2 Precision, Accuracy, Reliability, and Uncertainty

The terms precision, accuracy, reliability, and uncertainty are used in this book and by many others and can cause confusion. Formal definitions of these terms as related to estimated parameters and predictions are described here using an archery analogy and by relating them to the statistical terms bias and variance or standard error of the regression. (The archery analogy was suggested by Richard L. Cooley, retired from the U.S. Geological Survey, oral communication, 1988).

Precision: In archery, a set of shots is precise if the shots fall within a narrow range, regardless of whether they are near the bull's eye. A parameter estimate or prediction is more precise if associated coefficients of variation or confidence intervals are smaller. A model fits the observations more closely if the objective function is smaller, and this may indicate a more precise model depending on the measure used (see Chapter 6). More precise estimates or predictions are said to have lower variance. A precise parameter estimate results when the observations provide abundant information about the parameter, given the model construction. A precise prediction results when the parameters important to the prediction are precisely estimated.

Accuracy: In archery, a set of shots is accurate if the shots are distributed evenly about the bull's eye, though they may fall within a large radius around the bull's eye. Accurate estimates and predictions are, on average, close to the true, unknown value, but the range of values may be large. An accurate parameter estimate results when (1) the model is accurate and (2) the observations are unbiased. The observations may or may not provide abundant information about the parameter; abundant information would result in a parameter estimate that is both accurate and precise if points 1 and 2 were satisfied. An accurate prediction results when (1) the model is accurate and (2) parameter values important to the prediction are accurate. The observations may or may not provide much information about the parameters important to predictions. Accurate estimates and predictions are sometimes referred to as unbiased; inaccurate estimates and predictions are biased.

Reliability: In archery, a set of shots is reliable if the shots are distributed in a narrow range about the bull's eye. Reliable parameter estimates and predictions are both accurate and precisely determined. Reliable parameter estimates and predictions result when (1) the model accurately represents processes of importance to the observations and the predictions, and (2) the observations contain much information relevant to the predictions, so that the parameters important to the predictions are reliably estimated. From a probabilistic perspective, reliability is often defined as 1.0 minus the probability of failure.

Uncertainty: The direct inverse of reliability, so often defined as the probability of failure.

While these terms have distinct meanings, in practice, "accurate" often is used when "precise" is more applicable. In this book, we had to choose between always using these terms as defined here, or recognizing that many readers would proceed without having these definitions firmly in mind and would possibly be confused by proper usage. In some circumstances we chose more common usage to create what we thought would be an easier learning experience.

1.5 ADVANTAGEOUS EXPERTISE AND SUGGESTED READINGS

Most of this book requires little expertise in statistics and mathematics. Familiarity with basic statistics is useful, including definitions of the following terms: samples and populations; mean, standard deviation, variance, and coefficient of variation of samples and populations; normal probability distribution; log-normal probability distribution; confidence interval; and significance level. Familiarity with simple linear regression also is helpful. Good elementary references for these topics include Benjamin and Cornell (1970), Ott (1993), Davis (2002), and Helsel and Hirsch (2002). Useful advanced texts include Cook and Weisberg (1982), Seber and Wild (1989), and Draper and Smith (1998).

1.5 ADVANTAGEOUS EXPERTISE AND SUGGESTED READINGS

To use the exercises to learn the principles of sensitivity analysis, nonlinear regression, and associated evaluation of the regression, students will benefit from understanding groundwater flow problems well enough to follow the discussions of the physical problem considered. To perform the optional simulations of the groundwater model used in many of the exercises that accompany the methods, students will benefit from familiarity with the computer program MODFLOW-2000 (McDonald and Harbaugh, 1988; Harbaugh et al., 2000; Hill et al., 2000; Anderman and Hill, 2001).

When this book is used to teach a semester- or quarter-long academic course, it may be desirable to start with two to four weeks of instruction on statistics and linear regression. Recommended topics include graphical data analysis, hypothesis testing, simple linear regression, and multiple linear regression. If, for example, Helsel and Hirsch (2002) is used, the readings and exercises in Table 1.1 address the suggested material.

If Davis (2002) is used to learn basic statistics, the topics in Table 1.2 are suggested.

TABLE 1.1 Suggested Reading Assignments and Exercises in Helsel and Hirsch (2002)

Chapter	Topic	Reading Assignment	Exercise
2	Graphical data analysis	Introduction; Section 2.1.5	None
3	Uncertainty	Sections 3.1, 3.2, 3.4	3.1 (parametric interval)
4	Hypothesis testing	Introduction; Sections 4.1, 4.2, and 4.4	4.1 (for untransformed data)
5	t-Tests	Introduction; Section 5.2	5.2
8	Correlation coefficients	Introduction; Sections 8.1 and 8.4	None
9	Simple linear regression	All except Section 9.6	9.1. Use data subsets to show the effect of small data sets.
11	Multiple regression	All except Section 11.8	11.1

TABLE 1.2 Suggested Reading Assignments and Exercises in Davis (2002)

Chapter	Topic	Reading Assignment
2	Summary statistics	pp. 34–39
2	Joint variation of two variables	pp. 40–46
2	Comparing normal populations	pp. 55–58
2	Testing the mean, P-values, significance	pp. 60–66
2	Confidence limits, t-distribution	pp. 66–75
4	Runs tests	pp. 185–191
4	Simple linear regression	pp. 191–204, 227–228
6	Multiple regression	pp. 462–470

1.6 OVERVIEW OF CHAPTERS 2 THROUGH 15

The primary topics of this book are (1) methods for sensitivity analysis, data assessment, model calibration, and uncertainty analysis developed on the basis of inverse modeling theory; and (2) guidelines for the effective application of these methods. The methods are presented in Chapters 3 to 9 and the guidelines are presented in Chapters 10 to 14. Field applications and tests of the methods and guidelines are presented in Chapter 15. Chapter 2 presents an overview of the exercises and the computer programs used in this work. Three appendixes go into greater depth concerning several aspects of the nonlinear regression method used and one appendix presents selected statistical tables. Chapters 2 through 15 are described in more detail in the following paragraphs.

Chapter 2 presents an overview of (1) three computer codes for inverse modeling that are used throughout the book, (2) a hypothetical groundwater management problem to which the methods are applied, and (3) exercises that use this groundwater management problem to clearly demonstrate the methods.

Chapters 3 to 5 present methods for measuring model fit, initial model sensitivity analysis, and parameter estimation. Chapter 3 discusses how observations of the simulated system are compared to equivalent simulated values using objective functions. Terms of the objective functions are defined, and least-squares objective-function surfaces are introduced. Chapter 4 discusses sensitivity analysis methods for evaluating the information that the observations provide toward estimating a set of parameters and using such an analysis to design parameterizations and decide what parameters to estimate. Several statistics are presented that are independent of model fit and thus can be applied prior to having achieved a successful inversion. These are called fit-independent statistics. Chapter 5 presents the modified Gauss–Newton gradient method for estimating parameter values that produce the best fit to the observations by minimizing the least-squares objective function.

Chapters 6 to 8 present methods for evaluating model fit, parameter estimates, data needs, and prediction sensitivity and uncertainty. Most of these methods involve calculating and evaluating diagnostic and inferential statistics and conducting graphical analyses. Chapter 6 discusses methods for evaluating model fit, including using residuals (differences between observed and simulated values) and weighted residuals to calculate statistical measures of fit, and graphs that can be used to help detect model error and assess normality of weighted residuals. Chapter 7 presents methods for evaluating estimated parameters and their uncertainty, including confidence intervals and measures of the support that the observations provide for the estimated parameter values. Methods for assessing model linearity are also discussed. Chapter 8 discusses evaluation of model predictions and their sensitivity and uncertainty, and methods for identifying data that would improve model predictions. Topics include measures for assessing the importance to predictions and to confidence intervals on predictions of observations and prior information on parameters. Monte Carlo methods of evaluating uncertainty are discussed briefly.

Chapter 9 presents methods for calibrating transient and transport models, and for recalibrating and reevaluating existing models when new data become available.

1.6 OVERVIEW OF CHAPTERS 2 THROUGH 15

TABLE 1.3 Guidelines for Effective Model Calibration

Model Development (Chapter 11)

1. Apply the principle of parsimony (start very simple; build complexity slowly)
2. Use a broad range of system information (soft data) to constrain the problem
3. Maintain a well-posed, comprehensive regression problem
4. Include many kinds of observations (hard data) in the regression
5. Use prior information carefully
6. Assign weights that reflect errors
7. Encourage convergence by making the model more accurate and by evaluating the observations
8. Consider alternative models

Model Testing (Chapter 12)

9. Evaluate model fit
10. Evaluate optimized parameter values

Potential New Data (Chapter 13)

11. Identify new data to improve simulated processes, features, and properties
12. Identify new data to improve predictions

Prediction Uncertainty (Chapter 14)

13. Evaluate prediction uncertainty and accuracy using deterministic methods
14. Quantify prediction uncertainty using statistical methods

---

Exercises at the ends of Chapters 3 to 9 demonstrate the methods. Most of the exercises involve the simple hypothetical groundwater management problem mentioned in the beginning of this chapter.

Chapters 10 to 14 present fourteen guidelines that address using the methods presented in Chapters 3 to 9 to analyze, simulate, calibrate, and evaluate models of complex systems. The guidelines are grouped into four topics: (1) model development, (2) model testing, (3) potential new data, and (4) prediction uncertainty. Chapter 10 introduces the guidelines and Chapters 11 to 14 each focus on the guidelines that address one of the four topics.

Table 1.3 lists the guidelines to introduce the reader to the basic ideas they promote. For example, a fundamental aspect of the approach is to start simple and to build complexity slowly.

Chapter 15 addresses the use and testing of the methods and guidelines. First, issues of computer execution time, which are nearly always of concern when calibrating models, are discussed. Then, selected publications describing tests of the guidelines using synthetic test cases and use of the guidelines in field applications are listed. The remainder of Chapter 15 discusses a few aspects of two field cases to illustrate some of the methods and guidelines presented in the book.

2

COMPUTER SOFTWARE AND GROUNDWATER MANAGEMENT PROBLEM USED IN THE EXERCISES

This chapter briefly describes the computer programs and the groundwater management problem on which exercises presented in Chapters 2 through 9 are based. The exercises can be completed using results provided in figures and tables of the book, or hands-on computer exercises can be pursued.

The groundwater system can be simulated using the Ground-Water Flow Process of MODFLOW-2000 or MODFLOW_2005. Sensitivity analysis, parameter estimation, data needs assessment, predictions, and uncertainty evaluation can be performed using the Observation, Sensitivity, and Parameter-Estimation Processes of MODFLOW-2000 or the capabilities of UCODE_2005 or PEST. Explicit instructions for the MODFLOW-2000 and UCODE_2005 and possibly for other codes and graphical interfaces are provided on the web site listed in Chapter 1, Section 1.1 of this book. Performing the exercises using the computer programs or reviewing the instructions for doing so is expected to facilitate use of the methods in the simulation of other systems.

2.1 COMPUTER PROGRAMS MODFLOW-2000, UCODE_2005, AND PEST

The computer software used for the exercises was listed in Chapter 1, Section 1.1, and access through web sites is described there. The discussion here refers to MODFLOW-2000 Version 1.15, UCODE_2005 Version 1.0, and PEST Version 9.0. Later versions of these codes may have capabilities not discussed here.

Effective Groundwater Model Calibration: *With Analysis of Data, Sensitivities, Predictions, and Uncertainty.* By Mary C. Hill and Claire R. Tiedeman
Published 2007 by John Wiley & Sons, Inc.

MODFLOW-2000 is applicable only to solution of the transient, three-dimensional groundwater flow equation represented using a control-volume finite-difference numerical method. In MODFLOW-2000, groundwater flow is simulated using the Ground-Water Flow Process, and inverse modeling calculations are performed using the Observation, Sensitivity, and Parameter-Estimation Processes.

UCODE_2005 and PEST are universal inverse codes with broad applicability. They can be used with any simulation model that has ASCII input and output files and can be executed from a command prompt. Both programs use very similar template and instruction files to interact with the simulation model.

MODFLOW-2000, UCODE_2005, and PEST have many capabilities in common and have a few key differences. Table 2.1 lists and compares selected capabilities of each program for defining observations and parameters and lists some graphical user interfaces that support the programs.

All of these codes perform inverse modeling, posed as a parameter-estimation problem, by calculating parameter values that minimize a weighted least-squares objective function using nonlinear regression. The methods shared by MODFLOW-2000 and UCODE_2005 are described in Chapters 3 and 5. In addition, UCODE_2005 has a trust region option that is mentioned briefly in Chapter 5. The trust region approach can reduce the number of iterations required for difficult problems by as much as 50 percent (Mehl and Hill, 2002). For the problems considered in the exercises, PEST differs from the methods described in this book mostly in its definition of the Marquardt parameter and its use of a line search capability that improves regression performance in some circumstances. See Chapter 5, Section 5.1.1 for comments about the Marquardt parameter.

The method for calculating sensitivities in MODFLOW-2000 differs substantially from that in UCODE_2005 and PEST. In MODFLOW-2000, the Sensitivity Process calculates sensitivities using the sensitivity-equation method, which is the most accurate method available. Implementing the sensitivity-equation method requires extensive custom programming, which can easily double the size of a code. Any subsequent change to the capabilities of the forward simulation generally requires additional coding to accommodate sensitivity-equation sensitivities. The required substantial investment means that sensitivity-equation sensitivities probably will be available for only a very few codes and will rarely be available for all possible parameters, observations, or simulated dynamics.

Codes that calculate sensitivity-equation sensitivities can be produced using a program called ADIFOR (http://www.unix.mcs.anl.gov/autodiff/ADIFOR/). The resulting code tends to be difficult to develop further because the alterations created by the program are not modularly constructed and not clearly coded, but this option can be very useful.

UCODE_2005 and PEST can use sensitivities generated by programs such as MODFLOW-2000, or sensitivities can be calculated using perturbation methods. Perturbation sensitivities tend to be less accurate than sensitivity-equation sensitivities but require no custom programming. This is what allows these codes to be used with any process model. The less accurate sensitivities primarily can affect performance in two ways: (1) convergence of the nonlinear regression can be less

TABLE 2.1 Capabilities of MODFLOW-2000 Version 1.15, UCODE_2005 Version 1.0, and PEST Version 9.0

Capability	MODFLOW-2000[a]	UCODE_2005 and PEST[a]
OBSERVATION DEFINITION		
Heads and temporal changes in head, not necessarily at cell centers	Yes	Yes
Flows at head-dependent boundaries, not necessarily ending at a cell boundary	Yes	Yes
Flows at constant-head boundaries	Yes	Yes
Heads at constant-head boundaries	Yes	Yes
Advective transport	Yes, with the ADV Package	Yes
Any other observation	No	Yes
PARAMETER DEFINITION		
Zone arrays	Yes	Difficult[b]
Multiplication arrays	Yes	Difficult[b]
Pilot points interpolation method	Yes, using multiplication arrays	UCODE_2005, difficult[b]; PEST, efficient through regularization capability
Additive parameters[c]	Easy	Difficult[b]
Association of a parameter with more than one model characteristic (e.g., layer and riverbed hydraulic conductivity)	No	Yes
REGRESSION CAPABILITIES		
Trust region	No	UCODE_2005 only[d]
Line search	No	PEST only[d]

[a] MODFLOW-2000 (Harbaugh et al., 2000; Hill et al., 2000), UCODE_2005 (Poeter et al., 2005) and PEST (Doherty, 2005) are public domain, open source programs. For websites, see Section 1.1 or the reference list.

[b] Difficult if achieved using the capabilities of the listed code(s). If the process model, such as the Ground-Water Flow Process of MODFLOW-2000, performs these functions easily, these codes can take advantage of that.

[c] The additive parameter capability of MODFLOW-2000 is very general, allowing most interpolation methods to be applied to any characteristic that can be represented using parameters. This includes variations in streambed characteristics along the length of a river and hydraulic-conductivity variations caused by depositional processes.

[d] The trust region approach in UCODE_2005 can reduce iterations for difficult problems by 50 percent (Mehl and Hill, 2002). Performance of the line-search method has not been documented.

stable for poorly conditioned problems, as demonstrated by Mehl and Hill (2002), so that UCODE_2005 (without the trust region option) and PEST may not converge when MODFLOW-2000 does converge, and (2) parameter correlation coefficients calculated from the variance–covariance matrix on the parameter estimates can be inaccurate enough to be misleading (Hill and Østerby, 2003). Consequently, parameter correlation coefficients calculated by UCODE_2005 and PEST cannot be used as reliably as those calculated by MODFLOW-2000 to determine the existence of extreme parameter correlation. This issue is discussed in more detail in Chapter 4, Section 4.3; Exercise 4.1 clearly demonstrates this problem.

MODFLOW-2000 and UCODE_2005 or PEST can be used together to simplify the processes that UCODE_2005 and PEST use to define parameters and simulated equivalents of observations. This is advantageous when the MODFLOW-2000 Ground-Water Flow and Observation Process capabilities apply, but some other aspect of the problem, such as the estimation of a parameter of interest, is not supported by MODFLOW-2000. In this situation, the MODFLOW-2000 Ground-Water Flow Process capabilities are used as the process model for UCODE_2005 or PEST, and the MODFLOW-2000 Parameter and Observation capabilities are used to simplify the parameter substitution and extraction of simulated values in UCODE_2005 or PEST. As mentioned earlier, both UCODE_2005 and PEST can use sensitivities calculated by the process model; they can calculate other needed sensitivities using the perturbation method.

2.2 GROUNDWATER MANAGEMENT PROBLEM USED FOR THE EXERCISES

The exercises in this book focus on a groundwater management problem within the hypothetical geographic area depicted in Figure 2.1a. The groundwater system is of interest because pumping wells are being completed in aquifers 1 and 2 to supply local domestic and industrial water needs. In addition, a proposal has been submitted to local authorities for construction of a landfill (Figure 2.1a). The developers claim that the landfill is outside the capture zone of the proposed wells, and that any effluent from the landfill will reach the river sufficiently diluted to meet regulatory standards. Local authorities would like to investigate this claim.

Data on the flow system without pumpage are available for model development and are from a period of time that is consistent with long-term average conditions. Seasonal variations appear to be small. Upon completion of the water-supply wells, transient and steady-state data can be collected under pumping conditions. A key issue is whether the decision on the proposed landfill should be delayed until after the transient data are collected. The developers have requested a quick decision.

The flow system is complicated enough to require a numerical model for its simulation, but lacks some complexities typical of many field problems. Most notably, the subsurface material lacks local heterogeneity. The upper aquifer and confining bed are homogeneous, and the lower aquifer has a mild degree of regional heterogeneity. This is advantageous because the system is simple enough to clearly demonstrate the methods in the book. Also, using a synthetic test case means that

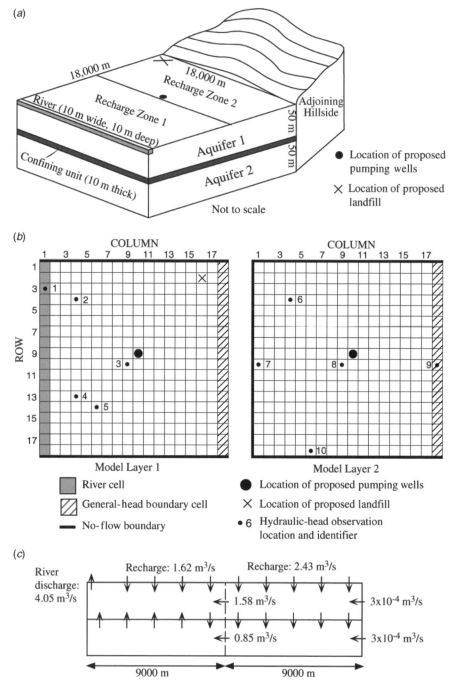

FIGURE 2.1 System and model used in exercises: (*a*) flow system; (*b*) finite-difference grid, boundary conditions, and locations of observation wells, proposed pumping wells and the landfill; (*c*) flows through a cross section; and (*d*) hydraulic heads. Parts (*c*) and (*d*) are produced using the true parameter values and no pumping.

2.2 GROUNDWATER MANAGEMENT PROBLEM USED FOR THE EXERCISES

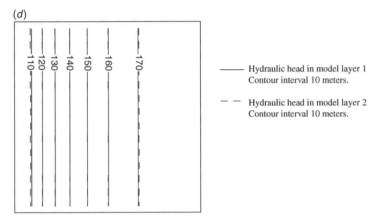

FIGURE 2.1 *Continued.*

results can be compared to "truth." Extension of the methods to realistic problems is discussed throughout the book, and especially in the guidelines and examples presented in Chapters 10 through 15.

2.2.1 Purpose and Strategy

A numerical model is needed to address this groundwater management problem because there are multiple aquifer layers and spatial variation in hydraulic conductivity and boundary conditions that are not conducive to analytic solution. To coordinate with data availability, a steady-state model without pumpage is developed first and used to produce a preliminary evaluation of effluent transport from the landfill when pumpage is applied. The effluent transport is simulated using particle tracking methods. The concern here is whether the effluent goes to the well—other issues like first arrival time are not of concern. If the particle goes to the well, there is no reason to use more computationally demanding transport simulations.

The developers of the landfill raise important questions about the steady-state model, so that additional data are needed. We use the steady-state model to evaluate potential new data that can be collected once the supply wells are completed, and we use the analysis in combination with field considerations to design a monitoring network. The data are collected, a transient model is produced that includes the pumping wells, and the model is recalibrated using the data from both steady-state and stressed conditions. Finally, the effluent transport issue is reevaluated using the recalibrated model.

2.2.2 Flow System Characteristics

The groundwater flow system used for most of the exercises is shown in Figure 2.1*a*. The flow system is comprised of two confined aquifers separated by a confining unit. Inflow occurs as areal recharge and as flow across the boundary with the adjoining hillside. At steady state without pumping, outflow occurs only as discharge to the river. In transient simulations and for steady-state simulations with pumpage, discharge is simulated from both of the model layers at the location shown in Figure 2.1a.

MODFLOW-2000 is used to simulate groundwater flow, and its ADV Package (Anderman and Hill, 2001) is used to simulate advective transport of effluent from the landfill. The domain is divided laterally into 18 rows and 18 columns (Figure 2.1b). Each confined aquifer is represented by one model layer. In layer 1, hydraulic conductivity is uniform. In layer 2, hydraulic conductivity increases linearly in steps with distance from the river, with each step consisting of a pair of model columns. The confining bed is not represented as a separate model layer, but as a vertical hydraulic conductivity that controls flow between the two model layers. Boundary conditions include two zones of areal recharge applied to model layer 1, one zone coincident with the 9 columns closest to the river, and the other coincident with the 9 columns closest to the hillside (Figure 2.1a). Inflow from the hillside to layers 1 and 2 and outflow from layer 1 to the river are simulated as head-dependent boundaries (Figure 2.1b). No-flow boundaries are specified on the bottom of the model domain and on all model sides except that adjacent to the hillside.

True steady-state simulated volumetric flows in the system without pumping are illustrated in Figure 2.1c, and simulated hydraulic heads are shown in Figure 2.1d. Without pumping, the flow system is actually two-dimensional because all stresses, boundary conditions, and subsurface properties and, therefore, all hydraulic heads and flows are the same for any cross section perpendicular to the river.

2.3 EXERCISES

Exercises are presented at the end of Chapters 2–9, and cover all of the methods and ideas included in those chapters. Most of the exercises involve the simple groundwater management problem described in Section 2.2. Through the development, calibration, and analysis of the steady-state and transient models that address this problem, the following steps are accomplished:

Steady-State Model

Simulate steady-state hydraulic heads (Exercises 2.1 and 2.2)
Define steady-state parameters and observations (Exercises 3.1 and 3.2)
Evaluate the initial steady-state model (Exercise 3.3)
Perform sensitivity analysis (Exercise 4.1)
Calibrate the steady-state model (Exercise 5.2)
Evaluate model fit to observations and prior information (Exercises 6.1 and 6.2)
Evaluate estimated parameter values (Exercises 7.1–7.3)
Make predictions using the calibrated steady-state model, perform sensitivity analysis, and evaluate potential new data (Exercise 8.1)
Evaluate prediction uncertainty (Exercise 8.2)

2.3 EXERCISES

Transient Model

Simulate hydraulic heads in the transient model (Exercises 9.1 and 9.2)
Define transient parameters and observations (Exercises 9.3 and 9.4)
Evaluate the initial transient model (Exercise 9.5)
Perform sensitivity analysis (Exercise 9.6)
Recalibrate the model using original steady-state observations and new transient observations (Exercise 9.7)
Evaluate the calibrated transient model (Exercises 9.8–9.11)
Make predictions using the recalibrated model (Exercise 9.12)

In groundwater model development, defining parameters often is difficult, as discussed in Chapter 1, Section 1.2.1. The exercises do not address this phase of model construction. Rather, the hypothetical flow system is designed so that its hydrogeologic and hydrologic characteristics can be accurately represented using only a few model parameters. Accurate representation of these aspects of the system allows the methods presented in the book to be illustrated more clearly. A more complicated problem might cause students to think that inaccurate parameterization is the problem when actually other issues are involved.

The exercises contain an explanation, followed by questions to be answered or issues to be explained. These questions and issues are listed under the heading *Problem* and usually involve examination and evaluation of results. The results are obtained as follows. The Ground-Water Flow Process capabilities of MODFLOW-2000 are used to simulate groundwater flow. The sensitivity and inverse modeling exercises can be performed using UCODE_2005, the Sensitivity and Parameter-Estimation Processes of MODFLOW-2000, or, in most situations, PEST. The results of these simulations are contained in figures and tables included in this book.

Students can complete all exercises in this book and thoroughly learn all methods presented without performing model simulations. To perform the model simulations, download the instructions, data, and codes as described in Chapter 1, Section 1.1.

Exercises marked *Optional* can be skipped without disturbing continuity. The exercises in Chapter 9 are all marked optional because they provide additional experience with methods already used in previous exercises. It can be advantageous to replace the Chapter 9 exercises with application of the methods to models related to other student investigations, possibly with class presentation of results.

Exercise 2.1: Simulate Steady-State Heads and Perform Preparatory Steps In these exercises, MODFLOW-2000 is used to simulate steady-state hydraulic heads for the flow system described in Section 2.2. Initial and final computer files, and instructions for modifying files, creating new files, and performing the simulations are available from the web site for this book; see Chapter 1, Section 1.1 for information about obtaining these files and instructions. Students who are not performing the simulations may skip these exercises.

3

COMPARING OBSERVED AND SIMULATED VALUES USING OBJECTIVE FUNCTIONS

The match of observed to simulated values is one of the most important indicators of how well a model represents an actual system. Objective functions measure this fit. Model calibration efforts largely involve attempting to construct a model that produces a good fit. Here, good fit means the objective function is as small as possible. Methods such as regression can determine parameter values that are optimal, meaning that they produce the best fit given the constructed model. The resulting parameter values are said to be optimal, optimized, or estimated by the regression. In later chapters of this book, we will see that a close fit is not the only goal of model calibration. However, methods that optimize parameter values are an important component of model calibration and can be used advantageously.

This chapter presents the objective functions used in this book to quantify the match between observed and simulated values and discusses alternative objective functions. It also lists the conditions needed for model results to be accurate when produced using regression methods, discusses quantities used in the objective functions, and introduces objective-function surfaces.

3.1 WEIGHTED LEAST-SQUARES OBJECTIVE FUNCTION

The weighted least-squares objective function is first presented with a commonly used diagonal weight matrix. This allows use of summations, which are easier for

---

Effective Groundwater Model Calibration: With Analysis of Data, Sensitivities, Predictions, and Uncertainty. By Mary C. Hill and Claire R. Tiedeman
Published 2007 by John Wiley & Sons, Inc.

3.1 WEIGHTED LEAST-SQUARES OBJECTIVE FUNCTION

many readers to understand than matrix and vector notation. The objective function is then presented with a full weight matrix.

Often the term weighted regression is applied to regression with a diagonal weight matrix and generalized regression is applied to regression with a full weight matrix (Draper and Smith, 1998, p. 223). In this book we refer to both as *weighted regression*. Regression without weighting is called *ordinary regression*.

3.1.1 With a Diagonal Weight Matrix

The objective function is first defined in the context of a groundwater model. Using hydraulic heads and flows as the observations, the weighted least-squares objective function, $S(b)$, can be expressed as

$$S(b) = \sum_{i=1}^{NH} \omega_{h_i}[y_{h_i} - y'_{h_i}(b)]^2 + \sum_{j=1}^{NQ} \omega_{q_j}[y_{q_j} - y'_{q_j}(b)]^2 + \sum_{k=1}^{NPR} \omega_{p_k}[y_{p_k} - y'_{p_k}(b)]^2 \quad (3.1a)$$

where b = a vector (which can be thought of as a list) containing values of each of the *NP* parameters being estimated;

NP = the number of estimated parameters;
NH = the number of hydraulic-head observations;
NQ = the number of flow observations;
NPR = the number of prior information values;
y_{h_i} = the *i*th observed hydraulic head being matched by the regression;
$y'_{h_i}(b)$ = the simulated hydraulic head that corresponds to the *i*th observed hydraulic head (a function of b);
y_{q_j} = the *j*th observed flow being matched by the regression;
$y'_{q_j}(b)$ = the simulated flow that corresponds to the *j*th observed flow (a function of b);
y_{p_k} = the *k*th prior estimate included in the regression;
$y'_{p_k}(b)$ = the *k*th simulated value (restricted to linear functions of b in UCODE_2005 and MODFLOW-2000);
ω_{h_i} = the weight for the *i*th head observation;
ω_{q_j} = the weight for the *j*th flow observation;
ω_{p_k} = the weight for the *k*th prior estimate.

For *NH* and *NQ*, multiple observations at the same location or reach are each counted. Using y to indicate a generic contribution of any kind and ω to indicate its weight, the objective function is more commonly expressed as

$$S(b) = \sum_{i=1}^{ND+NPR} \omega_i[y_i - y'_i(b)]^2 = \sum_{i=1}^{ND+NPR} \omega_i e_i^2 \quad (3.1b)$$

where ND = the number of observations;

y_i = the ith observation or prior information value being matched by the regression;

$y'_i(\mathbf{b})$ = the simulated equivalent, defined as the simulated value (a function of $\mathbf{b}$) that corresponds to y_i;

ω_i = the weight for the ith contribution to the objective function;

e_i = the ith weighted residual, equal to $[y_i - y'_i(\mathbf{b})]$.

Some of these terms are discussed further in Section 3.4.

3.1.2 With a Full Weight Matrix

In the simple diagonal weight matrix assumed in Eq. (3.1b), the diagonal entries are nonzero, and the off-diagonal terms equal zero. Each entry on the diagonal is the weight for a single observation or piece of prior information. More generally, the weighting requires a full weight matrix, in which one or more of the off-diagonal matrix entries are nonzero. These off-diagonal entries are needed to represent correlated observation errors. For a full weight matrix, the least-squares objective function of Eq. (3.1b) is written using vector and matrix notation as

$$S(\mathbf{b}) = [\mathbf{y} - \mathbf{y}'(\mathbf{b})]^T \boldsymbol{\omega} [\mathbf{y} - \mathbf{y}'(\mathbf{b})] = \mathbf{e}^T \boldsymbol{\omega} \mathbf{e} \qquad (3.2)$$

where $\boldsymbol{\omega}$ is the weight matrix and $\mathbf{y}$ is a vector of observations and prior information, $\mathbf{y}'(\mathbf{b})$ is a vector of simulated values, and $\mathbf{e}$ is a vector of residuals. The dimensions of the matrix and vectors are as follows: $\boldsymbol{\omega}$ is a square matrix dimensioned $(ND + NPR)$ by $(ND + NPR)$; all three vectors have $(ND + NPR)$ elements. ND and NPR were defined for Eq. (3.1). In Eq. (3.2), data for both observations and prior information are included in the weight matrix and in the vectors $\mathbf{y}, \mathbf{y}'(\mathbf{b})$, and $\mathbf{e}$. The structures of the weight matrix and the vectors are displayed in Appendix B, Eq. (B.1) and (B.2).

MODFLOW-2000 supports full weight matrices for all types of observations (except hydraulic head) and for prior information. MODFLOW-2000 can accommodate some common temporal correlations in the errors of hydraulic-head observations by differencing as discussed in Section 9.1.2 and in Hill et al. (2000, pp. 33–34). UCODE_2005 supports full weight matrices for all types of observations and for prior information and can accommodate any type of differencing.

Full weight matrices are discussed further in Guideline 6, Section G6.2, in Chapter 11.

3.2 ALTERNATIVE OBJECTIVE FUNCTIONS

Alternatives to the least-squares objective function described in this work are the maximum-likelihood objective function, the L_1 norm, and multiobjective optimization.

3.2 ALTERNATIVE OBJECTIVE FUNCTIONS

3.2.1 Maximum-Likelihood Objective Function

The maximum-likelihood objective function reduces to the least-squares objective function in most applications (as shown in Appendix A). The maximum-likelihood objective function is presented here and its value is calculated and printed by UCODE_2005 and MODFLOW-2000, because it can be used for model discrimination by itself or in the calculation of other statistics (e.g., see Carrera and Neuman, 1986; Loaiciga and Marino, 1986; Burnham and Anderson, 2002). The maximum-likelihood objective function is calculated as

$$S'(b) = (ND + NPR)\ln 2\pi - \ln|\omega| + e^T \omega e \qquad (3.3)$$

where $|\omega|$ is the determinant of the weight matrix, and, without loss of generality, it is assumed that the weight matrix is defined such that the common error variance σ^2 described in Appendixes A and C equals 1.0. Unlike the least-squares objective function, Eq. (3.3) can be negative. Appendix A presents the derivation of Eq. (3.3) and the assumptions required for the derivation and explains the equivalence of using Eq. (3.2) or (3.3) in practice. An alternative but, in practice, equivalent version is derived by Burnham and Anderson (2002, p. 12).

3.2.2 L_1 Norm Objective Function

The L_1 norm equals the sum of the absolute values of weighted residuals (Xiang et al., 1993; Menke, 1989). Minimizing the L_1 norm is often accomplished using the simplex method and does not require sensitivities or derivatives of the objective function. Inferential and diagnostic statistics that are derived from sensitivities and used for the sensitivity analysis, data assessment, and uncertainty evaluation described in this book can be obtained if the sensitivities are calculated separately. In this situation these statistics could be used as described here. L_1 norms are rarely used for nonlinear systems because they do not perform as well as nonlinear regression and provide less information.

3.2.3 Multiobjective Function

Multiobjective optimization uses multiple objective functions. The objective functions may be least-squares objective functions such as those considered in this work, or, for example, the objective function may be defined as the sum of costs for well installation and sampling that are to be kept as small as possible (e.g., see Deb, 2001; Reed et al., 2003; Vrugt et al., 2003). Objective functions also can include terms related to the smoothness of the estimated parameters (e.g., see Vasco et al., 1997).

Defining what to include in the different objective functions is an important part of multiobjective optimization. Some situations are clear, such as when one objective function represents the violation of established criteria, and another represents well installation and sampling costs. Other situations are not as clear, such as when one objective function represents the fit to some of the observations used in model calibration, and another represents the fit to other observations. For example, is it best to include all hydraulic heads in one objective function and all streamflow gain and loss observations in another? Should the data in different subbasins be included as separate objective functions, or as one combined objective function? In the former example, if the heads are all combined, that objective function is more likely to suffer from extreme parameter correlation. Is that advantageous to understanding system dynamics? Users who consider multiobjective optimization are encouraged to consider such questions and design their multiobjective optimizations carefully.

3.3 REQUIREMENTS FOR ACCURATE SIMULATED RESULTS

Theoretically, the least-squares objective function can be used to produce a model that accurately represents a system and provides accurate measures of model uncertainty only if three conditions are met. Two of these conditions relate to true errors, which equal the unknown amounts by which an observation or prior information equation differs from the value in the actual system. The conditions are: (1) Relevant processes, system geometry, and so on are adequately represented and simulated; (2) true errors of the observations and prior information are random and have a mean of zero; and (3) weighted true errors are independent, which means that the weighting needs to be proportional to the inverse of the variance-covariance matrix on the true observation errors (Draper and Smith, 1998, p. 34, 222). The true errors cannot be analyzed, so weighted residuals are investigated and the characteristics of the true errors are inferred. Tests for weighted residuals are described in Chapter 6.

To estimate parameter values with the least-squares objective function there is no requirement about the statistical distribution of the true errors (Helsel and Hirsch, 2002, Table 9.1). However, normality is often assumed, which allows calculation of observation error variances and covariances from field data and construction of linear confidence intervals. The first is discussed in Guideline 6 in Chapter 11; linear confidence intervals are discussed in Sections 7.5.1 and 8.4.2. Tests for normality are presented in Sections 6.4.5. Model linearity can be tested using measures discussed in Section 7.7.

3.3.1 Accurate Model

Many aspects of requirement 1 above are application specific, but some methods of sensitivity analysis and comparing observed and simulated values can be useful for achieving the requirement and/or for testing and demonstrating to what degree it is achieved. Much of this book presents such methods and shows how to use them.

3.3.2 Unbiased Observations and Prior Information

Requirement 2 is important because if an observation or prior information equation is biased—that is, the difference between observed and simulated values is expected to be consistently negative or positive—the model is likely to be biased. For

3.3 REQUIREMENTS FOR ACCURATE SIMULATED RESULTS

example, consider streamflow observations that are affected by a process that makes them higher than would be expected given the simulated processes (e.g., if baseflow contributes to streamflow but is not included in a rainfall-runoff model). Optimized parameter values may produce a good fit to the observations, but the system may not be simulated correctly. If the model is then used to simulate other circumstances the predictions are likely to be inaccurate. One consequence of requirement 2 is that bias cannot be accommodated by weighting. Instead, every effort needs to be made to eliminate bias in the observations. For the streamflow example, the noted bias is commonly eliminated by subtracting estimates of base flow to create the observations used in the regression. The importance of requirement 2 to the validity of regression methods is explained in Appendix C.

3.3.3 Weighting Reflects Errors

To understand requirement 3, consider that weighting performs two related functions. First, weighting needs to produce weighted residuals that have the same units so that they can be squared and summed using Eqs. (3.1) or (3.2). Obviously, summing numbers with different units produces nonsense. Second, weighting needs to reduce the influence of observations and prior information that are less accurate relative to those that are more accurate. These two functions relate directly to the theoretical requirement that the weight matrix be proportional to the inverse of the variance-covariance matrix of the true errors (requirement 2), which is derived in Appendix C. Errors are discussed in Chapter 11 under Guideline 6; examples are provided in Chapter 15. The assumptions implied by using a diagonal weight matrix are discussed in Appendix A.

Mathematically, requirements 2 and 3 can be expressed as:

$$E(\varepsilon) = 0$$

for a diagonal weight matrix (Eq. 3.1): $\omega_i \propto 1/\sigma_i^2$ (3.4)

for a full weight matrix (Eq. 3.2): $\omega^{1/2} \propto V(\varepsilon)^{-1}$

where $\propto$ means "proportional to," ε is a vector of true errors, σ_i^2 is the variance of the true error of observation i, and $V(\varepsilon)$ is the variance–covariance matrix of the true errors, with variances along the diagonal and covariances off the diagonal. The true errors in vector ε relate observed or prior information values, y_i, to true, unknown values, y_i^{true}, through the expressions:

$$y_i = y_i^{\text{true}} + \varepsilon_i, \quad i = 1, \text{ND} + \text{NPR} \quad \text{or, equivalently,}$$
$$\boldsymbol{y} = \boldsymbol{y}^{\text{true}} + \boldsymbol{\varepsilon} \tag{3.5}$$

Additive errors are assumed in Eq. (3.5). This is not a very restrictive assumption because errors often are additive or can be converted to being additive, as discussed in the following paragraphs.

For many observations, and especially groundwater flow and concentration observations, errors are typically thought to be proportional to the true value, so that

$$y = y^{true}(1 + \varepsilon) = y^{true} + y^{true}\varepsilon. \tag{3.6}$$

An appropriate weighting strategy can be achieved by specifying the coefficient of variation as the statistic from which the weight is calculated, and using observed or simulated values to estimate y^{true} (e.g., see Keidser and Rosbjerg, 1991). The variance is then calculated as $[(c.v.) \times a]^2$, where c.v. is the coefficient of variation and a is the observed or simulated value. The standard deviation equals $[(c.v.) \times a]$. Anderman and Hill (1999) show that using simulated, rather than observed, concentrations is needed to obtain unbiased parameter estimates in transport problems, and this conclusion is likely to be generally applicable. See Section 9.2 for more discussion of weighting concentrations. MODFLOW-2000 supports using observations to calculate weights; UCODE_2005 supports using either observations or simulated values.

Errors that can be made additive through a transformation include, for example, multiplicative errors for which $y_i = (y_i^{true}) \times (\varepsilon_i)$. This error model can be log-transformed to produce $\ln(y_i) = \ln(y_i^{true}) + \ln(\varepsilon_i)$, in which the errors are additive as in Eq. (3.5). Margulis et al. (2002) present a study in which errors are multiplicative. Observation transformations that convert multiplicative errors to be additive can be easily implemented in UCODE_2005 or PEST, though doing so can make model results harder to communicate to resource managers.

3.4 ADDITIONAL ISSUES

Issues related to prior information, weighting, and weighted residuals are discussed.

3.4.1 Prior Information

The linear prior information equations supported by MODFLOW-2000 and UCODE_2005 have the form

$$P'_p(b) = \sum_{j=1}^{NP}(a_{p,j}b_j) = a_{p,1}b_1 + a_{p,2}b_2 + \cdots + a_{p,NPR}b_{NP} \tag{3.7}$$

where p indicates the pth prior information equation, $a_{p,j}$ are coefficients, and b_j is the jth parameter value. In this book, the subscript p is sometimes replaced by a prior information name and j is replaced by the parameter name instead of a parameter number.

Often, prior information equations have one nonzero coefficient $a_{p,j}$ equal to 1.0, so they are of the form $P'_p = b_j$. In this case, the contribution to the objective function (Eq. (3.1) or (3.2)) is simply the weighted difference between the prior value of a parameter, P_p, and b_j.

3.4 ADDITIONAL ISSUES

More than one term is needed on the right side of Eq. (3.7) when the prior information relates to a linear function that includes more than one parameter value. Consider the following two groundwater examples.

Example 1: In each of two confined models, specific storage values are defined as parameters and are estimated by regression. The names of these parameters are SS1 and SS2. The combined storage coefficient of both layers has been measured from aquifer-test drawdown data that are not being used as observations in the calibration of a regional-scale model. In this situation, the prior estimate, P_p, equals the combined storage coefficient from the aquifer test; the simulated value, P'_p, equals the simulated combined storage coefficient; and the two parameters involved, b_{SS1} and b_{SS2}, are specific storage values for each model layer. In this situation, there are two nonzero coefficients in Eq. (3.4): $a_{p,SS1}$, the coefficient for b_{SS1}, equals the thickness of layer 1; $a_{p,SS2}$, the coefficient for b_{SS2}, equals the thickness of layer 2.

Example 2: The distribution of hydraulic conductivity is expected to be smooth on the basis of an evaluation of depositional environment and hydraulic gradient. This smooth distribution is simulated by interpolating from a number of locations at which parameters are defined. Smoothness is imposed by introducing prior estimates, P_p, that equal zero; simulated values, P'_p, that equal the difference between parameter values at neighboring locations; and two parameters, b_{j1} and b_{j2}, that are involved in each prior information equation. Each equation has two nonzero values of $a_{p,j}$, one equal to 1.0 and one equal to -1.0. The prior information equations are therefore of the form

$$P'_p = b_{j1} - b_{j2}$$

Contributions to the objective function (Eq. (3.1) or (3.2)) are weighted differences between P_p, which equals 0.0, and P'_p. The variance of the error could be derived from geostatistical arguments, but to the authors' knowledge this has not been investigated.

Prior information must be used carefully. Two issues related to the use of prior information are discussed briefly here, and are further discussed in Section 5.5 and in Chapter 11 under Guideline 5. First, prior information on sensitive parameters can obscure important information available from the regression. This occurs when prior information is used to restrict the parameter estimate from becoming unreasonable during regression. However, unreasonable parameter estimates can lead to important insight about problems with the model or with the observations.

Second, for insensitive parameters in models with long forward execution times, it can be advantageous to set the parameter value equal to its prior estimate during regression, rather than estimating the parameter. This can significantly reduce execution times without substantially affecting the results. For final model runs, including the prior information and estimating the parameter allows the modeler to (1) assess whether the parameter value remains close to the prior value as expected, and (2) include the uncertainty of the parameter in the calculation of diagnostic statistics used to evaluate the regression and uncertainty in predictions.

As for observations, the model can be used to identify new prior information for which the cost of measurement would likely be a good or bad investment. See Section 8.2 and Guideline 11 under Chapter 13.

3.4.2 Weighting

The purpose of weighting is described in Section 3.3.3. For a diagonal weight matrix, Eq. (3.4) presents the requirement that weights of Eq. (3.1) need to be proportional to 1.0 divided by the variance of the data measurement error; that is, $\omega_{ii} \propto 1/\sigma_i^2$. Specifying the weights on the basis of the inverse of the error variance achieves the goal of emphasizing observations and prior information that are thought to be accurate relative to those that are thought to be inaccurate. It is always important to analyze data error. Weighting provides a way for that analysis to be formally included in model development.

An approach that is consistent with $\omega_{ii} \propto 1/\sigma_i^2$ is to define the weighting in an attempt to achieve the stricter requirement that:

$$\omega_{ii} = 1/\sigma_i^2 \qquad (3.8)$$

For a full weight matrix, the equivalent expression is

$$\boldsymbol{\omega} = \mathbf{V}(\boldsymbol{\varepsilon})^{-1} \qquad (3.9)$$

where $\mathbf{V}(\boldsymbol{\varepsilon})$ is the variance–covariance matrix of the observation errors, with variances along the diagonal and covariances off the diagonal. Setting the weights to be equal to, rather than proportional to, the right-hand sides results in some very useful properties, as described in Chapter 6, Section 6.3.2 and Guideline 6 in Chapter 11. Eq. (3.8) and (3.9) are used extensively in this book.

Most modelers can envision standard deviations or coefficients of variation more easily than variances, and MODFLOW-2000 and UCODE_2005 allow the user to specify these statistics to characterize error; the codes then calculate the variance internally. Examples of converting judgments about errors to standard deviations and coefficients of variation are discussed under Guideline 6. As noted there, if more than one source of error exists, the variance of each source needs to be determined and the variances need to be summed to obtain the final variance of the observation or prior information.

If the statistic (e.g., variance, standard deviation, or coefficient of variation) used to weight observations and prior information accurately reflects the uncertainty in the estimate, as suggested above, then (1) the observation or prior information can be viewed in a Bayesian sense and (2) measures of uncertainty produced by the model may reflect the actual uncertainty of the observations and prior information. For prior information, this issue was mentioned in Section 1.3.2 and in Section 3.4.1, and is discussed in more detail in Guideline 5 (Chapter 11).

3.5 LEAST-SQUARES OBJECTIVE-FUNCTION SURFACES

In some situations, errors in observations are not independent. For example, errors in streamflow gain and loss observations calculated from streamflow measurements can be correlated as discussed in Chapter 11 in Section G6.1 under the heading "Determine Covariances for Weight Matrices." These correlations indicate that the information present in the observations is redundant. Correlations close to 0.0 indicate little redundancy; correlations close to 1.0 or -1.0 indicate extreme redundancy. Although experience to date indicates that including the correlations in the weight matrix often has a minor effect on estimated parameter values, using the full weight matrix may be important to calculated uncertainties (see Chapter 11, Section G6.1).

3.4.3 Residuals and Weighted Residuals

Residuals are calculated as

$$[y_i - y'_i(\boldsymbol{b})] \tag{3.10}$$

and represent the match of the simulated values to the observations or prior estimates. For a diagonal weight matrix, weighted residuals are calculated as

$$\omega_i^{1/2}[y_i - y'_i(\boldsymbol{b})] \tag{3.11}$$

and represent the fit of the regression relative to the weights.

For a full weight matrix, weighted residuals are calculated as

$$\boldsymbol{\omega}^{1/2}[\boldsymbol{y} - \boldsymbol{y}'(\boldsymbol{b})]. \tag{3.12}$$

The square-root of the weight matrix is calculated such that $\boldsymbol{\omega}^{1/2}$ is symmetric (S. Christensen, Univ. of Aarhus, Denmark, written commun., 1996).

For weighting as suggested by Eq. (3.8) and (3.9) and discussed in Chapter 11 under Guideline 6, weighted residuals represent the fit of the regression in the context of the expected accuracy of the observations or prior estimates. Those expected to be less accurate are de-emphasized when weighted residuals are considered; those expected to be more accurate are emphasized.

3.5 LEAST-SQUARES OBJECTIVE-FUNCTION SURFACES

For one or two parameters, it is possible to plot the objective function and to easily diagnose any problems with its minimization. Objective-function surfaces for two parameters can be constructed through the following steps: (1) vary the values of the two parameters over selected ranges, (2) calculate the simulated equivalents of the observations for each set of parameters, (3) calculate the sum of weighted squared residuals (Eq. (3.1) or (3.2)) for each set of parameters, (4) plot these objective-function values against the two parameter values, and (5) contour the plotted values.

The objective-function surfaces resemble topographic maps except that instead of elevation above sea level, the topography is created by areas with lower and higher values of the objective function. Also, instead of coordinate direction, the "location" is characterized by the values of the parameters. The goal of regression is to identify the parameter values for which the objective-function is the smallest, which is analogous to finding the location of the lowest point in the landscape.

Figure 3.1a shows a simple two-parameter model and the distribution of the hydraulic heads calculated using the true values of transmissivity parameters T1 and T2. Figure 3.1b shows its weighted least-squares objective-function surface (for plotting convenience, the logarithm of this surface is shown) plotted against the log of T1 and T2. For a linear problem, the objective-function contours would be concentric ellipses or parallel straight lines symmetrically spaced about a trough. The nonlinearity of Darcy's Law with respect to hydraulic conductivity results in the much different shape shown in Figure 3.1b.

In practice, most models have more than two parameters and it is not possible to visualize the entire objective function. However, objective-function surfaces can be useful in two ways.

1. The model can be redesigned to be represented with only two parameters. For example, for a groundwater model one parameter can be defined that multiplies all the hydraulic-conductivity values in the system and a second parameter can be defined that multiplies all the recharge values in the system. The resulting objective-function surface can reveal extreme parameter correlation or other problems with multiple minima that exist but are difficult to detect when the system is represented using more parameters. This procedure is illustrated in Exercise 5.1.
2. For a problem with many defined parameters, objective-function surfaces can be used to evaluate pairs of parameters that are difficult to estimate.

With UCODE_2005, it is easy to create the data sets for such plots through the Investigate Objective Function mode. Similar data sets can be produced using PEST with SENSAN. There is no simple method to produce such data sets with MODFLOW-2000.

Objective functions for three or four dimensions can be represented using more sophisticated methods, but this is not considered here.

3.6 EXERCISES

Exercise 3.1: Steady-State Parameter Definition This exercise stresses the importance of checking the simulated values resulting from defined parameter values and correcting any errors in how the parameters are defined.

This exercise involves defining and checking parameters of the steady-state flow system described in Section 2.2. The flow system properties, parameter names, and the starting parameter values are shown in Table 3.1. The conductance of the head-dependent boundary adjacent to the hillside (see Figure 2.1a and 2.1b) is not estimated because this property has a minor effect on the flow system, as shown by the small amount of flow that enters this model boundary (Figure 2.1c).

3.6 EXERCISES

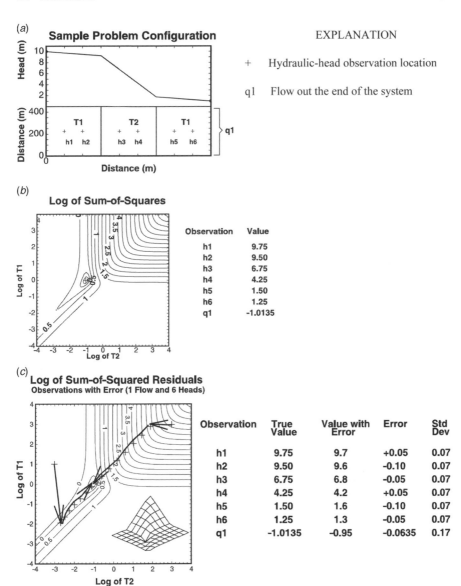

FIGURE 3.1 Objective-function surfaces for a simple model. (*a*) One-dimensional porous-media flow field bounded by constant heads on the left and right and consisting of three transmissivity zones and two transmissivity values T1 and T2. Hydraulic heads calculated using the true parameter values are shown. (*b*) Logarithm of the weighted least-squares objective function that includes observations of hydraulic heads h1 through h6, in meters, and flow q1, in cubic meters per second. The observations contain no error. (*c*) Logarithm of the weighted least-squares objective function using observations with error, and a three-dimensional portrayal of the objective-function surface. Sets of parameter values produced by modified Gauss–Newton nonlinear regression iterations are identified (+), starting from two sets of starting values and progressing as shown by the arrows. (From Poeter and Hill, 1997.)

TABLE 3.1 Parameter Name and Starting Value for Properties of the Steady-State Flow System for Which Parameters Are Estimated in Subsequent Exercises

Flow System Property	Parameter Name	Starting Value[a]
Horizontal hydraulic conductivity of layer 1, in m/s	HK_1	3.0×10^{-4}
Hydraulic conductivity of the riverbed, in m/s	K_RB	1.2×10^{-3}
Vertical hydraulic conductivity of confining bed, in m/s	VK_CB	1.0×10^{-7}
Horizontal hydraulic conductivity of layer 2 in columns 1 and 2, in m/s	HK_2	4.0×10^{-5}
Recharge in recharge zone 1, in cm/yr	RCH_1	63.072
Recharge in recharge zone 2, in cm/yr	RCH_2	31.536

[a] Five significant digits are used for recharge because of a units conversion.

All work for Exercise 3.1 involves modifying computer files and simulating the system. Instructions are available from the web site for this book described in Section 1.1. Students who are not performing the simulations may skip Exercise 3.1.

Exercise 3.2: Observations for the Steady-State Problem In this exercise, observations of the steady-state flow system described in Section 2.2 are defined and checked, and weights on the observations are defined and calculated.

The hydraulic-head observations used for the steady-state system are listed in Table 3.2. Their locations are shown in Figure 2.1b. All head observations are from wells located at the centers of model cells. In addition, there is one flow observation equal to the groundwater discharge to a river reach. The reach extends along the entire length of the river, and the gain in streamflow is 4.4 m³/s.

(a–b) Define observations in model input files.

Exercises 3.2a–b involve modifying computer files and simulating the system. Instructions are available from the web site for this book described in

TABLE 3.2 Hydraulic-Head Observations

Well Identifier	Observation Name	Layer	Row	Column	Observed Head (m)	Variance of Well Elevation Measurement Error (m²)	Variance of Water-Level Measurement Error (m²)	Variance of the Observation Error (m²)
1	hd01.ss	1	3	1	101.804	1.00	0.0025	1.0025
2	hd02.ss	1	4	4	128.117	1.00	0.0025	1.0025
3	hd03.ss	1	10	9	156.678	1.00	0.0025	1.0025
4	hd04.ss	1	13	4	124.893	1.00	0.0025	1.0025
5	hd05.ss	1	14	6	140.961	1.00	0.0025	1.0025
6	hd06.ss	2	4	4	126.537	1.00	0.0025	1.0025
7	hd07.ss	2	10	1	101.112	1.00	0.0025	1.0025
8	hd08.ss	2	10	9	158.135	1.00	0.0025	1.0025
9	hd09.ss	2	10	18	176.374	1.00	0.0025	1.0025
10	hd10.ss	2	18	6	142.020	1.00	0.0025	1.0025

3.6 EXERCISES

```
DATA AT HEAD LOCATIONS
                     OBER-   SIMUL.
        OBSERVATION  VATION  EQUIV.                            WEIGHTED
OBS#       NAME        *       *     RESIDUAL   WEIGHT**.5   RESIDUAL
  1      hd01.ss      102.    100.     1.58       0.999         1.58
  2      hd02.ss      128.    139.   -11.2        0.999       -11.2
  3      hd03.ss      157.    174.   -17.7        0.999       -17.7
  4      hd04.ss      125.    139.   -14.4        0.999       -14.4
  5      hd05.ss      141.    157.   -16.2        0.999       -16.2
  6      hd06.ss      127.    140.   -13.1        0.999       -13.1
  7      hd07.ss      101.    103.    -1.76       0.999        -1.75
  8      hd08.ss      158.    174.   -15.8        0.999       -15.8
  9      hd09.ss      176.    190.   -13.9        0.999       -13.9
 10      hd10.ss      142.    157.   -15.0        0.999       -15.0
-----------------

DATA FOR FLOWS REPRESENTED USING THE RIVER PACKAGE
        OBSERVATION  MEAS.   CALC.                             WEIGHTED
OBS#       NAME      FLOW    FLOW   RESIDUAL   WEIGHT**.5    RESIDUAL
 11      flow01.ss   -4.40   -4.86    0.461      2.27           1.05
```

FIGURE 3.2 Part of MODFLOW-2000 LIST output file showing initial model fit and weights for the head and flow observations.

Section 1.1. Students who are not performing the simulations may skip Exercises 3.2a–b.

(c) *Check simulated values.*

Observed hydraulic heads and flows and their simulated equivalents in the initially constructed model are shown in Figure 3.2.

Problem: Using Figure 3.2, does the model fit suggest data input error?

(d) *Calculate weights on hydraulic-head and flow observations.*

As discussed in Sections 3.3 and 3.4.2 and in Chapter 11 under Guideline 6, assignment of weights requires an analysis of the likely accuracy of the observations. In the simple model used for these exercises, this assignment is easier than usual because any deviation from the accurate simulated values has been added intentionally. More realistic situations are discussed elsewhere in this book, including in Guideline 6 and Chapter 15.

The observed heads of Table 3.2 were generated by simulating hydraulic head using the true model and adding randomly generated noise with known variance. The added noise has the following characteristics:

1. The elevation of each observation well has a mean error of 0.0 and a variance of 1.0, as shown in Table 3.2.

2. In addition, each head measurement has an error associated with the water-level measurement method. This error has a mean of zero and a variance of 0.0025, as shown in Table 3.2.
3. The flow has an error with a mean of zero and a coefficient of variation of 10 percent.

Problem

- Compute the weights for each observation from the values specified for the head observations in Table 3.2 and the flow observation in the text above. Calculate the weights as the inverse of the observation variance. The final variance equals the sum of the variances of the components.
- Check your calculations against the weights printed in output files from Exercise 3.2b or using the output shown in Figure 3.2.

Exercise 3.3: Evaluate Model Fit Using Starting Parameter Values This exercise involves assessing initial model fit. If the evaluation from Exercise 3.2 indicates no problems, the model fit resulting from the starting parameter values is worth evaluating. Use the tables of observed and simulated hydraulic heads and flows located in the output files from Exercise 3.2, which are shown in Figure 3.2, and are produced by students performing the exercises.

Problem

- Comment on the model fit achieved with the starting parameter values.
- How do the residuals compare to the weighted residuals?

For students performing the model simulations, do the following parts of this exercise.

- Attempt to achieve a better model fit by changing the parameter values manually, using your knowledge about the behavior of the groundwater flow system. Make changes three to twelve times. Each time, document the lack of fit being addressed, the reason the change attempted was expected to address that lack of fit, whether or not the change produced the expected results, and whether there were any unexpected, welcome, or unwelcome consequences.
- When finished, restore the starting parameter values.

4

DETERMINING THE INFORMATION THAT OBSERVATIONS PROVIDE ON PARAMETER VALUES USING FIT-INDEPENDENT STATISTICS

This chapter focuses on selected sensitivity analysis methods that measure the information that observations provide for defining parameters and estimating parameter values. The sensitivity analysis described in this chapter uses what are herein called *fit-independent statistics*. The statistics are fit independent in that residuals (Eq. (3.7)) are not used to calculate these statistics—only sensitivities and the weighting are used. Sensitivities are defined in Section 4.3.1.

Sensitivity analysis is a very broad field. This book includes some sensitivity methods that are not common in other text books, such as the fit-independent statistics, and there are many methods that are not presented in this book. Other sensitivity analysis methods are presented by, for example, Saltelli et al. (2000, 2004). The methods presented in this book are generally classified as local methods because they use sensitivities calculated for one set of parameter values. They are most useful if the model is not too nonlinear with respect to the parameter values. In most circumstances the methods presented have been found to be useful; models apparently have to be extremely nonlinear for the methods to fail completely.

This chapter focuses on sensitivity analysis methods using fit-independent statistics that measure the information provided by observations for parameter values. The methods are discussed again in Chapter 7 along with fit-dependent statistics that serve the same basic purpose. Chapter 8 introduces fit-independent statistics for evaluating the information observations provide on predictions and the importance of parameter values to predictions. Thus, fit-independent statistics can be used

---

Effective Groundwater Model Calibration: With Analysis of Data, Sensitivities, Predictions, and Uncertainty. By Mary C. Hill and Claire R. Tiedeman
Published 2007 by John Wiley & Sons, Inc.

to evaluate each link of the observation–parameter–prediction sequence connected quantitatively by the model, as discussed in Chapters 1 and 10.

This chapter begins by discussing how observations provide information about model construction and parameter definition, as well as providing information about parameter values. Then, sensitivities are defined mathematically and conceptually, the importance of scaling the sensitivities is discussed, and fit-independent statistics are presented. Finally, advantages and limitations of fit-independent statistics are discussed.

4.1 USING OBSERVATIONS

Observations are used to construct models, define parameters, and estimate parameter values. These roles are discussed briefly here and more in Chapter 11.

4.1.1 Model Construction and Parameter Definition

Observations provide information about model construction and parameter definition (also called parameterization) as well as about the value of model parameters. Observations provide information about what dynamics and features of a system are important. For example, consider the following circumstances.

1. Hydraulic-head observations indicate smooth spatial changes in hydraulic gradient in a groundwater system. Given the geologic history and hydraulic conditions of the system, it is suspected that the gradual changes in hydraulic gradient reflect a hydraulic-conductivity distribution that varies gradually. Such a distribution might be well represented in a model using an interpolation method in which the hydraulic-conductivity values at the interpolation points are defined as parameters.
2. Hydraulic-head observations indicate abrupt spatial changes in hydraulic gradient under natural conditions. Given the geologic history and hydraulic conditions of the system, it is suspected that the abrupt changes in hydraulic gradient reflect a hydraulic-conductivity distribution that varies abruptly. Such a distribution might be well represented using a zonation method in which the hydraulic-conductivity values in each of several hydrogeologic units are defined as parameters.
3. Concentration observations suggest that near its source, a plume in a groundwater system sinks significantly in a short distance and then sinks very slowly as it spreads and moves downstream toward the northeast. The mass of the plume appears to diminish with time. This situation is likely to result from density effects at high concentrations, the effects of areal recharge over time creating small downward vertical velocities, and the effects of advection, dispersion, and decay. The groundwater flow model boundary conditions, hydraulic-conductivity field, and so on need to reproduce these conditions as the plume moves.

The process by which observations are used to construct a model is dominated by professional judgment and often by trying different options in an ad hoc fashion. When the choices are discrete they do not lend themselves to gradient-based optimization methods. To the extent that the options involve parameters, the methods described in this chapter can be used as measures of the information provided by the observations on the parameter value. The information provided by the observations on, for example, system processes often can be inferred from the information provided on the parameter value. A complementary way of expressing this is that the importance indicated by a sensitivity analysis related to the parameters often can be used to infer the importance of a related process.

4.1.2 Parameter Values

For any given calibration, most of the effort generally is spent trying to use the information provided by the observations to adjust parameter values. To make this task more meaningful and manageable, this book suggests that nonlinear regression be used to estimate parameter values given a set of observations. The ability of the regression to precisely estimate a set of parameters is related to the information the observations provide on the parameter values. The statistics typically used in nonlinear regression to determine how well parameters are estimated are defined in linear regression textbooks such as Draper and Smith (1998). These include p-values, t-statistics, and so on. These statistics depend on model fit being optimized and are calculated after regression is successfully completed. For models with lengthy execution times, methods that do not require completion of regression can be very helpful, and the fit-independent statistics presented in this chapter are designed to serve this purpose. The fit-independent statistics are closely related to standard statistics, as noted in the subsequent discussion.

The information that the observations provide about model construction and parameter definition is difficult to quantify. One option is to construct and estimate parameters for a variety of plausible alternative models, as discussed under Guideline 8 in Chapter 11 and in the context of prediction uncertainty under Guideline 14 in Chapter 14. This approach has the advantage of accounting for model nonlinearity and the disadvantage of sometimes requiring unattainable computer resources.

An approximate approach to evaluating the information observations provide about model construction is to assume that if observations provide a large amount of information about a parameter value, then they also are likely to provide a large amount of information about the model construction related to that parameter, including how the parameter is defined. Often such a conclusion is valid and therefore focusing attention on model construction and parameter definition in areas of high parameter value sensitivity can help improve model fit to the observations. In nonlinear models, however, exceptions will occur. For example, if the material blocking groundwater flow in one part of the simulated system has an extremely small value of simulated hydraulic conductivity, most of the measures of importance discussed in this chapter will tend to be small. As more moderate values of hydraulic conductivity are simulated, the sensitivities can increase if the blockage is important to reproducing the dynamics represented by the observations. Sometimes such exceptions can be identified through understanding of the flow system.

Thus, a way of assessing the information provided by observations about different aspects of model construction and parameter definition is to define parameters that control those aspects. For example, parameters can be defined to control the thickness of hydrogeologic units, to position points used in interpolation, or to position zone boundaries. This has been done, for example, by Zheng and Wang (1996) and Tung and Chou (2002). The sensitivity analysis methods presented in this work are, therefore, limited in their generality only by the parameters the user chooses to define.

While parameters can be defined to represent any aspect of a simulated system, there are advantages to defining parameters frugally and carefully, as discussed in Guidelines 1 and 3 in Chapter 11. Generally, some types of parameters are not defined because these aspects of the system are better supported by independent data and/or are less important to fitting observations than other types of parameters. For example, in groundwater systems the product of hydraulic conductivity and hydrogeologic-unit thickness is important. However, it is more common to define parameters to represent hydraulic conductivity, which can vary over many orders of magnitude, than hydrogeologic-unit thickness, which is often known within 50 percent or less. Focusing on parameters that represent the least known and most important aspects of a system is a good strategy in most circumstances.

4.2 WHEN TO DETERMINE THE INFORMATION THAT OBSERVATIONS PROVIDE ABOUT PARAMETER VALUES

Determining the information that observations provide toward estimating parameter values is valuable throughout model development. This analysis can help make the most of every model run and, therefore, becomes increasingly important as models require greater execution time. Determining observation information with respect to parameter values is most commonly used to:

- Decide what observations to include
- Design the defined parameters
- Decide which of the defined parameters to estimate
- Evaluate which potential new observations are important to the parameters
- Evaluate how the analysis is affected by model nonlinearity

These issues are discussed briefly here to motivate the rest of the chapter and to provide perspective, and in more detail in Chapter 11 under Guidelines 3 and 4. Examples of these issues are presented in Chapter 15.

4.2 WHEN TO DETERMINE INFORMATION THAT OBSERVATIONS PROVIDE

Three issues need to be considered when determining which observations to include in a regression. The first is that some observations may be affected by processes that are not simulated. For example, hydraulic heads may reflect perched conditions which are typically not simulated by saturated groundwater flow models. Omission of this category of observations needs to be based on the relevance of the observations to the simulated processes and the importance of the omitted processes to the predictions of interest. This analysis can often be addressed using the sensitivity analysis methods described in this book. The second issue is that observations are often clustered, and culling of clustered observations is one mechanism to consider. The effects of clustering can be evaluated in part with sensitivity analysis, as discussed in Chapter 11 under Guideline 4. The third issue occurs when there is the opportunity to conduct additional field work and the information provided by a potential observation is important to prioritizing the field effort. The importance of the potential observation to parameter estimates can be evaluated using the methods described in this chapter; its importance to predictions can be evaluated using the methods described in Chapter 8.

When designing the defined parameters, observations commonly are more reliable indicators of system dynamics than other types of data. In many groundwater systems, for example, observations of hydraulic heads, flows, concentrations, and so on are more reliable indicators of system properties than are direct measurements of those properties. This is mostly because of problems with accessibility and scale. For example, these problems make it difficult to obtain accurate measurements of hydraulic conductivity and produce inconsistencies between the scale of most hydraulic-conductivity measurements and the scale of the model (e.g., see Barth et al., 2001; Barlebo et al., 2004). Thus, although hydraulic-conductivity measurements are valuable, it is important to consider them in the context of their likely errors and the errors of other available data. The fit-independent statistics help in this evaluation.

The fit-independent statistics described in this chapter can be used to determine the parameters that are well supported by the observations, which is important when designing defined parameters. For example, a groundwater modeler may be interested in the detail supported by the observations for the hydraulic-conductivity distribution of a system. In parts of the system where observations provide abundant information, more parameters generally can be supported; where observations provide little information, fewer parameters generally can be supported. When deciding whether to define additional parameters, it is important to know how much the new parameters depend on the observation data, accounting for the effects of observation error.

Deciding which of the defined parameters to estimate is important because for models with long execution times, regression runs can be lengthy. Execution times can be reduced by excluding from the regression insensitive parameters (those for which the observations provide very little information) or selected correlated parameters (those for which the observations do not provide unique information for each value). Exclusion of such parameters improves the performance of the regression, rarely affects regression results, and reduces execution times. When this strategy is followed for nonlinear models, it is important to recalculate the fit-independent statistics occasionally using updated parameter values, because, as noted at the end of this section, the value of the statistics will change as parameter values change.

Evaluating which potential new observations are important to the parameters is a valuable step for guiding collection of additional field data and can be done using the sensitivity statistics presented in this chapter. These statistics produce fit-independent measures of the information that individual potential observations provide about individual or sets of parameters. A thorough discussion of using the statistics in this context is given in Guideline 11 of Chapter 13.

Alternative methods for selecting new observations that improve the parameter estimates use criteria related to minimizing parameter uncertainty (e.g., Knopman and Voss, 1988, 1989; Nishikawa and Yeh, 1989). These methods often involve the design of observation networks and thus generally focus on identifying sets of observations, rather than on examining the information provided by individual observations. These methods use many of the same measures of parameter uncertainty that are used by the statistics presented here, so results are expected to be similar. Recent work on monitoring network design methodologies has tended to focus on minimizing prediction, rather than parameter, uncertainty. This topic is discussed in Chapter 8.

For nonlinear models, different sensitivities are calculated for different parameter values, as discussed in Section 1.4. If a model is too nonlinear, the sensitivities vary so much that fit-independent statistics calculated from them become useless for the purposes discussed here. However, experience to date has shown that for most nonlinear models of groundwater systems, the statistics presented here have been found to be very useful. Some examples are discussed in Guideline 3 (Chapter 11), in Guideline 11 (Chapter 13), and in Chapter 15.

4.3 FIT-INDEPENDENT STATISTICS FOR SENSITIVITY ANALYSIS

Fit-independent statistics are calculated using sensitivities, which are defined in the following section. Subsequent sections define scaling and five fit-independent statistics.

Fit-independent statistics are measures of *leverage*—the potential for an observation to make a difference based on the observation sensitivities. In contrast, *influence statistics* measure the actual difference. The actual effect depends on the observed value, and, therefore, influence statistics depend on model fit. Leverage and influence statistics are discussed in Chapter 7, Section 7.3. Many of the fit-independent statistics described here are compared to influence statistics and the results of cross-validation in Foglia et al. (in press). The issue of observation importance to parameters spans the first two components of the observation–parameter–prediction triad composed of entities that are directly connected by the model, as discussed in Chapters 1 and 10.

4.3 FIT-INDEPENDENT STATISTICS FOR SENSITIVITY ANALYSIS

4.3.1 Sensitivities

Sensitivities are calculated as the derivatives of simulated equivalents to observations (such as simulated hydraulic heads and flows) with respect to the model parameters. That is,

$$\left(\frac{\partial y'_i}{\partial b_j}\right)\bigg|_b \quad (4.1)$$

where y'_i is defined after Eq. (3.1b) as the simulated value that corresponds to an observation or item of prior information, b_j is the jth parameter, and the notation indicates that the sensitivities are calculated for the parameter values listed in vector b. The latter is important because for nonlinear problems the sensitivities are different when calculated for different parameter values. For this reason, the sensitivities of Eq. (4.1) are called local sensitivities by Saltelli et al. (2000). This issue is discussed in Section 4.4.

Some models, such as MODFLOW-2000, calculate the derivatives using sensitivity-equation sensitivities (Hill et al., 2000, pp. 67–71) or adjoint states (Thomas Clemo, Boise State University, written communication, 2004) which produce the most accurate sensitivities; other models, such as UCODE_2005 and PEST, approximate sensitivities using forward, backward, or central differences. For example, the forward-difference approximation to Eq. (4.1) is

$$\left(\frac{\partial y'_i}{\partial b_j}\right)\bigg|_b \approx \left(\frac{y'_i(b + \Delta b) - y'_i(b)}{\Delta b_j}\right) \quad (4.2)$$

where Δb is a vector of zeros except that the jth element equals Δb_j. Equation (4.2) is calculated by running the model once using the parameter values in b to obtain $y'_i(b)$, then again after changing the jth parameter value to obtain $y'_i(b + \Delta b)$, and finally taking the difference and dividing by the change in the jth parameter value. Execution time issues for calculating sensitivities are discussed in Chapter 15, Section 15.1. Accuracy issues are discussed in Sections 2.1, 4.4, and 7.4, and by Yager (2004)

The sensitivities indicate the slope of a plot of a simulated value y'_i relative to one parameter or, approximately, how much a simulated value would change if a parameter value were changed, divided by the change in the parameter value. The parameters are considered individually; sensitivities do not account for changes in multiple parameters.

Sensitivities can be used to indicate the importance of the observations to the estimation of parameter values. Observations are likely to be very valuable in estimating a parameter value if their simulated equivalents change substantially given a small change in the parameter value; observations contribute very little to estimating a parameter if their simulated equivalents change very little even with a large change in the parameter value.

4.3.2 Scaling

Generally, it is useful to compare the relative importance of different observations. A problem with making this comparison using sensitivities is that sensitivities are in the units of the simulated value divided by the units of the parameter, both of which can vary considerably. For example, for groundwater models the simulated values might be hydraulic heads measured in meters, flows measured in cubic meters per day, and concentrations measured in milligrams per liter; parameters might be hydraulic conductivity measured in meters per day and recharge measured in millimeters per year. The solution pursued here and by others is to scale the sensitivities to achieve quantities with the same units. The scaling used depends on the intended purpose of the resulting scaled sensitivities.

In both MODFLOW-2000 and UCODE_2005, scalings are used to produce dimensionless scaled sensitivities (dss) that are accumulated for each parameter to produce composite scaled sensitivities (css). Composite scaled sensitivities provide information about individual parameters, but cannot be used to evaluate whether a set of observations can estimate each parameter uniquely. Problems with uniqueness occur when coordinated changes in parameter values produce the same fit to the observations. Parameter correlation coefficients (pcc) indicate whether observations provide information for estimating parameters uniquely. Leverage statistics reflect the importance of observations on the basis of the effects measured by both css and pcc. Finally, one-percent scaled sensitivities ($1ss$) can be used to produce sensitivity maps, but there are difficulties with this scaling. These statistics are discussed below.

4.3.3 Dimensionless Scaled Sensitivities (dss)

When a diagonal weight matrix is used, dimensionless scaled sensitivities, dss_{ij}, are calculated as (Hill, 1992; Hill et al., 1998)

$$dss_{ij} = \left(\frac{\partial y'_i}{\partial b_j}\right)\bigg|_{b} |b_j| \omega_{ii}^{1/2} \qquad (4.3a)$$

where $y'_i =$ a simulated value. Here the notation indicates that the simulated value is associated with an observation (the ith observation), but similar scaling can be used with sensitivities of other quantities, such as potential observations. This is discussed in Guideline 12 in Chapter 13.

$b_j =$ the jth estimated parameter.

$\left(\frac{\partial y'_i}{\partial b_j}\right) =$ the derivative, or sensitivity, of the simulated value associated with the ith observation with respect to the jth parameter, evaluated at the set of parameter values in $\boldsymbol{b}$.

$\boldsymbol{b} =$ a vector that contains the parameter values at which the sensitivities are evaluated; for nonlinear models, sensitivities will be different for different values in $\boldsymbol{b}$.

$\omega_{ii} =$ the weight of the ith observation.

Similar scaling was used by Cooley et al. (1986) and Harvey et al. (1996).

4.3 FIT-INDEPENDENT STATISTICS FOR SENSITIVITY ANALYSIS

For log-transformed parameters, Eq. (4.3b) can be used to reflect the improved regression performance produced by log transformation. However, use of Eq. (4.3b) means that dss and css can vary considerably between model runs if the transformed parameters change. MODFLOW-2000 and UCODE_2005 use Eq. (4.3b).

$$dss_{ij} = \left(\frac{\partial y'_i}{\partial (\ln b_j)}\right)\bigg|_b |\ln(b_j)|\omega_{ii}^{1/2} = \left[\left(\frac{\partial y'_i}{\partial b_j}\right)\bigg|_b b_j\right]|\ln(b_j)|\omega_{ii}^{1/2} \qquad (4.3b)$$

To better understand the dimensionless scaled sensitivity, consider Eq. (4.3a) with the square root of the weight replaced by $1/\sigma$, where σ is the standard deviation of the observation error (as discussed in Chapter 3, Section 3.4.4 and in Guideline 6 in Chapter 11, it is advantageous to define the observation weights as $\omega_{ii} = 1/\sigma^2$). Also, divide and multiply the equation by 100, to achieve

$$dss_{ij} = \left(\frac{\partial y'_i}{\partial b_j}\right)\bigg|_b \left|\frac{b_j}{100}\right|\left(\frac{100}{\sigma}\right) \qquad (4.4)$$

By Eq. (4.4), the dimensionless scaled sensitivity indicates the amount the simulated value would change, expressed as a percent of the observation error standard deviation, given a one-percent increase in the parameter value. If $dss_{ij} = 1$, a one-percent change in the parameter value, b_j, would produce a change in the simulated value, y'_i, equivalent to one percent of the standard deviation of measurement error, σ. If $dss_{ij} = 10$, a one-percent change in the parameter value would produce a change in the simulated value equivalent to 10 percent of σ. Thus, dimensionless scaled sensitivities include the effects of sensitivity and of observation error. This is discussed further in Section 4.3.4 on composite scaled sensitivities.

The dimensionless scaled sensitivities can be used in two ways.

First, they can be used to compare the importance of different observations to the estimation of a single parameter b_j. Observations with large dss_{ij} are likely to provide more information about parameter b_j compared to observations associated with small dss_{ij} (large and small in absolute value). Also, observations with large dss_{ij} can be considered more important to the estimation of parameter b_j.

Second, the dimensionless scaled sensitivities can be used to compare the importance of different parameters to the calculation of a single simulated value y'_i. Parameters that are more important to the simulated value have dss_{ij} that are larger in absolute value. An example of using dimensionless scaled sensitivities is provided in Exercise 4.1b.

For a full weight matrix, dimensionless scaled sensitivities are calculated as

$$dss_{ij} = \sum_{k=1}^{ND}\left[\left(\frac{\partial y'_k}{\partial b_j}\right)\bigg|_b b_j(\omega^{1/2})_{ki}\right] \qquad (4.5)$$

where ND is the number of observations used in the regression, $\omega^{1/2}$ is the square root of the weight matrix determined such that $\omega^{1/2}$ is a symmetric matrix, and $(\omega^{1/2})_{ki}$ is the matrix element in row k and column i.

In Eq. (4.5), the dimensionless scaled sensitivity dss_{ij} for simulated value y'_i is a function of the sensitivity of y'_i with respect to b_j as well as a function of the sensitivities of other simulated values y'_k, $k \neq i$, with respect to b_j. For observations with errors that are correlated with the error of observation y_i, $(\omega^{1/2})_{ki} \neq 0.0$. For observations with errors that are not correlated with the error of observation y_i, $(\omega^{1/2})_{ki} = 0.0$. In practice, the off-diagonal terms of $(\omega^{1/2})_{ki}$ are likely be smaller than the diagonal terms, so the contribution of the sensitivity term $\partial y'_i/\partial b_j$ to dss_{ij} is likely to be greater than the contribution of $\partial y'_k/\partial b_j$, $k \neq i$.

The fit-dependent equivalent to the dimensionless scaled sensitivity is the statistic DFBETAS presented in Chapter 7, Section 7.5.2.

4.3.4 Composite Scaled Sensitivities (css)

Composite scaled sensitivities reflect the total amount of information provided by the observations for the estimation of one parameter. They are calculated for each parameter using dimensionless scaled sensitivities and can be calculated for some or all observations. The composite scaled sensitivity, css_j, for the jth parameter calculated for ND observations is (Hill, 1992; Anderman et al., 1996; Hill et al., 1998)

$$css_j = \sum_{i=1}^{ND} \left[(dss_{ij})^2|_b /ND \right]^{1/2} \qquad (4.6)$$

where the quantity in parentheses equals a dimensionless scaled sensitivity of Eq. (4.3) or (4.5). The composite scaled sensitivity is equal to a scaled version of the square root of the diagonal of $X^T\omega X$, which is the regression variance times the Fisher information matrix (Burnham and Anderson, 2002); X is a matrix of the sensitivities defined in detail following Eq. (5.2). Statistics that perform a similar function are the L_1 norm of sensitivities used by R. L. Cooley (U.S. Geological Survey, written communication, 1988) and the CTB statistic of Sun and Yeh (1990a) and Sun (1994). The CTB statistic is scaled using the weight on prior information for parameter b_j instead of the parameter value b_j as in Eq. (4.3) or (4.5).

Often composite scaled sensitivities are used in a comparative manner, whereby larger values indicate parameters for which the observations provide more information. If there are composite scaled sensitivities that are less than one percent of the largest value, regression often will have trouble converging. In this situation, the values of parameters with small composite scaled sensitivities may need to be assigned prior information or have the parameter specified rather than estimated by the regression (see Guideline 5 in Chapter 11).

Composite scaled sensitivities are also meaningful individually. By using Eq. (4.4), they can be interpreted as the average amount that the simulated values change, expressed as a percent of the standard deviation of the observation error, given a one-percent change in the parameter value. This interpretation of the composite scaled sensitivity shows clearly that a parameter can be estimated only if the information provided by the observations, as expressed through their sensitivities, dominates the effects of observation error (noise in the data). The information provided by the sensitivities is related to the observation types, locations, measurement times, and system conditions. If a css_j value is too small, the observation data may be too noisy relative to the sensitivity information provided, and the regression may not be able to estimate a value of b_j. An example of using composite scaled sensitivities is provided in Exercise 4.1b.

Linear regression can be used to illustrate the interaction between the noise in the data and the sensitivity information that the observations provide by virtue of their type, location, and time. In linear regression, values of the independent variable, X, play a role similar to the sensitivities in nonlinear regression. The effect of the sampled range of X values on linear regression behavior is analogous to the effect of the range of observation types, locations, times, and system conditions on nonlinear regression behavior. In linear regression, the amount of noise in the data that can occur while still enabling accurate estimation of the parameters depends on the range of X values sampled. As this range increases, the regression can detect a trend in the data (and estimate parameters of the linear model) in the presence of a greater amount of data error. Figure 4.1 illustrates this concept by showing a linear model plotted with three different sets of observation data. In Figure 4.1a, the noise in the data overwhelms the information provided by the data locations (the range of X values), and consequently it is difficult to discern a trend in the data. In Figure 4.1b, the data locations are the same, but there is less noise in the data, and the trend in the data is thus much more discernable. In Figure 4.1c, the noise level is the same as in Figure 4.1b, but the information content of the observations is reduced by reducing the range of X, and the noise level again overwhelms the information that the observations provide.

The interaction between the information content of the observations, as reflected in their sensitivities, and the noise in the observations suggests that there is some css_j value below which the observations provide insufficient information to estimate parameter b_j. Although experience to date has not clearly identified this critical value, we suggest a value of 1.0. A css_j value of 1.0 means that a one-percent change in the parameter value produces, on average, a change in simulated values that is equivalent to one percent of the measurement error standard deviation. Parameter values with composite scaled sensitivities less than 1.0 are more likely to be poorly estimated, in that confidence intervals are large and regression convergence problems are persistent.

4.3.5 Parameter Correlation Coefficients (pcc)

Parameter correlation coefficients (pcc) used in conjunction with composite scaled sensitivities produce a useful sensitivity analysis. Parameter correlation coefficients are calculated as $\text{Cov}\{\boldsymbol{b}\}_{jk}/[\text{Var}\{\boldsymbol{b}\}_{jj} \text{Var}\{\boldsymbol{b}\}_{kk}]$, where $\text{Cov}\{\boldsymbol{b}\}_{jk}$ is the covariance between two parameters and $\text{Var}\{\boldsymbol{b}\}_{jj}$ and $\text{Var}\{\boldsymbol{b}\}_{kk}$ are the variances of each of

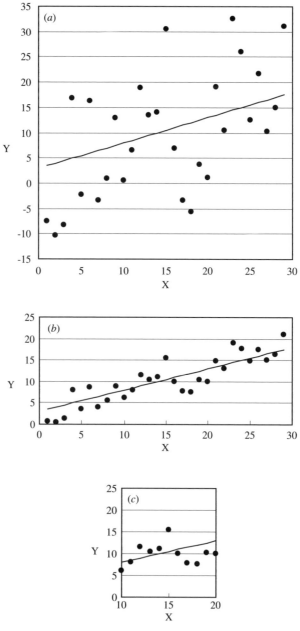

FIGURE 4.1 A single linear model, $y = b_0 + b_1 X$, with three sets of data. The true parameter values are $b_0 = 3$ and $b_1 = 0.5$. (a) The noise in the data has a standard deviation of $\sigma = 15$. (b) The data are at the same X values, and $\sigma = 5$. (c) The noise level is the same as in (b), but the range of X values is reduced.

4.3 FIT-INDEPENDENT STATISTICS FOR SENSITIVITY ANALYSIS

the parameters. Further discussion of *pcc* is presented in Chapter 7, Section 7.2.1, because it is closely associated with the parameter statistics discussed there. Limitations of *pcc* are discussed in Section 4.4.2. Here we present comments needed to support wise use of parameter correlation coefficients in sensitivity analysis.

The *pcc* are calculated for each possible pair of model parameters. They indicate whether parameter values can be estimated uniquely by regression, given the constructed model and the observations and prior information provided. The *pcc* values can vary from -1.00 to $+1.00$. The *pcc* for a parameter with itself is always 1.00. If the *pcc* for a pair of parameters is equal to or very close to -1.00 or $+1.00$, the two parameters generally cannot be estimated uniquely. Extreme correlation between more than two parameters is indicated if *pcc* values for all pairs of the parameters involved are near -1.00 and $+1.00$ and indicates that the parameters involved generally cannot be estimated uniquely. If the absolute values of all *pcc* are less than about 0.95, then it is likely that all parameter values can be estimated uniquely. However, this is a rule of thumb; experience has shown that unique estimates sometimes can be obtained even with absolute values of *pcc* that are very close to 1.00.

Correlation coefficients are typically displayed as a matrix. This matrix is always symmetric, with diagonal elements equal to 1.00. For example,

	PAR1	PAR2	PAR3
PAR1	1.00	0.96	0.05
PAR2	0.96	1.00	0.98
PAR3	0.05	0.98	1.00

Here, PAR1, PAR2, and PAR3 are parameter names. Alternatively, the large values can be listed in a table, such as

Parameter Pair	Correlation Coefficient
PAR1–PAR2	0.96
PAR2–PAR3	0.98

This table lists all *pcc* values greater than 0.95

In the global output file for MODFLOW-2000 and the main output file for UCODE_2005, the full parameter correlation coefficient matrix is printed, followed by a list of values larger in absolute value than 0.85. Moderate correlations between 0.85 and 0.95 are included because parameter correlation coefficients can change substantially during calibration and it is useful to know whether a previously high correlation has become just moderately high or low. In addition, moderate parameter correlations can contribute to large confidence intervals in some circumstances.

An example of evaluating *pcc* as part of a sensitivity analysis is provided in Exercise 4.1c. Additional exercises on parameter correlation include

Exercise 5.1a, which uses objective-function surfaces for a two-parameter problem to investigate the performance of regression in the presence of extreme parameter correlation, and Exercise 7.1f, which uses regression from different starting parameter values to test the uniqueness of parameter estimates with correlation coefficients very close to 1.00.

Parameter correlation can also be evaluated using eigenanalysis and singular value decomposition. These alternatives are discussed in Chapter 7, Section 7.2.5. In this book, we focus on parameter correlation coefficients because in most cases they are easier for most modelers and resource managers to understand and critical values for identifying nonunique estimates are clearer. The alternative methods do not offer enough advantage to overcome these considerations.

It is important to assess the *pcc* in conjunction with the scaled sensitivities and leverage statistics before proceeding with regression. Limitations of *pcc* are discussed in Section 4.4.2. If the *pcc* for one or more parameter pairs is very close to +1.00 or −1.00, the regression may be unable to uniquely estimate the extremely correlated parameters. Options for addressing this situation are discussed in Chapter 7, Section 7.4.

4.3.6 Leverage Statistics

Leverage statistics combine the information provided by the *dss*, *css*, and *pcc* to identify observations able to dominate the regression. Leverage statistics are calculated using only sensitivities and weights and so are independent of model fit. They are introduced here because of their utility when used in conjunction with *dss*, *css*, and *pcc*. The equation for the leverage statistic is presented in Chapter 7, Section 7.3 because it is closely associated with the parameter statistics discussed there.

One leverage statistic is calculated for each observation. Observations with large values of leverage could dramatically affect one or more of the estimated parameter values, depending on the value of the observation. Often, but not always, observations with greater leverage have large absolute values of *dss* for one or more parameters and large *css*. An observation with small *dss* and *css* values can attain a large value of leverage if it is instrumental in reducing the correlation between two or more parameters.

4.3.7 One-Percent Scaled Sensitivities

A final scaling considered here produces one-percent scaled sensitivities, denoted $1ss_{ij}$, which are calculated as

$$1ss_{ij} = \left(\frac{\partial y'_i}{\partial b_j}\right)\bigg|_b \frac{b_j}{100} \qquad (4.7)$$

Commonly, one-percent scaled sensitivities are calculated for simulated values at every node of a model grid instead of just at observation locations, so the subscript *i* would be used to identify every grid node. Sensitivities at every node of a grid are

4.3 FIT-INDEPENDENT STATISTICS FOR SENSITIVITY ANALYSIS

readily available when sensitivities are calculated using the sensitivity-equation method using, for example, MODFLOW-2000.

One-percent scaled sensitivities maintain the units of the simulated values. They approximately equal the amount that the simulated value would change if the parameter value increased by one percent. When calculated for simulated values of the same type, larger values of $1ss_{ij}$ indicate greater sensitivity to b_j, which indicates that an observed equivalent to simulated value i could be important to the estimation of the parameter value. Because they have dimensions, one-percent scaled sensitivities cannot be used to compare the importance of different types of observations to the estimation of parameter values and generally cannot be used to form a composite statistic. However, retaining units of the simulated values allows the one-percent scaled sensitivities to more effectively communicate results in some circumstances.

The omission of weights from Eq. (4.7) means that the one-percent scaled sensitivities do not reflect the importance of the observations on the regression as effectively as do the dss or leverage statistics. The omission of the weighting does, however, make the statistic easier to calculate.

One-percent scaled sensitivities can be used to create contoured sensitivity maps the same way heads at every node are used to create contoured head maps. Similar maps can be produced using UCODE_2005 or PEST by defining enough model locations as observations so that accurate maps are created, but this is an arduous undertaking. Maps of one-percent scaled sensitivities can be used to identify where additional observations would be most important to the estimation of different parameters. For example, if composite scaled sensitivities show that existing observations do not provide ample information to estimate a parameter, large absolute values on maps of one-percent scaled sensitivities for this parameter can help show where in the model domain a new observation would provide the most information about the parameter. However, as noted later, significant limitations exist.

There are three disadvantages that limit the use of one-percent scaled sensitivity maps in practice. First, there are potentially a large number of maps to evaluate. For each parameter, there is a map for each model layer and, in transient models, for each time step. Searching these maps for the largest values of one-percent scaled sensitivities can be cumbersome. Furthermore, the largest values of one-percent scaled sensitivity often occur at different locations and times for each of the parameters. Additional criteria are needed to determine important locations, such as the potential effect of the observation on simulated predictions and their uncertainty. Second, conclusions drawn from one-percent sensitivity maps can be difficult to justify to resource managers because the many maps can be overwhelming, and different conclusions might be drawn from different maps. Third, the maps can only be produced for an observation type for which the simulated equivalent can be calculated over the entire model domain, such as hydraulic heads in groundwater systems. The maps do not provide information about other types of observations, such as flows or advective-transport observations in groundwater systems. The opr statistic presented in Chapter 8 generally is a better method of identifying important new observations.

Despite their practical limitations, an excellent use of one-percent scaled sensitivities is for instructional purposes. For relatively simple problems, using knowledge of the physics and other processes that control the simulated system to explain the patterns and magnitudes of one-percent sensitivities helps modelers understand what sensitivities mean and provide to regression calculations. This is illustrated in Exercise 4.1d. For more complex problems, one-percent sensitivity maps can help the modeler better understand the processes controlling the simulated system.

4.4 ADVANTAGES AND LIMITATIONS OF FIT-INDEPENDENT STATISTICS FOR SENSITIVITY ANALYSIS

The fit-independent statistics presented in this chapter have the advantage that in many circumstances they provide a good evaluation of the information provided by the observations for estimating parameters without first having to complete a successful regression. For models with long execution times, using fit-independent statistics to design the parameterization and decide which parameters to estimate in a given regression run can be advantageous.

Limitations of fit-independent statistics generally are related to the scaling, inaccurate sensitivities, or the nonlinearity of the sensitivities, as discussed in the following sections. Additional comments and guidance for addressing difficulties are provided in Guideline 3 of Chapter 11, Section G3.2.

4.4.1 Scaled Sensitivities

Three issues related to scaled sensitivities are discussed: (1) they do not account for parameter correlations, (2) the scaling by parameter values defined in Eqs. (4.3) to (4.7) works well for some circumstances but not for others, and (3) though relatively robust in the presence of inaccurate sensitivities and model nonlinearity, they fail to perform well if the model is extremely nonlinear.

Scaled sensitivities are limited in that they do not account for the possibility that while the observations may provide substantial information about individual parameters, coordinated changes in the parameter values may produce the same model fit. Thus, it cannot be determined if the observations can be used to estimate each parameter uniquely. This occurs when parameters are highly correlated and can be detected by calculating the parameter correlation coefficients and leverage statistics defined in Sections 4.3.5 and 4.3.6 and discussed further in Sections 4.4.2 and 4.4.3.

The scaling by the parameter value used in the definitions of dimensionless, composite, and one-percent scaled sensitivities is useful when the effect of changing parameter values by a multiplicative factor is of interest. For example, in groundwater models it is common to think of errors in flow parameters, such as recharge, as some percentage of the flow (such as 5 or 10 percent), rather than as plus or minus a particular flow value. Similarly, potential changes in hydraulic conductivity commonly are thought of as a multiplicative factor such as plus and minus an

4.4 ADVANTAGES AND LIMITATIONS OF FIT-INDEPENDENT STATISTICS

order of magnitude (multiplying and dividing by 10), rather than plus and minus a particular hydraulic-conductivity value. The utility of the scaling results from the underlying physics.

For some types of parameters, scaling by the parameter value can produce misleading results. For example, in groundwater models, parameters that represent hydraulic head at constant-head boundaries pose a special problem. By using Eqs. (4.3) to (4.5) and (4.7), a flow system at sea level would have different dimensionless and one-percent scaled sensitivities than the identical system at 100 meters above sea level, which indicates that these scaled sensitivities provide misleading results. In other types of models, the test of whether a change of datum would change the *dss* values can be used to determine when scaling by the parameter value is problematic.

In MODFLOW-2000, the scaling difficulty affects Constant-Head Boundary (CHD) parameters (Harbaugh et al., 2000, pp. 78–79). For CHD parameters, modified versions of Eqs. (4.3) to (4.5) and (4.7) are used in which the scaled sensitivities are not multiplied by the parameter value. Thus, the values printed in the table of dimensionless scaled sensitivities (in the MODFLOW-2000 output file) for these parameters are not dimensionless; they have units of 1.0 divided by length. They can be thought of as the amount that the dependent variable would change if the CHD parameter changed by 1.0 unit, where the unit depends on how the parameter is defined and is commonly foot or meter.

UCODE_2005 is generally applicable, so presentation of scaled sensitivities cannot be tailored to particular types of parameters. Modelers need to be aware that scaled sensitivities of parameters for which a change of datum would change the *dss* values are misleading and should not be used.

An alternative scaling that may be useful in some circumstances was proposed by Tiedeman et al. (2003) and suggests that the parameter value be replaced by the parameter standard deviation (s_{bi}). This scaling can be achieved by multiplying one-percent scaled sensitivities by $100/s_{bi}$. As discussed in Chapter 7, Section 7.2.2, s_{bi} depends on model fit, making such scaled sensitivities fit-dependent.

Finally, the effects of both nonlinearity and scaling by the parameter value cause scaled sensitivities to be different for different sets of parameter values. If the differences that occur for a reasonable range of parameter values are too extreme, such that different parameters are rated as important when calculated at one set of parameter values and not important when calculated at another set, the scaled sensitivities are inadequate for the purposes they serve in the guidelines discussed in Chapters 10–14. Their utility can be tested by calculating values for several sets of parameter values.

In practice, the sensitivity analysis suggested in this book has proved to be useful even for highly nonlinear problems, as gauged by the modified Beale's measure discussed in Chapter 7, Section 7.7. For example, their utility is demonstrated in many groundwater flow and transport problems (Anderman et al., 1996; Barlebo et al., 1996; D'Agnese et al., 1997, 1999, 2002; Poeter and Hill, 1997; Hill et al., 1998). Problems that are too nonlinear for the sensitivity analysis to be useful also may be too nonlinear for the gradient optimization methods described in Chapter 5 to be useful, but this has not been tested. Problems that are too nonlinear for gradient optimization methods need to be addressed using global search methods such as simulated annealing and genetic algorithm (see Chapter 5, Section 5.2), which are much more computationally intensive than gradient methods.

4.4.2 Parameter Correlation Coefficients

Parameter correlation coefficients (pcc) have two advantages. First, they are easier to understand than alternatives such as eigenvector analysis or the closely related singular value decomposition, as noted in Section 4.3.5. Second, except for problems related to accuracy of the sensitivities, the degree of correlation can be determined easily by comparing the absolute value of the pcc to the value 1.00.

There are three limitations associated with the pcc.

First, the nonlinearity of inverse problems can cause correlation coefficients to be quite different for different sets of parameter values, as shown in Figure 4.2. In Figure 4.2, the objective-function surface has a distinct minimum, indicating that the parameters can be uniquely estimated. The absolute values of pcc calculated at many of the parameter values are significantly less than 1.00, correctly indicating the existence of a unique minimum. However, pcc with absolute values very close to 1.00 are calculated for some sets of parameter values. These large pcc values could lead to the incorrect conclusion that a unique minimum does not exist. In practice, nonuniqueness can only be clearly concluded if supported by an analysis of the simulated processes and available data, by using pcc values calculated for a range of parameter values, or by using regression to investigate uniqueness as discussed in Chapter 7.

A second concern about pcc is that they can be inaccurate when calculated using sensitivities with an inadequate number of correct significant digits (Hill and

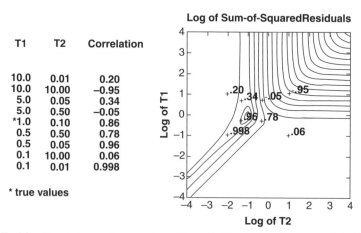

FIGURE 4.2 Correlation of parameters T1 and T2 of the simple model shown in Figure 3.1a. Correlation coefficients are calculated at different parameter values and are plotted on the $\log_{10}$ weighted least-squares objective-function surface shown in Figure 3.1c. T1 and T2 are in square meters per day. (From Poeter and Hill, 1997.)

Østerby, 2003). The accuracy of perturbation sensitivities suffers if the perturbation amount is too large for nonlinear parameters or too small for insensitive parameters, or if simulated values have an insufficient number of significant digits. The latter can occur because the process model lacks numerical precision or does not print numbers with sufficient precision, or the template or instruction input files for codes such as UCODE_2005 or PEST have not been set up to include enough significant digits. See Poeter et al. (2005) for further discussion. The accuracy of sensitivity-equation sensitivities can suffer if the convergence criteria for the solver are too large or the numerics of the model are inadequate. Following suggestions to enhance sensitivity precision is always important when calculating *pcc* values.

The third issue related to *pcc* is that as parameter sensitivity decreases, greater sensitivity precision is required for the *pcc* to be accurate (Hill and Østerby, 2003). In general, as more parameters are defined, parameter sensitivity is reduced. Composite scaled sensitivities (*css*) can be used to identify insensitive parameters. Combining parameters can be used to identify existing correlation that is obscured by having many defined parameters.

For example, consider that the hydraulic conductivity K_{rock} of a fractured rock aquifer in an initial groundwater model has been divided into parameters $K_{granite}$ and K_{schist}, corresponding to different rock types within the aquifer, in a subsequent model. Commonly, the css_j for $K_{granite}$ and for K_{schist} will be smaller than that for K_{rock}. Even if the sensitivities in the two models are precise to the same number of significant digits, the *pcc* in the first model may be more accurate than those in the second model because the parameter sensitivity has decreased. As the number of defined parameters increases, in any problem a point will be reached at which the *pcc* are no longer reliable indicators of parameter correlation.

The reverse also is true—as parameter sensitivity increases, less precision is required for the *pcc* to be accurate. This characteristic can be used to advantage. For example, in groundwater models, the correlation that commonly occurs between recharge and hydraulic conductivity may not be revealed by *pcc* when many parameters are defined. By combining all the recharge and all the hydraulic-conductivity parameters into a few parameters (using multiplication arrays to preserve the original spatial distribution of values), a more definitive test of parameter correlation can be achieved. Any extreme correlation that occurs for the few parameters also is present in the set of many parameters, it just cannot be identified using the *pcc* values calculated for the large set of parameters.

4.4.3 Leverage Statistics

Leverage statistics have the advantage of reflecting the importance of observations produced by the effects measured by both scaled sensitivities and correlation coefficients. They also have the advantage of not needing to be scaled and therefore do not inherit the difficulties of scaling discussed in Section 4.4.1. One difficulty of leverage statistics is that they do not reveal why observations are important; scaled sensitivities and parameter correlation coefficients can be used to gain insight. In addition, nonlinearity is likely to produce the same types of changes in leverage values that occur for *pcc* values. To the authors' knowledge, this has not been tested.

4.5 EXERCISES

Exercise 4.1: Sensitivity Analysis for the Steady-State Model with Starting Parameter Values In this exercise, sensitivities, parameter correlations, and leverage statistics for the steady-state flow system described in Chapter 2, Section 2.2 are calculated and evaluated.

(a) *Calculate sensitivities for the steady-state flow system.*

This exercise involves modifying computer files and simulating the system. Instructions are available from the web site for this book described in Chapter 1, Section 1.1. Students who are not performing the simulations may skip this exercise.

(b) *Use dimensionless and composite scaled sensitivities (dss and css) to evaluate observations and defined parameters.*

Dimensionless and composite scaled sensitivities are presented in Table 4.1. These statistics are discussed in Sections 4.3.3 and 4.3.4, Guideline 3 in Chapter 11, and Guideline 11 in Chapter 13. Plotting the composite scaled sensitivities on a bar graph as shown in Figure 4.3 is an effective method for showing how much information the observations likely provide for each parameter.

Problem
- Use the dimensionless scaled sensitivities of Table 4.1 and the discussion in Section 4.3 to identify which observations are most important to estimation of parameter HK_1. Use information about the flow system to explain why the *dss* for observations hd01.ss, hd07.ss, and flow01.ss are much smaller than the *dss* for the other observations.
- Use the composite scaled sensitivities of Table 4.1 and Figure 4.3 and the discussion in Section 4.3 to assess whether it is likely that all of the parameters for this model can be estimated with the available head and flow observations.

(c) *Evaluate parameter correlation coefficients (pcc) to assess parameter uniqueness.*

Use the parameter correlation coefficients shown in Tables 4.2 and 4.3 and the criterion presented in Section 4.3.5 to identify parameter values that might be difficult to estimate uniquely with the 10 head and one flow observations. In these tables, results calculated by (a) MODFLOW-2000, with the more accurate sensitivity-equation sensitivities, and (b) UCODE_2005, using less accurate perturbation sensitivities, are presented to show the effects of sensitivity inaccuracy.

TABLE 4.1 Dimensionless and Composite Scaled Sensitivities Calculated for Exercise 4.1a Using MODFLOW-2000 (Values Calculated by UCODE_2005 Are Similar)

Observation		Parameter Labels[a]					
Number	ID	HK_1	K_RB	VK_CB	HK_2	RCH_1	RCH_2
1	hd01.ss	0.110×10^{-4}	−0.225	0.105×10^{-6}	0.383×10^{-5}	0.150	0.0749
2	hd02.ss	−33.3	−0.225	−0.284	−5.47	24.0	15.3
3	hd03.ss	−57.9	−0.225	−0.493	−15.7	38.3	35.9
4	hd04.ss	−33.3	−0.225	−0.284	−5.47	24.0	15.3
5	hd05.ss	−46.5	−0.225	−0.394	−9.95	32.9	24.1
6	hd06.ss	−33.4	−0.225	−0.635	−5.35	24.0	15.6
7	hd07.ss	−2.34	−0.225	−2.38	2.08	1.82	1.04
8	hd08.ss	−57.5	−0.225	−0.133	−16.0	37.8	36.1
9	hd09.ss	−66.6	−0.225	$−0.580 \times 10^{-1}$	−23.3	38.1	52.1
10	hd10.ss	−46.3	−0.225	−0.330	−10.1	32.6	24.4
11	flow01.ss	$−0.547 \times 10^{-3}$	$−0.663 \times 10^{-4}$	$−0.260 \times 10^{-5}$	$−0.190 \times 10^{-3}$	−7.36	−3.68
Composite scaled sensitivity		41.3	0.214	0.783	11.0	27.4	25.6

[a] See Table 3.1 for parameter label definitions.

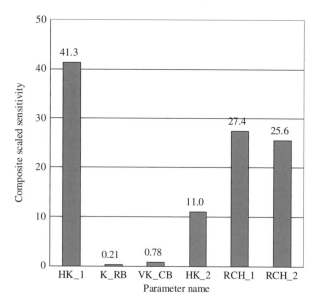

FIGURE 4.3 Composite scaled sensitivities (*css*) for the steady-state simulation calculated using starting parameter values.

TABLE 4.2 Parameter Correlation Coefficient (*pcc*) Matrix[a] Calculated by (a) MODFLOW-2000 and (b) UCODE_2005 with Central-Difference Perturbation, Using the Starting Parameter Values for the Steady-State Problem with 10 Hydraulic-Head Observations and One Streamflow Gain Observation

	HK_1	K_RB	VK_CB	HK_2	RCH_1	RCH_2
			(a) MODFLOW-2000			
HK_1	1.00	−0.37	−0.57	−0.75	**0.95**	−0.63
K_RB		1.00	−0.11	0.31	−0.22	0.25
VK_CB			1.00	0.82	−0.68	0.81
HK_2		Symmetric		1.00	−0.83	**0.98**
RCH_1					1.00	−0.76
RCH_2						1.00
			(b) UCODE_2005			
HK_1	1.00	−0.39	−0.57	−0.76	**0.95**	−0.63
K_RB		1.00	−0.10	0.32	−0.24	0.25
VK_CB			1.00	0.82	−0.68	0.81
HK_2		Symmetric		1.00	−0.83	**0.98**
RCH_1					1.00	−0.76
RCH_2						1.00

[a] Cells are in bold type for $pcc \geq 0.95$.

4.5 EXERCISES

TABLE 4.3 Parameter Correlation Coefficient (*pcc*) Matrix[a] Calculated by (a) MODFLOW-2000 and (b) UCODE_2005 for Starting Parameter Values for the Steady-State Problem Using Only the 10 Hydraulic-Head Observations

	HK_1	K_RB	VK_CB	HK_2	RCH_1	RCH_2
			(a) MODFLOW-2000[b]			
HK_1	1.00	**1.00**	**1.00**	**1.00**	**1.00**	**1.00**
K_RB		1.00	**1.00**	**1.00**	**1.00**	**1.00**
VK_CB			1.00	**1.00**	**1.00**	**1.00**
HK_2		Symmetric		1.00	**1.00**	**1.00**
RCH_1					1.00	**1.00**
RCH_2						1.00
			(b) UCODE-2005[c]			
HK_1	1.00	**0.97**	**1.00**	**1.00**	**1.00**	**1.00**
K_RB		1.00	**0.97**	**0.97**	**0.97**	**0.97**
VK_CB			1.00	**1.00**	**1.00**	**1.00**
HK_2		Symmetric		1.00	**1.00**	**1.00**
RCH_1					1.00	**1.00**
RCH_2						1.00

[a]Cells are in bold type for $pcc \geq 0.95$.
[b]The correct values of 1.00 calculated by MODFLOW-2000 use the more accurate sensitivity-equation sensitivities.
[c]The incorrect values calculated by UCODE_2005 are caused by the less accurate central-difference perturbation sensitivities.

Problem
- When the flow observation is included, are any of the *pcc* calculated by MODFLOW-2000 (Table 4.2a) above 0.90? Above 0.95? Although parameters with *pcc* values up to nearly 1.00 can probably be estimated uniquely, it is useful to be aware of which parameters have these relatively high correlations.
- Based on the comments above, what do the *pcc* values indicate about the likelihood of being able to estimate all of the parameters independently using the head and flow data?
- When only hydraulic-head observations are included, why are all the parameters extremely correlated, as indicated by the results from MODFLOW-2000 (Table 4.3a)?
- Why are the correlation coefficients calculated by UCODE_2005 unable to capture fully the extreme parameter correlation of all parameters when using only hydraulic-head observations (Table 4.3b)?

(d) Use contour maps of one-percent sensitivities for the steady-state flow system.

Contour maps of one-percent scaled sensitivities for the steady-state system, calculated for the starting parameter values, are shown in Figure 4.4. Each map is

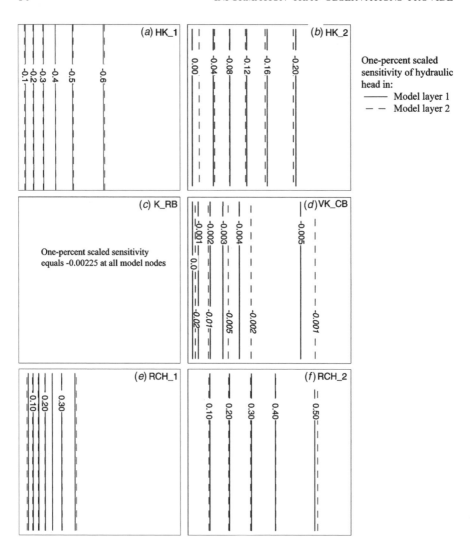

FIGURE 4.4 Contour maps of one-percent scaled sensitivities of hydraulic head for the steady-state model, calculated using Eq. (4.7), where y' is hydraulic head evaluated at each cell in the model grid and b is one of the six model parameters: (*a*) HK_1, (*b*) HK_2, (*c*) K_RB, (*d*) VK_CB, (*e*) RCH_1, and (*f*) RCH_2. The sensitivities are calculated using the starting parameter values. Contour labels apply to sensitivities in both layers for all maps except that for VK_CB.

related to one parameter and can be used to identify areas with relatively large and small absolute values of scaled sensitivity. Areas with large absolute values indicate where hydraulic-head measurements are likely to be most important for estimating the parameter. Because the sensitivities are scaled by the parameter values and all sensitivities are for the same observation type, the one-percent scaled sensitivity maps also can be compared with each other.

4.5 EXERCISES

The simple system considered here provides the opportunity to (1) identify how sensitivities reflect system dynamics, and (2) demonstrate the utility of sensitivity analysis and the role sensitivities play in regression. This exercise can form a frame of reference for considering sensitivities calculated in more complicated systems.

Problem: Explain the one-percent sensitivity maps from the steady-state system (Figure 4.4), basing your analysis on characteristics of fluxes into, out of, and within the flow system. The cell-by-cell fluxes and the boundary fluxes of the steady-state flow system along a cross section perpendicular to the river (along a row) are shown in Figure 4.5. The steady-state flow system is two-dimensional, because all features are the same for any row; thus, all information about the system is portrayed in a cross section along any row of the model. The total fluxes through the entire model are obtained by multiplying the values in Figure 4.5 by 18, which is the number of rows. In explaining the sensitivities, answer the following questions:

- Why are the one-percent scaled sensitivities negative for hydraulic-conductivity parameters HK_1 and HK_2, and positive for the recharge parameters RCH_1 and RCH_2?
- Why are the magnitudes of the one-percent scaled sensitivities larger for HK_1 than for HK_2?

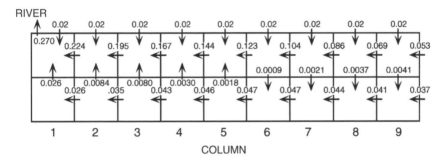

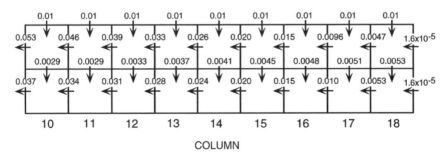

FIGURE 4.5 Cell-by-cell fluxes, in m³/s, along any model row of the steady-state flow system with the true parameter values.

- Why do the one-percent scaled sensitivities for RCH_1 vary only over the left half of the system, whereas those for RCH_2 vary over the entire domain?
- Why is VK_CB the only parameter for which there are substantial differences in the one-percent scaled sensitivities for model layers 1 and 2?
- Why are the one-percent scaled sensitivities for K_RB the same throughout the system?

(e) *Evaluate leverage statistics.*

For the initial model, four observations have leverage statistics that are larger than 0.90:

flow01.ss	1.00
hd01.ss	0.99
hd07.ss	0.97
hd09.ss	0.94

Problem: Leverage statistics reflect the combined effects of sensitivity and correlation. Use the leverage statistics, the discussions of Sections 4.3 and 4.4, and Tables 4.1 and 4.2 to address the following questions.

- For each of the high leverage observations, which parameters have the largest dimensionless scaled sensitivities?
- Evaluate whether the high leverage observations are dominated by sensitivity or correlation considerations. Use the system dynamics that contribute to the importance of each observation.

5

ESTIMATING PARAMETER VALUES

As part of model calibration, it is often useful to determine parameter values that produce the smallest possible value of the objective function in Eq. (3.1) or (3.2). The process of calculating such parameter values is called optimization. If more than one set of parameter values produce the same small objective-function value, the resulting parameter values define multiple minima; if only one set of parameter values produces the smallest objective-function value, the resulting parameter values define a unique minimum. If the optimization problem has a unique minimum and the objective function is smooth enough, as in Figure 3.1b, c, optimization methods that use calculated sensitivities are very advantageous in that they are computationally efficient. These are called *gradient methods* because they generally use the gradient of the objective-function surface to determine how to proceed toward the minimum. They are also called *regression methods*.

Nonlinear regression, instead of the simpler linear regression, is needed when simulated values are nonlinear functions of the parameters being estimated. This is common in groundwater models, as discussed in Chapter 1, Section 1.4.1. Model nonlinearity produces important complications to regression and has been the topic of considerable investigation in several fields. Seber and Wild (1989) and Dennis and Schnabel (1996) are excellent upper-level texts on nonlinear regression. The discussion in this book is the most accessible nonlinear regression presentation known to the authors.

This book uses a modified Gauss–Newton nonlinear regression method. The method uses an iterative form of standard linear-regression equations and works

---

Effective Groundwater Model Calibration: *With Analysis of Data, Sensitivities, Predictions, and Uncertainty.* By Mary C. Hill and Claire R. Tiedeman
Published 2007 by John Wiley & Sons, Inc.

well only with modifications. This chapter describes the difficulties of the method and most of the modifications used by MODFLOW-2000 and UCODE_2005. Appendix B describes additional modifications, including quasi-Newton updating.

The modified Gauss–Newton method presented here is an extension of the method presented by Cooley and Naff (1990, Chap. 3), which is similar to methods presented by Seber and Wild (1989), Sun (1994), Tarantola (2005), Dennis and Schnabel (1996), and other texts on nonlinear regression. The modified Gauss–Newton method presented in this book also can be categorized as a Marquardt or Levenberg–Marquardt method. The approach forms the basis of most multicriteria gradient optimization methods (Ehrgott, 2000).

The modified Gauss–Newton method presented here has performed well relative to alternatives in that fewer or an equivalent number of total model evaluations are required and it is at least as robust as the alternatives. Cooley (1985), Hill (1990), and Cooley and Hill (1992) compare the modified Gauss–Newton method to quasi-linearization, quasi-Newton, Fletcher–Reeves, a combined Fetcher–Reeves/quasi-Newton, a modified Gauss–Newton/full-Newton hybrid, and the modified Gauss–Newton method with the quasi-Newton updating described in Appendix B. They considered problems of steady-state and transient groundwater flow in which relatively few parameters are estimated. Results presented by Mehl and Hill (2003) suggest that the double-dogleg trust region approach of Dennis and Schnabel (1996) can substantially reduce execution times for difficult problems. This method is available in UCODE_2005; it is not described in this book.

5.1 THE MODIFIED GAUSS–NEWTON GRADIENT METHOD

Parameter values that minimize the least-squares objective function (Eqs. (3.1) and (3.2) in Chapter 3) are calculated using normal equations. Section 5.1.1 presents the normal equations for the modified Gauss–Newton method used in this work and uses a one-parameter problem to illustrate aspects of the method. Section 5.1.2 presents a two-parameter example problem that demonstrates the iterations required to solve nonlinear regression using the normal equations. Finally, Section 5.1.3 discusses the convergence criteria that govern when to stop the iterative process.

5.1.1 Normal Equations

Normal equations are derived by taking the derivative of the objective function with respect to the parameters and setting the derivative equal to zero. By using Eq. (3.2), this becomes

$$\frac{\partial}{\partial \boldsymbol{b}}\left[[\boldsymbol{y} - \boldsymbol{y}'(\boldsymbol{b})]^T \boldsymbol{\omega}[\boldsymbol{y} - \boldsymbol{y}'(\boldsymbol{b})]\right] = \boldsymbol{0} \tag{5.1}$$

where $\boldsymbol{0}$ is a vector of NP values that all equal zero, and NP is the number of estimated parameters.

5.1 THE MODIFIED GAUSS–NEWTON GRADIENT METHOD

When $y'(b)$ is nonlinear, Eq. (5.1) is solved by approximating $y'(b)$ as a linear function using two terms of a Taylor series expansion, so that

$$y'(b) \cong y^\ell(b) = y'(b_0) + \left.\frac{\partial y'(b)}{\partial b}\right|_{b=b_0}(b - b_0) \qquad (5.2a)$$

where $y^\ell(b)$ = the linearized form of $y'(b)$;
b_0 = the vector of parameter values about which $y'(b)$ is linearized;
$\left.\dfrac{\partial y'(b)}{\partial b}\right|_{b=b_0}$ = the sensitivity matrix calculated using the parameter values listed in vector b_0.

The vector $y'(b)$ has $ND + NPR$ elements, where ND is the number of observations and NPR is the number of prior information equations; the vector b has NP elements, where NP is the number of estimated parameters.

If the sensitivities are expressed as the matrix X, Eq. (5.2a) can be written

$$y^\ell(b) = y'(b_0) + X|_{b=b_0}(b - b_0) \qquad (5.2b)$$

where X is the sensitivity matrix (also called the Jacobian matrix), with elements equal to $\partial y'_i/\partial b_j$. X has $ND + NPR$ rows and NP columns, so $i = 1, ND + NPR$ and $j = 1, NP$ as shown in Appendix B.

To understand what linearizing $y'(b)$ means, it is useful to consider a model that has only one parameter. Here we consider the Theim equation, which describes the shape of a steady-state cone of depression around a pumping well given a homogeneous groundwater system and a constant head at radial distance r_0. Figure 5.1a shows the nonlinear function linearized about $b_0 = 0.005$, which is a starting guess for the value of the transmissivity parameter (T). At b_0, the linearized approximation equals the nonlinear function; away from b_0 the linearized approximation generally differs from the nonlinear function. The function can be represented by the notation $y'(b)$, where b is not bold because there is only one parameter and y is not bold because it represents the function, not a vector of values simulated for a set of observations.

The Gauss–Newton normal equations are developed by substituting the linearized approximation of $y'(b)$ into the objective function. Using the expression of the least-squares objective function from Eq. (3.2) gives

$$S^\ell(b) = [y - y^\ell(b)]^T \omega [y - y^\ell(b)] \qquad (5.3)$$

Again using the one-parameter problem to understand what this equation represents, Figure 5.1b shows the shapes of the least-squares objective function calculated using the nonlinear model and the model linearized about b_0. The figure shows that the linearized objective function reaches a minimum value at $T = 0.007$, which is closer to the minima of the nonlinear objective function than is the starting guess of $T = 0.005$. Starting at b_0, the Gauss–Newton method uses the objective function

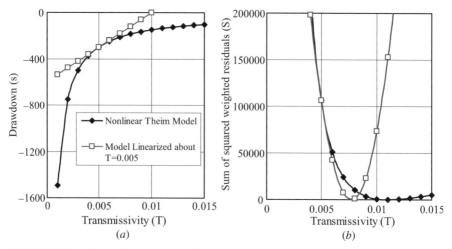

FIGURE 5.1 (a) Nonlinear model $y'(b)$ and linearized approximation of $y'(b)$, linearized about point $T = b = b_0 = 0.005$. (b) The objective function calculated using the nonlinear function $y'(b)$ and the linearized objective function calculated using the linear approximation of $y'(b)$. The nonlinear model is the Theim equation, $s = [Q/(2\pi T)] \ln(r/r_0)$, with Q (pumpage) = 1, r_0 (distance of zero drawdown) = 1000, and generated "observations" at r (distance to the observation well) = 1, 2, 4, 6, 10, 40, 80, 120, 200, 300, 450, 600. No noise was added to the "observations".

formed using the linear model to determine how the parameter value should be changed. Ideally, moving from b_0 to the minimum of the linearized objective function will result in an estimated parameter value that is closer to the minimum of the nonlinear objective function. This is indeed the case for the objective functions shown in Figure 5.1b.

One difference between linear and nonlinear regression is that in linear regression, parameter values are estimated by solving the normal equations once. In contrast, nonlinear regression is iterative in that a sequence of parameter updates is calculated, solving linearized normal equations once for each update. Thus, in nonlinear regression there are parameter-estimation iterations.

The iterative form of the normal equations needed to solve nonlinear regression problems is produced by minimizing the objective function of Eq. (5.3). This is accomplished by taking the derivative with respect to the parameter values and setting it to zero. In addition, the superscript 0 (shown in Eq. (5.2b)) is replaced by r, which identifies the parameter-estimation iteration. The resulting Gauss–Newton nonlinear regression normal equations are

$$(X_r^T \omega X_r) d_r = X_r^T \omega (y - y'(b_r)) \tag{5.4}$$

where r = the parameter-estimation iteration number;

X_r = the sensitivity matrix calculated for the parameter values in b_r;

5.1 THE MODIFIED GAUSS–NEWTON GRADIENT METHOD

ω = the weight matrix;

$(X_r^T \omega X_r)$ = a symmetric, square matrix of dimension NP by NP (as noted in Chapter 4, Section 4.3.4, $X^T \omega X$ is related to the Fisher information matrix);

d_r = an NP-dimensional vector used to update the parameter estimates (called the parameter change vector in this book);

b_r = the vector of parameter estimates at the start of iteration r.

The sensitivity matrix X_r appears in Eq. (5.4) because taking the derivative of Eq. (5.3) produces sensitivities $\partial y'/\partial b$. When calculated at parameter values b_r, these sensitivities can be expressed as matrix X_r.

For the first parameter-estimation iteration, the model is linearized about starting parameter values defined by the modeler. In each subsequent iteration, the model is linearized about parameter values estimated in the previous iteration. For each parameter-estimation iteration, Eq. (5.4) is solved for d_r, and then d_r is used to update the parameter values for the start of iteration $r+1$, using the equation $b_{r+1} = b_r + d_r$. In practice, a modified form of this equation is used, as described later in this chapter. Figure 5.2 shows how Eq. (5.4) relates to the geometry of a linearized objective-function surface for a hypothetical two-parameter problem. The right side of Eq. (5.4) is proportional to the gradient of the linearized objective function. Without the $(X_r^T \omega X_r)$ term on the left side of Eq. (5.4), the parameter change vector d_r would point directly down the gradient of the linearized objective-function surface, as shown by arrow A in Figure 5.2. This is called the steepest descent direction. The $(X_r^T \omega X_r)$ term modifies the direction of d_r to point toward the minimum of the linearized objective-function surface, as shown by arrow B in Figure 5.2.

The basic Gauss–Newton method presented in Eq. (5.4) is prone to difficulties such as oscillations due to overshooting the optimal parameter values. It only

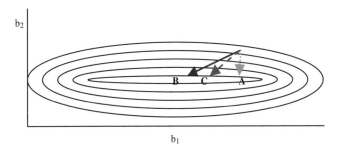

FIGURE 5.2 A linearized objective-function surface for a hypothetical two-parameter problem, illustrating the geometry of the normal equations. The arrows represent the direction relevant to the parameter change vector d_r. Arrow A points down gradient in a direction defined by the right-hand side of Eq. (5.4). Arrow B points in the direction of d_r solved for using Eq. (5.4) or (5.5). Arrow C shows that the direction of d_r, solved for using a nonzero Marquardt parameter in Eq. (5.6), is between arrows A and B.

works well when modified. Three important modifications are scaling, the Marquardt parameter, and damping. These are discussed in the following paragraphs.

Scaling Often the parameter values, and thus the sensitivities, have values that differ by many orders of magnitude. This can cause great difficulties with obtaining an accurate solution of Eq. (5.4). The accuracy of the parameter change vector d_r can be improved by scaling Eq. (5.4). The scaling is implemented as

$$(C^T X_r^T \omega X_r C) C^{-1} d_r = C^T X_r^T \omega (y - y'(b_r)) \tag{5.5a}$$

where C is a diagonal scaling matrix with element c_{jj} equal to $[(X^T \omega X)_{jj}]^{-1/2}$.

Marquardt Parameter The resulting scaled matrix has the smallest possible condition number (Forsythe and Strauss, 1955; Hill, 1990). Scaling with C changes the magnitude but not the direction of d_r. Therefore, in Figure 5.2 the parameter change vector d_r still points in the direction of arrow B after scaling has been implemented.

In some circumstances, the direction of the change vector d_r is nearly parallel to the contours of the objective-function surface and changing the parameter values using d_r yields little progress toward estimating optimal parameter values. In this case, changing the direction of d_r can be advantageous. The second modification involves introduction of a term that causes the direction of vector d_r to move toward the steepest-descent direction. The term is called the Marquardt parameter (Marquardt, 1963; Theil, 1963; Seber and Wild, 1989; Cooley and Naff, 1990). In Figure 5.2 a nonzero Marquardt parameter moves the direction of d_r from the direction of arrow B to the direction of arrow C. The Marquardt parameter is included in the scaled objective function of Eq. 5.5a as

$$(C^T X_r^T \omega X_r C + I m_r) C^{-1} d_r = C^T X_r^T \omega (y - y'(b_r)) \tag{5.5b}$$

where I is an $NP \times NP$ identity matrix and m_r is the Marquardt parameter. The procedure for determining the Marquardt parameter is discussed in the next section.

Damping Overshoot is a common problem with the Gauss–Newton method, so damping is introduced. Overshoot occurs when the parameter change vector points toward locations on the objective-function surface that are closer to the minimum of the nonlinear objective-function surface, but then extends beyond these locations to larger objective-function values. Damping helps prevent overshoot by allowing the parameters to change less than the full amount calculated by d_r. This can significantly improve regression performance. Damping is applied when updating the parameter values using the parameter change vector d_r.

Including damping in Eq. (5.5b) produces

$$(C^T X_r^T \omega X_r C + I m_r) C^{-1} d_r = C^T X_r^T \omega (y - y'(b_r)) \tag{5.6a}$$

$$b_{r+1} = \rho_r d_r + b_r \tag{5.6b}$$

where ρ_r is the damping parameter. Together, Eqs. (5.6a) and (5.6b) almost express the normal equations and the iterative process for the modified Gauss–Newton

5.1 THE MODIFIED GAUSS–NEWTON GRADIENT METHOD

optimization method used in UCODE_2005 and MODFLOW-2000. Additional modifications include an iteration control mechanism and a quasi-Newton modification that are discussed in Appendix B. In addition, UCODE_2005 provides a trust region approach not described here. Calculation of the Marquardt and damping parameters is discussed next.

Calculate the Marquardt Parameter   The Marquardt parameter is used to change the direction of and shorten d_r. These modifications improve regression performance for ill-posed problems (Marquardt, 1963). In Eq. (5.6a), m_r initially equals 0.0 for each parameter-estimation iteration r. If d_r is nearly orthogonal to the steepest descent direction, the resulting parameter changes are unlikely to reduce the value of the objective function, and m_r is changed to a nonzero value. The modification to the direction and length of d_r caused by $m_r > 0.0$ is illustrated in Figure 5.2 as the change from arrow B to arrow C.

In MODFLOW-2000 and UCODE_2005, the value of the Marquardt parameter is determined as suggested by Cooley and Naff (1990, pp. 71–72). If the cosine of the angle between the vector d_r and the vector orthogonal to the steepest descent direction is less than a threshold value (commonly 0.08), m_r is increased using the relation $m_r^{\text{new}} = a \times m_r^{\text{old}} + b$. Commonly, $a = 1.5$ and $b = 0.001$. The threshold value for the cosine of the angle and a and b can be specified by the user. PEST handles the Marquardt parameter somewhat differently in that it is applied to the unscaled matrix. The results obtained by PEST have been similar to those achieved by MODFLOW-2000 and UCODE_2005 in tests conducted by the authors. John Doherty (oral communication, 2003), author of PEST, suggested that PEST converged in one less parameter-estimation iteration in some circumstances, but the specifics of his numerical experiments are unknown.

Calculate the Damping Parameter   The damping parameter, ρ_r, shortens d_r and can vary in value from 0.0 to 1.0. This parameter modifies all values in the parameter change vector d_r by the same factor. Thus, in vector terminology, the direction of d_r is preserved. For each parameter-estimation iteration, the damping parameter initially equals 1.0 but is changed to a smaller value for either of two reasons:

1. To ensure that the absolute values of fractional parameter value changes are all less than a value specified by the user. This value is the input variable MaxChange of UCODE_2005 and MAX-CHANGE of MODFLOW-2000. In this book, this value is referred to as *max-allowed-change*.
2. To damp oscillations that occur when elements in d_r and d_{r-1} define opposite directions (Cooley, 1993), implemented as described in Appendix B.

To evaluate whether damping needs to be implemented for reason 1, fractional parameter value changes are calculated for each native parameter value as

$$(b_j^{r+1}|_{\rho_r=1.0} - b_j^r)/|b_j^r| = d_j^r/|b_j^r|, \quad j = 1, NP \tag{5.7}$$

where b_j^r is the jth element of vector $\boldsymbol{b}_r$, that is, the value of the jth parameter at parameter estimation iteration r. If b_j^r equals 0.0, 1.0 is used in the denominator. The value of b_j^{r+1} in Eq. (5.7) is calculated using Eq. (5.6b) with $\rho_r = 1.0$. That is, the value is calculated assuming no damping. In this book the absolute value of the largest fractional parameter value change calculated using Eq. (5.7) is referred to as *max-calculated-change*.

If *max-calculated-change* is greater than *max-allowed-change*, ρ_r is calculated as follows unless oscillation concerns (reason 2 above) result in an even smaller value:

$$\rho_r = \frac{\text{max-allowed-change}}{\text{max-calculated-change}} \tag{5.8}$$

Following computation of ρ_r by Eq. (5.8), $\boldsymbol{b}_{r+1}$ is calculated by Eq. (5.6b) and contains the parameter values for starting the next parameter-estimation iteration. A somewhat different procedure is used for calculating the damping parameter for model parameters that are log-transformed in the regression. This procedure is described in Section 5.4 and Appendix B.

Typically, *max-allowed-change* has been the same for all parameters. UCODE_2005 and PEST, however, allow different values of *max-allowed-change* to be assigned to different parameters. This is likely to be used to allow insensitive parameters to change more than sensitive parameters so that the insensitive parameters do not produce tiny damping parameters that can restrict updates of sensitive parameters to the point where no progress can be made.

5.1.2 An Example

To understand more clearly how the modified Gauss–Newton method works, consider its performance for the two-parameter model shown in Figure 5.3. The data shown in Figure 5.3a are transient groundwater level drawdowns caused by pumpage from a single well. The model used is the Theis equation, in which drawdown is a nonlinear function of two parameters: the transmissivity (T) and the storage coefficient (S). Both parameters are estimated. The observations are the drawdowns listed in Figure 5.3a.

The nonlinear objective-function surface is shown in Figure 5.3b. Conceptually, this is analogous to the objective function in Figure 5.1b produced using the nonlinear function of Figure 5.1b. Figure 5.3c and Figure 5.3d show approximations of the objective-function surface produced by linearizing the Theis equation about the parameter values marked by X_1 and X_2. The problem is linearized by replacing the Theis equation with the first two terms of a Taylor series expansion (Eq. (5.2)) in which $\boldsymbol{b}_0$ includes the parameter values at X_1 or X_2, and using this linearized model to replace $y'(\boldsymbol{b})$ in Eq. (3.2) to obtain Eq. (5.3). As in Figure 5.1b, the linearized objective-function surfaces approximate the nonlinear surface well near $\boldsymbol{b}_0$ and less well further away.

In Figure 5.3c, the objective function is linearized about a point (X_1) far from the minimum (•) of the nonlinear objective function. Moving from this point all the

5.1 THE MODIFIED GAUSS–NEWTON GRADIENT METHOD

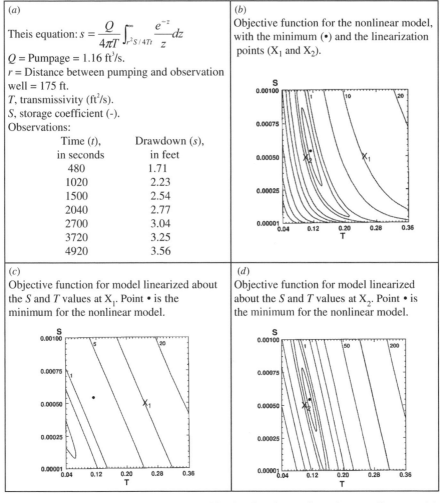

(a)
Theis equation: $s = \dfrac{Q}{4\pi T}\int_{r^2 S/4Tt}^{\infty} \dfrac{e^{-z}}{z}\,dz$

Q = Pumpage = 1.16 ft^3/s.
r = Distance between pumping and observation well = 175 ft.
T, transmissivity (ft^2/s).
S, storage coefficient (-).
Observations:

Time (t), in seconds	Drawdown (s), in feet
480	1.71
1020	2.23
1500	2.54
2040	2.77
2700	3.04
3720	3.25
4920	3.56

(b) Objective function for the nonlinear model, with the minimum (•) and the linearization points (X_1 and X_2).

(c) Objective function for model linearized about the S and T values at X_1. Point • is the minimum for the nonlinear model.

(d) Objective function for model linearized about the S and T values at X_2. Point • is the minimum for the nonlinear model.

FIGURE 5.3 Model equation, data, and objective-function surfaces for a nonlinear model. (Example from Cooley and Naff, 1990, p. 66.)

way to the minimum of the linearized objective-function surface (a point to the left of the plot) would overshoot the nonlinear objective-function minimum. As mentioned previously, this is a common problem with unmodified Gauss–Newton methods. Here, proceeding to the minimum of the linearized surface would produce a negative value of transmissivity, which is computationally infeasible. This is a situation in which more advantageous results can be obtained by limiting the parameter value changes using the damping parameter ρ_r of Eq. (5.6b). With damping, the regression moves only part of the way from X_1 to the minimum of the linearized surface.

In Figure 5.3d, the objective function is linearized about a point near the minimum of the nonlinear objective function. In this case, moving to the minimum of the linearized objective-function involves small changes in the parameter values, and damping is not needed. Moving to this minimum produces parameter values near the minimum of the nonlinear objective-function, which is the goal of the regression.

Figure 5.3d also shows that the linearized model closely replicates the objective-function surface near the minimum. This has consequences for the applicability of linear inferential statistics, such as linear confidence intervals, as discussed in Chapter 7, Section 7.5.1 and Chapter 8, Section 8.4.2. The figures of the objective-function surfaces also can be used to better understand nonlinear confidence intervals, as discussed in Chapter 7, Section 7.5.1 and Chapter 8, Section 8.4.3.

5.1.3 Convergence Criteria

Convergence criteria are needed to determine when to stop the modified Gauss–Newton iterative process. In UCODE_2005 and MODFLOW-2000, parameter estimation converges if either one of two convergence criteria are satisfied. By the first criterion, convergence is achieved when the parameter values change only a small amount from one parameter-estimation iteration to the next. This indicates that at the current regression iteration, the parameter values lie in a relatively flat area that is a minimum in the objective-function space. For untransformed parameters, this condition is satisfied if, for all parameters, *max-calculated-change* in Eq. (5.8) is less than *max-allowed-change* (user-specified variable TolPar of UCODE_2005 and TOL of MODFLOW-2000). That is, using the UCODE_2005 variable name,

$$\text{max-calculated-change} < \text{TolPar} \quad \text{for all } j = 1, NP \quad (5.9)$$

Preferably, this convergence is achieved in the final calibrated model with a criterion value no larger than 0.01. For log-transformed parameters, a modified form of Eq. (5.9) is used, as described in Section 5.4 and Appendix B.

TolPar typically is 0.01 or 0.001 for final regressions, indicating that convergence is reached when parameter values are changing between parameter-estimation iterations no more than 1 or 0.1 percent. There are situations in which it is advantageous for larger values of TolPar to be specified, especially for preliminary regressions.

Typically, TolPar has been the same for all parameters. UCODE_2005 and PEST, however, allow different values of TolPar to be assigned to different parameters. This is likely to allow inclusion of parameters that are too insensitive to achieve the small convergence criteria imposed on most parameters, but not so insensitive that the instabilities are very large. There has been little experience so far with this option.

By the second convergence criterion, the nonlinear regression converges if the model fit changes little over the course of two parameter-estimation iterations. If three consecutive values of the least squares objective function (Eq. (3.1) or (3.2)) change less than a user-defined amount (TolSOSC of UCODE_2005 and SOSC of MODFLOW-2000), nonlinear regression converges. The model-fit criterion often is useful early in the calibration process to avoid lengthy simulations that fail to improve model fit. However, satisfying this criterion does not provide as strong an indication that a minimum has been reached as the parameter-value criterion. Therefore, for final regression runs, it is preferable that the parameter-value criterion be satisfied.

As discussed by Cooley and Naff (1990, p.70), modified Gauss–Newton optimization typically converges within "a number of iterations equal to five or twice the number of parameters, whichever is greater." Well-conditioned problems (commonly those with large css values and little correlation) tend to converge in fewer iterations than poorly conditioned problems. It is rarely fruitful to increase the number of iterations to more than twice the number of parameters, and the resulting runs can take large amounts of computer time. It is generally more productive to consider alternative models (see Guideline 8, Chapter 11).

5.2 ALTERNATIVE OPTIMIZATION METHODS

Alternative algorithms for the minimization of the least-squared objective function with respect to parameter values include methods that use the gradient of the objective function and not the full sensitivity matrix (as used by, e.g., Carrera and Neuman, 1986; Hill, 1992; Xiang et al., 1993; Tarantola, 2005), and global optimization methods such as simulated annealing, genetic algorithms, tabu search, and shuffled complex evolution (SCE) (e.g., Zheng and Wang, 1996; Solomatine et al., 1999; Tsai et al., 2003b; Vrugt et al., 2003; Fazal et al., 2005).

For the first set of methods, the steepest descent direction, which equals the derivative of the objective function with respect to the parameter values, generally is calculated efficiently using adjoint states (Hill, 1992; Townley and Wilson, 1985). Scaled derivative of the objective function might be able to replace the composite scaled sensitivities in the guidelines, but this has not been tested. There are no replacements for the one-percent and dimensionless scaled sensitivities, the parameter correlations, and leverage statistics. However, adjoint states themselves can be useful, as discussed by Sykes et al. (1985). In addition, adjoint-state algorithms are often programmed to calculate the sensitivities and the parameter variance–covariance matrix to provide analyses that need them after convergence is reached. In this case, the methods suggested in this book could be used.

Global-search methods operate quite differently than gradient methods such as modified Gauss–Newton. Global-search methods do not use sensitivities. Instead, they proceed to the next set of parameters using a long history of the model fit produced by previous sets of parameters. The methods differ in how the previous sets are used. The advantage of global-search methods is their ability to identify parameter values that produce the best fit to observed values and prior information regardless of the degree of model nonlinearity and the presence of local minima. The disadvantage is that they are much more computationally intensive, often requiring execution times that are tens or hundreds of times as long as the execution times required by gradient-search methods.

Global-search methods are most useful for problems with very irregular objective-function surfaces that are not amenable to the much more numerically efficient gradient-search methods. For problems with such irregular objective functions, scaled sensitivities, parameter correlation coefficients, and leverage statistics are likely to change dramatically as parameter values change, and thus they are not useful. If the irregularity is local, the methods presented in this book may be useful in part of the solution space. For example, biological processes can be very nonlinear with regard to pH because outside some range of pH the organism dies. Within a certain range, however, and often the range of most interest, the process may be linear enough for the methods presented in this book to be useful.

Public-domain programs are available for implementing some common global-search methods. For example, MGO (Zheng and Wang, 2003) provides global-search capabilities using genetic algorithms, simulated annealing, and tabu search.

5.3 MULTIOBJECTIVE OPTIMIZATION

When developing models of many natural systems, data are scarce and it often is useful for all data relevant to model outputs to be considered simultaneously using a single objective function. Weighting is used to include many kinds of data. This book focuses on this approach to model calibration.

Alternatively, as mentioned in Chapter 3, Section 3.2.3, regression can be performed using subsets of the observations and prior information, whereby each subset is used to define a different objective function. This is called multiobjective optimization. A short description of multiobjective optimization can be found at http://www.fp.mcs.anl.gov/otc/Guide/OptWeb/multiobj/. Recent books on these methods include Statnikov and Matusov (1995) and Ehrgott (2000).

In multiobjective optimization, trade-offs between the different objective functions are an integral part of the evaluation. The trade-offs are obtained by weighting different objective functions differently. This has consequences in the implied relative accuracy of the data contained in each of the objective functions. This issue needs to be considered when determining feasible solutions using multiobjective function optimization. For example, solutions with weights that result in one set of data dominating or being ignored may be of interest as part of the analysis but generally are not viable solutions.

While for any one combination of weights the regression methods discussed here could be used, in recent applications multiobjective optimization has been accomplished with Shuffle Complex Evolution (SCE) (Nunoo and Mrawira, 2004).

5.4 LOG-TRANSFORMED PARAMETERS

Log-transformed parameters are often useful because the uncertainty of many parameters is best represented by a log-normal probability distribution. When the parameter is log-transformed, the uncertainty is then best represented by a normal

5.4 LOG-TRANSFORMED PARAMETERS

distribution, which is convenient to use. Log-transforming dependent variables was discussed in Chapter 3, Section 3.3.3, and is not addressed further here.

Log-transformation involves taking the logarithm of selected parameters. Thus, the parameters in vector b of Eq. (3.1) or (3.2) can be either native values or the log-transform of the native values. Log-transforming parameters can produce an inverse problem that converges more easily and prevents the native parameter values from becoming negative (Carrera and Neuman, 1986).

Log transformation can be defined using base e or base 10, where base e is also called the natural logarithm. Base 10 is easier for most modelers to use because a log-transformed value of 1 indicates a native value of 10, a log-transformed value of 2 indicates a native value of 100, and so on. Conversion between natural and base 10 logarithms involves multiplying by a factor of 2.3. In UCODE_2005 and MODFLOW-2000, the log-transform is implemented internally using natural logarithms ($\log e$); the input and output use base 10 logarithms as much as possible.

Even when some parameters are log-transformed, allowing modelers to consider native values has the advantage of emphasizing the connection between model results and field data. For example, even for log-transformed parameters, it is useful to define starting parameter values for regression runs as native values and to report final estimates as native values. UCODE_2005 and MODFLOW-2000 are constructed so that the user can consider native values as much as possible.

There are four special circumstances, one related to model input and three related to model output, in which the modeler has to deal more directly with log-transformed values.

The one model input situation occurs when there is prior information on the log-transformed parameter. In this case, only one parameter can be included in the prior information equation (one term in the summation presented after Eq. (3.2)). For UCODE_2005, the specified statistic needs to be related to the base 10 log of the parameter. MODFLOW-2000 can read statistics related to the native value and calculate the statistic related to the log-transformed parameter value. The value of the statistic specified can be determined using the methods described under Guideline 6 in Chapter 11.

The first model output situation is fairly subtle and will not be noticed by most modelers. It involves calculation of the damping parameter and the convergence criteria, which are used to control or measure the change in the parameter values. For native parameter values, Eqs. (5.8) and (5.9) are used. Calculation of these quantities is different for log-transformed parameters and is described in Appendix B.

The second model output situation is that log-transformed parameter estimates, standard deviations, coefficients of variation, and confidence interval limits appear in the MODFLOW-2000 and UCODE_2005 output files along with analogous statistics applicable to the native parameter values. In most circumstances, the user can ignore the statistics related to the log-transformed parameter values and instead use the statistics related to the native values. Related issues are discussed in Chapter 7, Section 7.2.4.

The third model output situation occurs when there is prior information defined for a log-transformed parameter. In this situation the associated residual, weight, and weighted residual are reported as the natural logarithms of the actual values, because the regression calculations use these values.

5.5 USE OF LIMITS ON ESTIMATED PARAMETER VALUES

Upper and lower limits on parameters that constrain possible estimated values are commonly available in inverse models and are suggested, for example, by Sun (1994, p. 35). Such limiting constraints on parameter values may appear to be necessary given the unrealistic parameter values that can be estimated through inverse modeling. However, this practice can disguise more fundamental modeling errors, as demonstrated by Poeter and Hill (1997) using a simple synthetic test case and Hill et al. (1998) using a complex synthetic test case. Anderman et al. (1996) show how unrealistic optimized values of recharge in a field problem revealed important model construction inaccuracies.

As discussed in Guideline 5 in Chapter 11, unrealistic estimated parameter values are likely to indicate either that (1) the data do not contain enough information to estimate the parameters or (2) there is a more fundamental model error. In the first circumstance, the best response is to use prior information or regularization on the parameter value, which tends to produce an estimate that is close to the most likely value, instead of at the less likely values that generally constitute the imposed upper and lower limits. In the second circumstance, the best response is to find and resolve the error. In the authors' opinions, the only circumstance in which it is advantageous to use limits on parameter estimates is to prohibit values that would make the process model fail.

To prevent the regression from calculating parameter values that would cause the process model to fail, UCODE_2005 supports limits. MODFLOW-2000 does not support limits because the required limits are imposed internally. For example, if a negative value of hydraulic conductivity is calculated by the regression, the value is changed to two orders of magnitude smaller than its starting value.

5.6 EXERCISES

Exercise 5.1 uses a two-parameter version of the test case to demonstrate the effects of extreme parameter correlation and the performance of the modified Gauss–Newton method and ends with an exercise asking students to derive the Gauss–Newton equation. In Exercise 5.2 the modified Gauss–Newton method is used to estimate the six parameters of the steady-state model.

Exercise 5.1: Modified Gauss–Newton Method and Application to a Two-Parameter Problem This exercise involves objective-function surfaces for a two-parameter version of the steady-state model described in Chapter 2, Section 2.2. Objective-function surfaces were discussed in Chapter 3, Section 3.5 and examples were shown in Figure 3.1. Objective-function surfaces constructed for the two combined parameters are used to show the effects of the different types of observations (hydraulic heads and flows) on objective-function surfaces and on nonlinear regression.

The two-parameter version of the model is developed by combining the six defined parameters into two parameters. One of the combined parameters multiplies hydraulic conductivities of the system and the other combined parameter multiplies recharge rates.

The combined hydraulic-conductivity parameter is defined so that if the parameter value equals 1.0, all the hydraulic-conductivity values equal their starting values. As the combined parameter value changes, all the hydraulic-conductivity values change proportionately. In UCODE_2005 or PEST, defining the combined hydraulic-conductivity parameter is straightforward. In MODFLOW-2000, the hydraulic conductivities controlled by the K_RB parameter (Table 3.1) cannot be combined with the other hydraulic conductivities and the value is fixed in the simulations. This does not compromise the analysis, because the observations are much less sensitive to K_RB than to most other parameters. The combined recharge parameter is defined so that if its value equals 1.0, both recharge parameters equal their starting values. As its value changes, the recharge values of zones 1 and 2 change proportionately. Combining the parameters in MODFLOW-2000 and UCODE_2005 is described in more detail in the computer instructions available from the web site described in Chapter 1, in Section 1.1.

Once a two-parameter model is constructed, UCODE_2005 or PEST can easily be used to produce data sets for constructing objective-function surfaces. There is no simple method of constructing such data sets with MODFLOW-2000. The objective-function values resulting from the UCODE_2005 and MODFLOW-2000 simulations are nearly identical. Objective-function surfaces using only hydraulic-head observations and including the flow observation with different weights are shown in Figure 5.4.

(a) Assess relation of objective-function surfaces to parameter correlation coefficients.

The objective-function surfaces from the two-parameter model are used in this exercise to investigate parameter correlations.

With hydraulic-head observations alone (Figure 5.4a), the objective-function surface is composed of parallel lines, and no unique minimum exists. In this situation, the two parameters are completely correlated, meaning that the correlation coefficients for all parameter pairs equal positive or negative 1.00 (here, +1.00). Thus, the hydraulic-head data cannot be used to estimate both parameters uniquely. Nonuniqueness would occur for any weighting, any combination of hydraulic-head observations in this system, and any number and configuration of hydraulic-conductivity and recharge parameters, as long as all parameters are estimated. Table 4.3a shows that with six parameters all correlation coefficients equal 1.00.

With the addition of the flow data weighted using a coefficient of variation of 10 percent (Figure 5.4b), which is a reasonable level of precision for such a measurement,

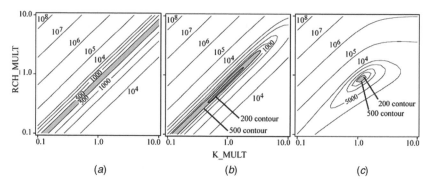

FIGURE 5.4 Objective-function surfaces for the two-parameter version of the simple test case, using (a) only head data, (b) head data and flow data weighted using a reasonable coefficient of variation of 10 percent, and (c) head data and flow data weighted using an unrealistically small coefficient of variation of 1 percent. In the regression, parameter K_MULT is log-transformed and parameter RCH_MULT is not. Logarithmic scales are used for both parameters so that the objective function values can be shown for a wide range of parameter values.

the objective-function surface indicates that a minimum exists, but that it covers a fairly broad area. When the flow is weighted using a coefficient of variation of 1 percent (Figure 5.4c), which indicates a generally unachievable level of precision, a clear minimum is apparent in the objective-function surface. Objective-function surfaces that contain a minimum indicate that the parameters are not completely correlated, and at the minimum the parameter correlation coefficients lie between the extreme values of −1.0 and 1.0. As shown in Figure 4.2, the correlation coefficient can be different for different sets of parameter values in nonlinear problems.

Problem
- Use Darcy's Law (Eq. (1.1)) to explain why the parameters are completely correlated when only hydraulic-head observations are used. In the equation, equate the recharge parameter, RCH_MULT, to q and the hydraulic-conductivity parameter, K_MULT, to K.
- Why does adding the flow measurement make such a difference in the objective-function surface?
- If adding one observation prevents the parameters from being completely correlated, what effect do you expect any error in that observation to have on the regression results?
- If the lines were all parallel to one of the axes, would the problem be correlation or insensitivity?

(b) *Examine the performance of the modified Gauss–Newton method.*

Parts of Exercise 5.1b involve modifying computer files and simulating the system. Instructions are available from the web site for this book described in

5.6 EXERCISES

Chapter 1, Section 1.1. Selected results are provided for students not performing the simulations.

First, perform nonlinear regression using the problem with two combined parameters for the situation in which only hydraulic-head observations are used. Perform regression in the four situations listed below. Check whether parameter estimation converged. Plot the progression of the parameter values produced by the modified Gauss–Newton method on the objective-function surface in Figure 5.4a. The parameter values are listed in Table 5.1.

1. Set MAX-CHANGE (see definition of this variable in Section 5.1.1, preceding Eq. (5.7)) to a large number, such as 10,000, and set the starting parameters to values near those that produce the best fit. The large value of MAX-CHANGE would never be used in practice; here it causes the regression to perform as if there were no damping in the modified Gauss–Newton method.
2. Keep MAX-CHANGE large, and set the starting parameter values to values located in the lower right corner of the objective-function surface of Figure 5.4a, where the surface is relatively flat.
3. Keep the starting values as in run 2, but decrease MAX-CHANGE to 0.5.
4. Keep MAX-CHANGE small, but set the starting parameter values to values near the upper central part of the objective-function surface. Compare estimates achieved in this run to those from run 3.

Second, perform nonlinear regression in the same four situations as described above, but include the flow observation weighted using a coefficient of variation of 10 percent. Plot the progression of the parameter values produced by the modified Gauss–Newton method on the objective-function surface in Figure 5.4b. The parameter values are listed in Table 5.2.

Third, perform nonlinear regression in the same four situations as described above, but include the flow observation and increase its weight by decreasing its coefficient of variation to 1 percent. Plot the progression of the parameter values produced by the modified Gauss–Newton method on the objective-function surface in Figure 5.4c. The parameter values are listed in Table 5.3.

Discuss the following questions related to the regression runs.

Problem
- Do the regression runs converge to optimal parameter values? How do the estimated parameter values compare among the different regression runs? Explain these results. Explain the difference in the progression of parameter values during these regression runs.
- Based on the results shown in Table 5.1, how can parameter correlation be detected if the correlation coefficients are not reliable? Is success of the modified Gauss–Newton method a reliable indicator?

TABLE 5.1 Parameter Values Obtained for the Four Runs Using Only Hydraulic-Head Observations

Iteration	Run 1		Run 2		Run 3		Run 4	
	K_MULT	RCH_MULT	K_MULT	RCH_MULT	K_MULT	RCH_MULT	K_MULT	RCH_MULT
1	1.00	1.00	9.0	0.20	9.0	0.20	1.0	9.0
2	1.09	0.86	1×10^{-14}	−12	4.5	0.11	0.74	4.5
3	1.06	0.81	2×10^{-14}	−6.0	2.4	0.056	0.51	2.25
4	1.05	0.81	3×10^{-14}	−3.0	1.2	0.079	0.76	1.3
5			4×10^{-14}	−1.5	0.60	0.12	0.99	0.94
6			7×10^{-14}	−0.75	0.32	0.18	1.06	0.82
7			1×10^{-13}	−0.37	0.26	0.21	1.03	0.79
8			2×10^{-13}	−0.19	0.26	0.20	1.02	0.78
9			3×10^{-13}	−0.09	0.26	0.20		
10			8×10^{-13}	−0.02				
	Converged		Did not converge		Converged		Converged	

TABLE 5.2 Parameter Values Obtained for the Four Runs Using Hydraulic-Head Observations and a Flow Observation Weighted Using a Reasonable Coefficient of Variation Value of 10 percent

Iteration	Run 1 K_MULT	Run 1 RCH_MULT	Run 2 K_MULT	Run 2 RCH_MULT	Run 3 K_MULT	Run 3 RCH_MULT	Run 4 K_MULT	Run 4 RCH_MULT
1	1.0	1.0	9.0	0.20	9.0	0.20	1.0	9.0
2	1.1	0.89	8×10^{-13}	0.89	4.5	0.22	1.0	4.5
3	1.2	0.89	1×10^{-12}	0.45	2.25	0.26	1.0	2.25
4	1.2	0.89	2×10^{-12}	0.22	1.2	0.38	1.1	0.89
5			4×10^{-12}	0.11	1.2	0.57	1.2	0.89
6			6×10^{-12}	0.056	1.2	0.86	1.2	0.89
7			1×10^{-11}	0.028	1.2	0.89		
8			2×10^{-11}	0.014	1.2	0.89		
9			3×10^{-11}	0.007				
10			4×10^{-11}	−0.0002				
	Converged		Did not converge		Converged		Converged	

TABLE 5.3 Parameter Values Obtained for the Four Runs Using Hydraulic-Head Observations and a Flow Observation Weighted Using an Unreasonable Coefficient of Variation Value of 1 percent

	Run 1		Run 2		Run 3		Run 4	
Iteration	K_MULT	RCH_MULT	K_MULT	RCH_MULT	K_MULT	RCH_MULT	K_MULT	RCH_MULT
1	1.0	1.0	9.0	0.20	9.0	0.20	1.0	9.0
2	1.1	0.91	9×10^{-13}	0.90	4.5	0.22	1.0	4.5
3	1.2	0.91	1×10^{-12}	0.45	2.25	0.26	1.0	2.25
4	1.2	0.91	2×10^{-12}	0.23	1.2	0.39	1.1	0.91
5			4×10^{-12}	0.11	1.2	0.58	1.2	0.91
6			7×10^{-12}	0.057	1.2	0.87	1.2	0.91
7			1×10^{-11}	0.028	1.2	0.91		
8			2×10^{-11}	0.014	1.2	0.91		
9			3×10^{-11}	0.007				
10			5×10^{-11}	−0.0002				
	Converged		Did not converge		Converged		Converged	

5.6 EXERCISES

(c) *Derive the Gauss–Newton normal equations (optional).*

Problem: As shown in Eq. (5.4), the unmodified Gauss–Newton equations can be expressed as

$$(X_r^T \omega X_r) d_r = X_r^T \omega (y - y'(b_r)) \tag{5.10}$$

Derive this equation by minimizing the objective function of Eq. (3.2) after substituting in the linearized version of $y'(b)$, which equals

$$y'(b) \approx y'(b_r) + X_r(b - b_r) \tag{5.11}$$

Exercise 5.2: Estimate the Parameters of the Steady-State Model In this exercise, a range of reasonable values is assigned to each of the six parameters of the steady-state flow system model described in Chapter 2, Section 2.2, and nonlinear regression is used to estimate the parameter values. Nonlinear regression is attempted without and then with prior information on two of the parameters.

Parts of this exercise involve modifying computer files and simulating the system. Instructions are available from the web site for this book described in Chapter 1, Section 1.1. Students not performing the simulations can skip those parts of the exercise.

(a) *Define a range of reasonable values for each parameter.*

In MODFLOW-2000 and UCODE_2005, a reasonable range specified by the user is compared with each parameter estimate as discussed in Section 5.5. This approach allows for a powerful check on likely model accuracy. In this exercise, ranges of reasonable values are defined for each steady-state model parameter. For students performing the simulations, files and instructions on the web site for this book can be used to complete the exercise.

(b) *First attempts at estimating parameters by nonlinear regression.*

In this exercise, first attempt to estimate the native values of all parameters, and then attempt to estimate the native values of the recharge parameters and the log-transformed values of the hydraulic-conductivity parameters. As discussed in Section 5.4, advantages of estimating the log of some parameter values instead of the native values are that (1) convergence problems can sometimes be alleviated, and (2) estimating the log-transform of a parameter prevents its native value from becoming negative. In this exercise, however, similar results are obtained whether or not the parameters are log-transformed. After the runs have been completed, consider the following questions. Selected results from the run without log-transformed parameters are presented in Tables 5.4, 5.5,

TABLE 5.4 Parameter Values for Each Parameter-Estimation Iteration of the Regression Run Without Log-Transformed Parameters in Exercise 5.2b

Iteration	HK_1	K_RB	VK_CB	HK_2	RCH_1	RCH_2
Start	3.00×10^{-4}	1.20×10^{-3}	1.00×10^{-7}	4.00×10^{-5}	63.07	31.54
1	3.40×10^{-4}	1.20×10^{-5a}	9.79×10^{-8}	2.82×10^{-5}	61.28	30.94
2	4.91×10^{-4}	2.16×10^{-5}	1.73×10^{-8}	4.00×10^{-7a}	63.90	22.10
3	4.92×10^{-4}	2.17×10^{-5}	1.00×10^{-9a}	4.00×10^{-7a}	64.06	21.94
4	4.92×10^{-4}	2.28×10^{-5}	3.00×10^{-9}	1.16×10^{-6}	62.84	23.13
5	4.94×10^{-4}	2.32×10^{-5}	1.00×10^{-9a}	4.00×10^{-7a}	63.20	22.76
6	4.94×10^{-4}	2.44×10^{-5}	3.00×10^{-9}	1.16×10^{-6}	62.03	23.92
7	4.97×10^{-4}	2.50×10^{-5}	1.00×10^{-9a}	4.00×10^{-7a}	62.40	23.54
8	4.97×10^{-4}	2.62×10^{-5}	3.00×10^{-9}	1.16×10^{-6}	61.27	24.65
9	4.99×10^{-4}	2.69×10^{-5}	1.00×10^{-9a}	4.00×10^{-7a}	61.63	24.28
10	4.99×10^{-4}	2.82×10^{-5}	3.00×10^{-9}	1.16×10^{-6}	60.54	25.36
			Did not converge			

[a]If hydraulic conductivities are assigned negative values by the regression, MODFLOW-2000 assigns them to be equal to the starting parameter value divided by 100.

and 5.6. For students performing the computer exercises, these results can be found in the model output files.

Problem

- In Table 5.4 and Figure 5.5, examine the changing values of the parameters and *max-calculated-change* (column 3 in the top part of Figure 5.5) to diagnose why the regressions did not converge. *Max-calculated-change* is the largest fractional parameter change that would occur if the damping parameter were equal to 1.0 (see Eq. (5.7)). A value of 0.50 indicates that the largest change (in absolute value) is a 50-percent increase in the parameter value, and a

TABLE 5.5 Composite Scaled Sensitivities for Each Parameter-Estimation Iteration of the Regression Run Without Log-Transformed Parameters in Exercise 5.2b

Iteration	HK_1	K_RB	VK_CB	HK_2	RCH_1	RCH_2
Start	41.3	0.214	0.783	11.0	27.4	25.6
1	41.6	20.9	0.515	7.45	37.9	30.4
2	35.2	10.8	0.033	0.085	28.2	17.2
3	35.0	10.8	0.473	0.499	28.1	17.0
4	35.1	10.3	0.464	0.543	27.2	17.8
5	35.1	10.1	0.472	0.499	27.2	17.4
6	35.2	9.58	0.464	0.543	26.3	18.1
7	35.2	9.35	0.471	0.499	26.2	17.8
8	35.3	8.92	0.463	0.544	25.4	18.4
9	35.2	8.68	0.471	0.499	25.4	18.1
10	35.4	8.28	0.464	0.545	24.6	18.8

5.6 EXERCISES

value of -1.00 indicates that the largest change is a 100-percent decrease. Also examine the sums of squared weighted residuals in Figure 5.5.

- In diagnosing why the regressions did not converge, also consider the composite scaled sensitivities (css) calculated for the starting parameter values (shown in Table 4.1 and Figure 4.3) and for the parameter values calculated at each iteration of the regression (shown in Table 5.5). What are the differences in the magnitudes of the css calculated at the starting parameter values? What might this indicate about the likelihood of estimating all six parameter values? How do the css calculated at iterations 2 through 10 differ from those calculated at the starting parameter values? How does this additional css information help explain the results shown in Table 5.4 and Figure 5.5?

SELECTED STATISTICS FROM MODIFIED GAUSS-NEWTON ITERATIONS

ITER.	MAX. PARAMETER PARNAM	CALC. CHANGE MAX. CHANGE	MAX. CHANGE ALLOWED	DAMPING PARAMETER
1	K_RB	-7.53194	2.00000	0.26554
2	HK_2	-2.11493	2.00000	0.94566
3	HK_2	-377.849	2.00000	0.52931E-02
4	VK_CB	33.7467	2.00000	0.59265E-01
5	HK_2	-79.4868	2.00000	0.25161E-01
6	VK_CB	33.8172	2.00000	0.59141E-01
7	HK_2	-71.9800	2.00000	0.27785E-01
8	VK_CB	33.8676	2.00000	0.59054E-01
9	HK_2	-62.8221	2.00000	0.31836E-01
10	VK_CB	33.9119	2.00000	0.58976E-01

SUMS OF SQUARED WEIGHTED RESIDUALS FOR EACH ITERATION

ITER.	SUMS OF SQUARED WEIGHTED RESIDUALS		
	OBSERVATIONS	PRIOR INFO.	TOTAL
1	1752.2	0.0000	1752.2
2	9286.4	0.0000	9286.4
3	650.03	0.0000	650.03
4	674.36	0.0000	674.36
5	603.16	0.0000	603.16
6	563.63	0.0000	563.63
7	504.43	0.0000	504.43
8	469.75	0.0000	469.75
9	420.64	0.0000	420.64
10	389.73	0.0000	389.73

PARAMETER ESTIMATION DID NOT CONVERGE IN THE ALLOTTED NUMBER OF ITERATIONS

FIGURE 5.5 Selected statistics from the modified Gauss–Newton iterations of the regression run without log-transformed parameters in Exercise 5.2b. This is a fragment from the global output file of MODFLOW-2000.

(c) *Assign prior information on parameters.*

The analysis from Exercise 4.1 and the performance of the regression in Exercise 5.2b suggested that prior information on parameters VK_CB and K_RB may be needed for the regression to converge. In this exercise, define the starting values of these two parameters as prior estimates. The prior information needs to be weighted in the same manner as observations need to be weighted. For both VK_CB and K_RB, assign a coefficient of variation of 0.3 to the prior estimates. Then, perform nonlinear regression.

TABLE 5.6 Parameter Values for Each Parameter-Estimation Iteration for Exercise 5.2c

Iteration	HK_1	K_RB	VK_CB	HK_2	RCH_1	RCH_2
Start	3.00×10^{-4}	1.20×10^{-3}	1.00×10^{-7}	4.00×10^{-5}	63.07	31.54
1	4.14×10^{-4}	1.16×10^{-3}	9.77×10^{-8}	2.12×10^{-5}	49.36	36.77
2	4.61×10^{-4}	1.17×10^{-3}	9.80×10^{-8}	1.37×10^{-5}	48.45	37.65
3	4.62×10^{-4}	1.17×10^{-3}	9.88×10^{-8}	1.49×10^{-5}	47.71	38.28
4	4.62×10^{-4}	1.17×10^{-3}	9.90×10^{-8}	1.53×10^{-5}	47.47	38.50
5	4.62×10^{-4}	1.17×10^{-3}	9.90×10^{-8}	1.54×10^{-5}	47.45	38.53
			Converged			

```
SELECTED STATISTICS FROM MODIFIED GAUSS-NEWTON ITERATIONS
         MAX. PARAMETER   CALC. CHANGE    MAX. CHANGE     DAMPING
 ITER.      PARNAM         MAX. CHANGE     ALLOWED       PARAMETER

   1        HK_2           -0.470616       2.00000        1.0000
   2        HK_2           -0.353595       2.00000        1.0000
   3        HK_2            0.106790       2.00000        0.81707
   4        HK_2            0.302890E-01   2.00000        1.0000
   5        HK_2            0.330029E-02   2.00000        1.0000

SUMS OF SQUARED WEIGHTED RESIDUALS FOR EACH ITERATION

               SUMS OF SQUARED WEIGHTED RESIDUALS
 ITER.    OBSERVATIONS    PRIOR INFO.      TOTAL
   1        1752.2         0.0000          1752.2
   2          81.454       0.19343E-01       81.473
   3          10.954       0.13860E-01       10.968
   4          10.562       0.93029E-02       10.571
   5          10.548       0.84770E-02       10.556
 FINAL       10.548        0.84769E-02       10.556

*** PARAMETER ESTIMATION CONVERGED BY SATISFYING THE TOL
CRITERION ***
```

FIGURE 5.6 Selected statistics from the modified Gauss–Newton iterations from Exercise 5.2c. This is a fragment from the global output file of MODFLOW-2000.

5.6 EXERCISES

Problem
- Compare the regression performance (Figure 5.6) with the results of Exercise 5.2b (Figure 5.5). Consider the values of *max-calculated-change* (column 3 in the top parts of Figures 5.5 and 5.6), and the sums of squared weighted residuals and the parameter values in Tables 5.5 and 5.6. For students who perform the computer exercises, these values are listed in the model output files.
- The two parameters with prior information have estimates that are nearly identical to the respective prior value. Why? If execution times are long, under what circumstances would you suggest including prior information and estimating these parameters? Explain.
- The statistic used to determine the weighting is important to whether the prior information can really be regarded as prior information or as regularization. For this problem, what would you conclude from the weighting used?

(d) *Parameter estimates and objective-function values.*

The starting, estimated, and true parameter values are shown in Table 5.7, and values of the objective function calculated for each of these parameter sets are shown in Table 5.8.

Problem
- Why do the estimated parameter values differ from the true parameter values?
- Comment on the objective-function values for the different parameter sets.

TABLE 5.7 Starting, Estimated, and True Values of the Parameters of the Steady-State Flow-System

Parameter Name	Starting Value	Estimated Value	True Value
HK_1	3.0×10^{-4}	4.62×10^{-4}	4.0×10^{-4}
K_RB	1.2×10^{-3}	1.17×10^{-3}	1.0×10^{-3}
VK_CB	1.0×10^{-7}	9.90×10^{-8}	2.0×10^{-7}
HK_2	4.0×10^{-5}	1.54×10^{-5}	4.4×10^{-5}
RCH_1	63.072	47.45	31.536
RCH_2	31.536	38.53	47.304

TABLE 5.8 Objective-Function Values Calculated Using the Starting, Estimated, and True Parameters

	Starting Parameters	Estimated Parameters	True Parameters
Objective-function value (heads and flows only)	1752.2	10.55	11.71
Objective-function value (heads, flows, and prior)	1752.2	10.56	23.13

(e) *Using objective-function surfaces to explore regression performance (Optional).*

As discussed in Chapter 3, Section 3.5, objective-function surfaces can be plotted for any two parameters, by systematically changing the values of only those two model parameters. In this exercise, use objective-function surfaces for selected parameter pairs to investigate the performance of the regression in Exercises 5.2b, 5.2c, and 5.2d. This is easily accomplished with UCODE_2005, as discussed in the instructions for this exercise on the web site described in Chapter 1, Section 1.1.

Problem: What insight is gained beyond what was provided by the sensitivity analysis using composite scaled sensitivities and parameter correlation coefficients?

6

EVALUATING MODEL FIT

The fit of model simulated values to the observations and prior information tests the ability of the model to realistically represent the simulated system. This chapter presents methods of evaluating model fit in the order they generally are used in a modeling project.

6.1 MAGNITUDE OF RESIDUALS AND WEIGHTED RESIDUALS

The first step in evaluating model fit generally involves determining the largest (in absolute value) residuals and weighted residuals, which were defined in Chapter 3, Section 3.4.3. Weighted residuals have the advantage of including the effects of errors. Large absolute values of weighted residuals indicate unexpectedly poor model fit more reliably than do large absolute values of unweighted residuals.

In initial model runs, the largest weighted residuals often indicate gross errors in the model, the observation data, the simulated equivalents of the observations, and/or the weighting. For example, some observations might be misrepresented in the simulation, suffer from incorrect data interpretation, or simply have been entered incorrectly in model input files. To help detect such problems, UCODE_2005 and MODFLOW-2000 output lists the five largest positive weighted residuals and the five largest negative weighted residuals (Hill et al., 2000, Table 14; Poeter et al., 2005, Table 28). In addition, the programs print the percent contribution of these individual weighted residuals to the objective function. Weighted residuals that

---

Effective Groundwater Model Calibration: With Analysis of Data, Sensitivities, Predictions, and Uncertainty. By Mary C. Hill and Claire R. Tiedeman
Published 2007 by John Wiley & Sons, Inc.

individually account for a large percent of the objective-function value are suspect and should be checked carefully. In subsequent model runs, after the major problems contributing to very large weighted residuals have been corrected, analysis of systematic misfit and statistics that measure overall model fit become increasingly important, as noted in Exercise 3.2.

6.2 IDENTIFY SYSTEMATIC MISFIT

Systematic model misfit can reveal problems with the model and/or the data. For example, a groundwater model may provide a good match to hydraulic heads but a very poor match to streamflow gains and losses. This indicates that the model poorly represents the dynamics of the simulated flow system. Alternatively, the model may fit all observations well, but only with parameter values that differ substantially and systematically from prior information. For example, in groundwater models, estimated hydraulic conductivities often are smaller than hydraulic conductivities measured by aquifer tests. One possibility is that aquifer tests use wells that commonly are screened in the subsurface materials with the highest hydraulic conductivities, and the volumes of these materials are small relative to the volumes represented by model hydraulic conductivities. In this situation, the problem is that the prior information is defined using data that are not representative of much of the subsurface included in the model.

Systematic misfit can be detected through application of the methods presented in this chapter. Often it is useful to apply the methods to subsets of the residuals and weighted residuals. Subsets generally are defined on the basis of observation or prior information type, location, time, and so on. Subset definition is problem dependent, and useful subsets often are determined only after some experimentation. For example, in a groundwater model it may be important for subsets to be defined based on well depth, model layer, distance from some types of boundaries, and so on. It is important to calculate the overall measures of model fit presented in Section 6.3 using subsets of observations and prior information, and the entire data set. For the graphical analyses of model fit presented in Section 6.4, it is important to use different symbols to represent different sets of observation and prior information.

6.3 MEASURES OF OVERALL MODEL FIT

Measures of overall model fit are single values that provide a quick evaluation of how well a model matches all or subsets of the observations and prior information. Measures calculated for alternative models of the same system often are used to judge how well the different models perform. The measures described here can be used in this way as long as the number of observations and prior information and their weighting do not vary between the models being compared.

6.3 MEASURES OF OVERALL MODEL FIT

The measures can be thought of as representing two competing goals—obtaining as good a fit as possible to observations and prior information and using as few parameters as possible. A better fit can always be obtained by increasing the number of parameters, but, as discussed in Chapter 11, Guideline 1, too many parameters can degrade the predictive ability of the model. Thus, all but the first measure presented here include a penalty for additional parameters. For the statistic to have a smaller value when a parameter is added, the model fit needs to be improved enough to overwhelm the increase in the penalty.

We do not provide an extensive list of overall measures of model fit. Measures not mentioned here include Kashyap's measure (Kashyap, 1982) and GCV (Craven and Whaba, 1979). These measures can be calculated using, for example, MMA (Poeter and Hill, in press).

6.3.1 Objective-Function Value

The value of the weighted least-squares or maximum-likelihood objective function (Eq. (3.1) to (3.3)) often is used informally to indicate model fit. Objective functions are rarely used for more formal comparisons because their values nearly always decrease as additional parameters are defined in the model and included in parameter estimation.

6.3.2 Calculated Error Variance and Standard Error

A common indicator of the overall magnitude of the weighted residuals is the calculated error variance, s^2, which equals (Cooley and Naff, 1990, p. 166; Ott, 1993)

$$s^2 = \frac{S(b)}{(ND + NPR - NP)} \quad (6.1)$$

where $S(b)$ is the weighted least-squares objective-function value (Eq. (3.1) or (3.2)) and the other variables are defined for Eq. (3.1). s^2 is dimensionless if the weighting is defined as suggested in this book. The square root of the calculated error variance, s, is called the standard error of the regression. Smaller values of both the calculated error variance and the standard error indicate a closer overall fit to the observations, and smaller values are preferred as long as the weighted residuals do not indicate model error (discussed in Section 6.4).

Overall Fitted Error Statistics A disadvantage to using s^2 and s directly as measures of model fit is that they have little intuitive appeal because they are dimensionless.

To obtain dimensional values that more effectively reflect the fit, s can be used to multiply the standard deviations and coefficients of variation used to calculate the weights for any group of observations. The resulting statistics are defined by Hill (1998, pp. 19, 53) as fitted error statistics, of which the fitted standard deviation and the fitted coefficient of variation are examples. These statistics express the average fit to different types of observations. For example, if a standard deviation of measurement error equal to 0.3 m is used to calculate the weights for most of the hydraulic-head observations and the calculated standard error is 3.0, the fitted standard deviation of 0.3 m × 3.0 = 0.9 m represents the overall fit achieved for these hydraulic heads. If a coefficient of variation of 0.25 (25 percent) is used to calculate weights for a set of spring-flow observations and the calculated standard error is 2.0, the fitted coefficient of variation of 0.25 × 2.0 = 0.50 (50 percent) represents the overall fit achieved to these spring flows. The standard deviation or coefficient of variation used to calculate the weighting reflects knowledge about observation or prior information error, and the fitted standard deviation or fitted coefficient of variation reflects both model fit and knowledge about error represented in the weighting. Although the fitted error statistic is not standard statistical terminology, in the authors' experience, it provides a meaningful way of communicating model fit.

Generally, this approach applies only if the fitted error statistic summarizes the fit to a fairly large number of observations or prior information. One or a few values can be evaluated more effectively by considering their residuals and weighted residuals directly.

Interpret the Calculated Error Variance The interpretation of the calculated error variance, s^2, or standard error, s, is related to the weighting used in the regression. If the weight matrix is defined as suggested in Eq. (3.8) or (3.9) and if the fit achieved by regression is consistent with the data accuracy as reflected in the weighting, the expected value of both the calculated error variance and the standard error is 1.0. This can be proved by substituting Eq. (3.2) into Eq. (6.1) and taking the expected value. It can be demonstrated using generated random numbers instead of residuals, as described in Exercise 6.1b.

If the calculated error variance or the standard error is significantly different from a value of 1.0, this indicates that the model fit is inconsistent with the weighting. A value of s or s^2 that is significantly greater than 1.0 indicates that the residuals are larger, on average, than is consistent with the statistics used to calculate the weighting. That is, the model fit is worse than would be expected based on the analysis of error used to determine the weighting. A value of s or s^2 that is significantly less than 1.0 indicates that the residuals are smaller, on average, than is consistent with the statistics used to calculate the weights. That is, the model fits the observations better than would be expected based on the analysis of observation error used to determine the weights.

For the calculated error variance, significant deviations from 1.0 can be evaluated by constructing a confidence interval. The confidence interval limits can be calculated as (Ott, 1993, p. 332)

$$\frac{ns^2}{\chi_U^2}; \frac{ns^2}{\chi_L^2} \qquad (6.2)$$

where n is the degrees of freedom, here equal to $ND + NPR - NP$ (see Eq. (3.1) for definitions); χ_U^2 is the upper tail value of a chi-square distribution (Appendix D,

Table D.5) with n degrees of freedom, with the area to the right equal to one-half the significance level of the confidence interval (the significance level, α, is 0.05 for a 95-percent interval); and χ_L^2 is the lower tail value of a chi-square distribution with n degrees of freedom with the area to the left equal to one-half the significance level.

Significant deviations from 1.0 also can be evaluated using a χ^2 test statistic, (Ott, 1993, p. 334). To consider the standard error instead of the calculated error variance, take the square root of each limit in Eq. (6.2). The confidence intervals are used to evaluate significant deviations of s or s^2 from 1.0 as described below.

If the confidence interval on s^2 includes the value 1.0, $\alpha = 0.05$, and the weighted residuals are random, then s^2 does not significantly deviate from 1.0 at a 5-percent significance level and model fit is consistent with the statistics used to calculate the weights on the observations and prior information. Expressed in terms of probability, there is only a 5-percent chance that the model fits the data in a way that contradicts the following assumptions: (1) the model is reasonably accurate and (2) the statistics used to calculate the weights correctly reflect the observation and prior information errors.

If the confidence interval does not include 1.0, the model fit is inconsistent with the statistics used to calculate the weighting. Of interest is whether statistics that are consistent with the model fit are realistic measures of error in the observations and prior information. For example, if the standard error of the regression is 2.0, statistics that would be consistent with the model fit would be 2.0 times the standard deviations or coefficients of variation used to determine the weighting. If a streamflow observation was thought to have a 5-percent coefficient of variation, would an increase by a factor of 2.0 to 10 percent be unreasonable? If so, unaccounted for observation error could not explain the large standard error, and model error would be suspected. Here, we refer to the adjusted statistics as *individual fit-consistent statistics*. They differ from overall fitted error statistics in that individual observations and prior information can be considered and are often important. For individual fitted error statistics, the variances, standard deviations, and coefficients of variation used to calculate the weights are adjusted. New weights that are consistent with the model fit are obtained by multiplying the variances by s^2 and the standard deviations and coefficients of variation by s. If the regression were carried out with these new weights, the same parameter estimates would be obtained and the residuals would be the same, but the weighted residuals would be different and s^2 would equal 1.0.

If the entire confidence interval on s^2 is less than 1.0 and the weighted residuals are randomly distributed, the model fit is better than anticipated based on the statistics used to calculate the weights. This is not necessarily an indication of the overfitting discussed in Guideline 1, but the possibility should be considered. Hill et al. (1998), obtained a small s^2 value because the actual observation error for a synthetic test case was smaller than expected. In this unusual case, the individual fitted error statistics were much smaller than the statistics used to determine the weighting and more accurately reflected the observation error.

If the entire confidence interval on s^2 is greater than 1.0, which is common, then the model fit is worse than anticipated based on the statistics used to calculate the weights. In this situation, the resulting interpretation depends on whether (1) the weighted residuals are randomly distributed, and (2) the individual fitted error statistics are so large that they could not reasonably be caused by observation and prior information errors. Randomness of the weighted residuals can be evaluated as discussed in Section 6.4 and in Exercise 6.2.

After the randomness of the weighted residuals has been evaluated and individual fitted error statistics have been calculated, the analysis depends on which of the following three situations apply.

1. The weighted residuals are randomly distributed and individual fitted error statistics can be justified (meaning that the observations and prior information error actually could be sufficiently larger than originally assumed). In this case, the analysis indicates that the model fit is consistent with the model being a reasonably accurate representation of the true system.
2. The weighted residuals are randomly distributed but individual fitted error statistics reflect unreasonable levels of observation and prior information error. In this case, the results of Hill et al. (1998) suggest that model error is significant but many sources of model error probably contribute to the lack of model fit. A few sources of model error do not dominate the model. This situation is not uncommon, and if the results of Hill et al. (1998) are valid, model predictions and measures of uncertainty can be accurate. Future studies are needed to test this conclusion.
3. The weighted residuals are not randomly distributed. In this case, the analysis suggests that there may be substantial and problematic model error. The best approach is to evaluate the model to determine the cause of the non-random residuals, and to evaluate the cause of any very large weighted residuals.

6.3.3 AIC, AIC_c, and BIC Statistics

The calculated error variance and standard error are sometimes criticized for not sufficiently representing the drawbacks associated with increasing the number of estimated parameters. The AIC, AIC_c, and BIC statistics were developed in the time-series literature to address this criticism (Brockwell and Davis, 1987; Burnham and Anderson, 2002). These statistics are calculated as the sum of the maximum-likelihood objective function (Eq. (3.3)) evaluated at the optimal parameter values, $S'(b')$, and terms that become large as more parameters are added. Although these statistics were developed for time-series problems, Carrera and Neuman (1986) successfully used them to discriminate between different parameterizations of a groundwater flow model. The references cited below for these statistics provide statistic derivations and additional discussion.

The AIC statistic was developed by Akaike (1973, 1974) and was corrected by Sugira (1978) to obtain AIC_c as described by Burnham and Anderson (2002, p. 66)

AIC$_c$ and AIC are calculated as:

$$\text{AIC}_c(b') = S'(b') + NP \times 2 + \frac{2 \times NP \times (NP+1)}{(NOBS + NPR - NP - 1)} \quad (6.3a)$$

$$\text{AIC}(b') = S'(b') + NP \times 2 \quad (6.3b)$$

where $S'(b')$ is the maximum-likelihood objective function of Eq. (3.3), NP is the number of estimated parameters, $NOBS$ is the number of observations used in the regression, and NPR is the number of prior estimate equations used in the regression. Often, $S'(b')$ is replaced in these equations by $n \times \log(S(b')/n))$, where $S(b')$ is defined in Eq. (3.2). AIC$_c$ is needed if $NOBS/NP < 40$ for any model considered.

The statistic BIC was developed by Akaike (1978) as a response to concern that AIC sometimes promoted use of more parameters than was required. The version of this statistic used by Carrera and Neuman (1986) is:

$$\text{BIC}(b') = S'(b') + NP \times \ln(NOBS + NPR) \quad (6.4)$$

For these statistics, smaller values generally indicate a more accurate model. However, if the statistics for a model with fewer parameters are only slightly larger than the statistics of another model with more parameters, it may be preferable to select the model with fewer parameters, unless the investigator has other information indicating the validity of the more complicated model. Burnham and Anderson (2002) suggest that of the three statistics, AIC$_c$ has distinct advantages.

These statistics can be cited in addition to s^2 or s; it is common to present all of these values in a table and/or graphically, for the models considered. MODFLOW-2000 prints AIC and BIC, UCODE_2005 prints AIC, AIC$_c$, and BIC.

6.4 ANALYZING MODEL FIT GRAPHICALLY AND RELATED STATISTICS

In addition to overall measures of model fit, several graphical analyses and related statistics can be used to assess whether the match of simulated values to observed values contradicts the requirements of Section 3.3 and thus indicates that the regression is not valid. The graphical methods were developed for groundwater inverse modeling by Cooley and Naff (1990), using the work of Draper and Smith (1981, 1998), and were slightly modified by Hill (1992, 1994). Required data files are produced by UCODE_2005 and MODFLOW-2000. The graphical methods are described in the following sections. Examples are presented here and in Exercise 6.2. In Chapter 10, Table 10.2 lists these graphs with questions they are likely to address and guidelines they are likely to support.

6.4.1 Using Graphical Analysis of Weighted Residuals to Detect Model Error

The graphical analyses of model fit presented here focus on weighted residuals. Regression results are valid (basically, model error is not indicated) only if (1) the weighted residuals from all types of observations and prior information appear to be statistically consistent (they all look like they have the same variance and a mean of zero) or (2) any statistical inconsistency can be explained by the correlation of weighted residuals expected through the fitting process imposed by the regression. The statistical consistency is evaluated using graphs of the weighted residuals with respect to: weighted or unweighted simulated values, independent variables such as space and time, and normal order statistics. This chapter focuses on using graphical analyses to detect model error. If model error is detected, see Guideline 9 in Chapter 12 for a discussion of how to proceed.

6.4.2 Weighted Residuals Versus Weighted or Unweighted Simulated Values and Minimum, Maximum, and Average Weighted Residuals

Graphs of weighted residuals can be plotted against either weighted or unweighted simulated values. The need to plot weighted instead of unweighted residuals and the advantages and disadvantages of weighting the simulated values are discussed here.

From an intuitive perspective, it makes sense that a model that fits the data well should not demonstrate a distinctively different fit to similar observations. In groundwater models, for example, simulated hydraulic heads would be expected to match observed hydraulic heads equally well in areas of high and low hydraulic head, all else being equal. Consider, for example, one area where the hydraulic heads are five meters, on average, above the heads in the other area. Residuals that are all negative in one area and all positive in the other would indicate model error.

Yet all observations are not similar. Weighted residuals need to be considered instead of unweighted residuals when errors associated with observations or prior information have different variances and/or are correlated for the analysis to detect model error. In the groundwater example, if the average depth to water in the observation wells and/or the methods used to determine the elevation of the wells differed in the two areas, larger residuals might be expected in one area. Use of weighted residuals eliminates the effects of this expected difference in model fit to observations in the two areas, allowing the graphs to be used more easily to detect model error.

Weighted simulated values are suggested for these graphs by Draper and Smith (1998, p. 63–64) because, in most circumstances, weighted residuals and weighted simulated values are statistically independent. However, Hill (1994) shows that three problems can occur. In some situations, plotting against unweighted residuals is advantageous. Thus, graphs of weighted residuals versus weighted and unweighted simulated values are considered here. The problems are discussed after describing how the graphs are constructed.

6.4 ANALYZING MODEL FIT GRAPHICALLY AND RELATED STATISTICS

Example graphs are shown in Figure 6.1. Ideally, weighted residuals are scattered evenly about 0.0 for the entire range of values on the horizontal axis, as in Figure 6.1. Figure 6.2 shows examples of graphs for which the weighted residuals are not random with respect to the weighted simulated values. When using MODFLOW-2000, the data needed to produce graphs of weighted residuals against weighted simulated values are listed in the output file with filename extension _ws. For UCODE-2005, the file with extension _ws includes weighted residuals and unweighted simulated values; weighted simulated values are listed in the output file with extension _ww.

The importance of testing for systematic misfit to subsets of the observations and prior information was discussed in Section 6.2. To identify systematic misfit, plot the weighted residuals for different subsets with different symbols. For example, in the exercises at the end of this chapter, hydraulic heads, flows, and prior information are plotted using different symbols. MODFLOW-2000 and UCODE_2005 facilitate this by allowing the user to specify a plot-symbol variable for each observation and piece of prior information. The plot-symbol variables are integers; plotting routines can use the integers to control the symbols used in graphs.

The statistics that summarize the distribution of the weighted residuals are the minimum, maximum, and average weighted residuals. The minimum and maximum weighted residuals display the range of weighted residuals at a glance. In practice, especially in the initial stages of calibration, the minimum and maximum weighted residuals often identify problems with the model or the observation data, as discussed in Section 6.1. The average weighted residual is a simple arithmetic average of the weighted residuals and ideally equals zero. In linear regression, the average always equals zero for the optimized parameter values; in nonlinear regression, the value of the average weighted residual generally approaches zero as calibration proceeds. In MODFLOW-2000, these statistics are printed in the LIST file, which is defined in the Name File; in UCODE_2005 they are printed in the main output file for the Forward, Sensitivity-Analysis, and Parameter-Estimation modes. That output file has filename extension #uout.

The three problems that can occur with graphs of weighted residuals versus weighted or unweighted simulated values and solutions for each are described next.

The first problem occurs when the weighted or unweighted simulated values extend over a wide range so that it is not possible to scale the associated axis to obtain a useful graph. The problem is illustrated in Exercise 6.2a. This problem can sometimes be resolved by using weighted simulated values or by log-transforming the axis for the weighted or unweighted simulated values. Another possibility is to multiply the weighted or unweighted simulated values of extreme points by a factor so that they plot closer to the other values. To ensure that the graph can still be used to test whether the weighted residuals vary systematically for any one type of data, this adjustment needs to be applied carefully. It is usually a good idea to apply the same factor to all weighted or unweighted simulated values for a given data type.

The second problem occurs when weights are calculated using coefficients of variation, as suggested after Eq. (3.5) and discussed in Chapter 11 under Guideline 6. In this case, the weight for an observed value y_i equals $1/(c.v.y_i)^2$ where $(c.v.)_i$

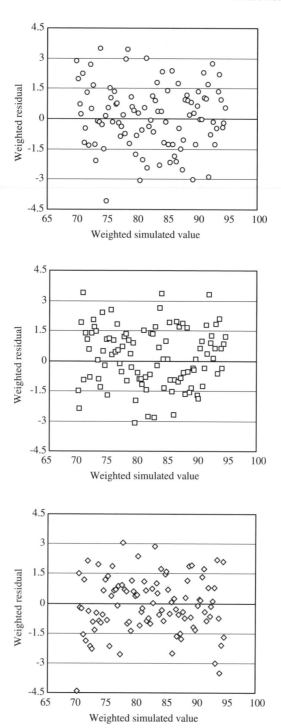

FIGURE 6.1 Example graphs of weighted residuals and weighted simulated values with no model bias. The values of weighted residuals plotted here are three different realizations of 100 generated normally distributed numbers with mean 0.0 and standard deviation 1.5. The standard deviation is used to define grid lines for the weighted residuals.

6.4 ANALYZING MODEL FIT GRAPHICALLY AND RELATED STATISTICS

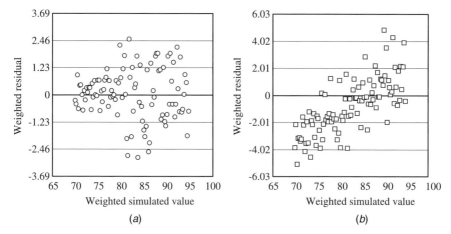

FIGURE 6.2 Example graphs of weighted residuals and weighted simulated values showing evidence of model bias for two different data sets. (*a*) The weighted residuals associated with smaller weighted simulated values vary less (have smaller variance). (*b*) The weighted residuals increase with increasing weighted simulated value. The standard deviations are of the weighted residuals (1.23 and 2.01) are used to define grid lines for the weighted residuals.

is the coefficient of variation. Then the weighted residual and the associated weighted simulated value are calculated as

$$\omega_i^{1/2} \times (y_i - y_i') = \frac{1}{(c.v.)_i y_i} \times (y_i - y_i') = \frac{1}{(c.v.)_i} - \frac{y_i'}{(c.v.)_i y_i} \quad (6.5a)$$

$$\omega_i^{1/2} \times y_i' = \frac{y_i'}{(c.v.)_i y_i} \quad (6.5b)$$

If the weight is calculated using the simulated value, as can be done using UCODE_2005, the weighted residual and associated weighted simulated value are calculated as

$$\omega_i^{1/2} \times (y_i - y_i') = \frac{1}{(c.v.)_i y_i'} \times (y_i - y_i') = \frac{y_i}{(c.v.)_i y_i'} - \frac{1}{(c.v.)_i} \quad (6.6a)$$

$$\omega_i^{1/2} \times y_i' = \frac{1}{(c.v.)_i} \quad (6.6b)$$

The second term of Eq. (6.5a) equals Eq. (6.5b). If $(c.v.)_i$ is the same for multiple observations, then the weighted residuals for these observations plot on a straight line with a slope of -1. Equation (6.6b) is the same for all simulated values with the same coefficient of variation, so the placement on the horizontal axis completely ignores the simulated value. In both circumstances, the vertical distribution of the weighted residuals still can be used to test their independence, but the purpose of the graph described in this section has largely been circumvented.

The most straightforward way to rectify this situation is to use the unweighted instead of weighted simulated values. Alternatively, Hill (1994) multiplies the weighted simulated values of Eq. (6.5b) by the observed values, y_i. Then, the modified weighted simulated values are

$$\omega_i^{1/2} \times y_i' \times y_i = \frac{y_i'}{(c.v.)_i} \tag{6.7a}$$

Similarly, multiplying Eq. (6.6b) by the simulated value yields

$$\omega_i^{1/2} \times y_i' \times y_i' = \frac{y_i'}{(c.v.)_i} \tag{6.7b}$$

Both modifications resolve the second problem but can worsen the first problem discussed earlier. The first option of plotting against simulated values is probably the most useful in many circumstances.

The third problem occurs when estimated parameters that have prior information are scaled using the value of the prior information. Such scaling is sometimes convenient because it produces prior information values that equal 1.0. Then, during regression, the percent change between the estimate and the prior value is obvious. For example, with this type of scaling, a parameter estimate of 1.5 indicates that the estimate is 50 percent larger than the prior value. In this circumstance, the weighted residuals for the prior information are calculated as

$$\omega_p^{1/2} \times (1.0 - P_p') \tag{6.8}$$

and the weighted simulated values are calculated as

$$\omega_p^{1/2} \times P_p' \tag{6.9}$$

If $\omega_p^{1/2}$ is equal to 1 divided by the coefficient of variation, which is common, and the coefficient of variation is the same for multiple prior parameters, which also is common, a graph of the weighted residuals and weighted simulated values forms a straight line with a slope of -1.0. As for the second problem, the graph would not indicate whether the weighted residuals vary systematically with the size of the simulated value. A meaningful graph can be obtained by plotting against unweighted simulated values or by calculating a modified weighted simulated value as

$$\omega_p^{1/2} \times P_p' \times P_p \tag{6.10}$$

Caution should be taken when altering the values used to create the graphs discussed in this section to ensure that the resulting graph serves the intended purpose of testing whether weighted residuals show systematic patterns of model fit when compared to simulated values.

6.4.3 Weighted or Unweighted Observations Versus Simulated Values and Correlation Coefficient R

Ideally, simulated values are close to observed values, so that graphs of observations against simulated values fall along a straight line with slope equal to 1.0 and an intercept of zero. Correspondingly, graphs constructed using weighted observations and weighted simulated values would have the same characteristics and can be useful when the weighting results in the values having a more condensed range. They also have the advantage that variations in expected error variance are accounted for already, making it easier to detect model error using the graph.

Comparing Figure 6.3 with Figure 6.2 shows that, all else being equal, plotting weighted residuals provides a better test of model bias than plotting weighted or unweighted observations. This is because the typically large range in magnitudes of the weighted or unweighted observations and simulated values can obscure trends in the differences between them. The greater the range, the smaller the same difference looks. This limitation is eliminated when weighted residuals are considered instead. Graphs of weighted residuals are less commonly used. Perhaps some modelers prefer to disguise model error.

When using MODFLOW-2000 or UCODE_2005, graphs of observed versus simulated values can be produced using data from the output file with filename extension _os; graphs of weighted observed values versus weighted simulated values can be produced using data listed in output file with filename extension _ww.

The correlation coefficient between the weighted observations and the weighted simulated values measures how well the trends in the weighted simulated values match those of the weighted observed values and, therefore, how closely the

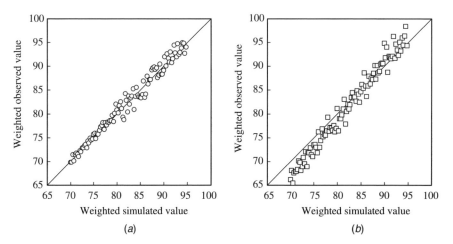

FIGURE 6.3 Example graphs of weighted observed and simulated values. The data plotted are the same data shown in the graphs of Figure 6.2. This display of the data does not reveal problems as clearly as do the graphs of weighted residuals and weighted simulated values in Figure 6.2.

points on a graph such as that shown in Figure 6.3 fall on the line. This correlation coefficient, R, can be calculated for a diagonal weight matrix as (Cooley and Naff, 1990, p. 166)

$$R = \frac{\sum_{i=1}^{ND}(\omega_i^{1/2}y_i - m_y)(\omega_i^{1/2}y_i' - m_{y'})}{\left[\sum_{i=1}^{ND}(\omega_i^{1/2}y_i - m_y)^2\right]^{1/2} \times \left[\sum_{i=1}^{ND}(\omega_i^{1/2}y_i' - m_{y'})^2\right]^{1/2}} \quad (6.11\text{a})$$

where y_i and y_i' are observed and simulated values, ω_i is the weight for the ith observation, and m_y and $m_{y'}$ are the means of the weighted observations and simulated values. For a full weight matrix the equation is

$$R = \frac{(\omega^{1/2}y - m_y)^T(\omega^{1/2}y' - m_{y'})}{\left[(\omega^{1/2}y - m_y)^T(\omega^{1/2}y - m_y)\right]^{1/2}\left[(\omega^{1/2}y' - m_{y'})^T(\omega^{1/2}y' - m_{y'})\right]^{1/2}} \quad (6.11\text{b})$$

where y, y', and ω were defined for Eq. (3.2). m_y and $m_{y'}$ are vectors with all ND elements equal to

$$m_y = \sum_{q=1}^{ND}(\omega^{1/2}y)_q/ND \quad (6.12)$$

$$m_{y'} = \sum_{q=1}^{ND}(\omega^{1/2}y'(b))_q/ND \quad (6.13)$$

Thus, m_y is a vector with each component equal to the average of the weighted dependent-variable observations, and $m_{y'}$ is an analogous vector using the weighted simulated values. Generally, a value of R that is greater than 0.90 indicates that the trends in the weighted simulated values closely match those of the weighted observations. However, R depends on the range of values and wide ranges are common when using different types of data. Use care when interpreting R.

When there is prior information, R also is calculated with y, $y'(b)$, and ω augmented as in Appendix A, in which case $ND + NPR$ replaces ND when calculating m_y and $m_{y'}$. In MODFLOW-2000, these statistics are printed in the LIST file; in UCODE_2005, they are printed in the main output file (filename extension #uout).

6.4.4 Graphs and Maps Using Independent Variables and the Runs Statistic

It is very important to evaluate weighted and unweighted residuals, observations, and simulated values with respect to the independent variables of a problem, such as space and time. Ideally, the signs and magnitudes of the weighted residuals plotted spatially on maps or temporally on graphs such as hydrographs show no discernible patterns and appear random. Distinct patterns, such as the presence of only positive weighted residuals in a particular model layer or region, can indicate substantial model error that may cause simulated predictions to be incorrect and misleading. Distinct patterns often are present, however, especially in temporal graphs. It is crucial for the modeler to understand the cause of such patterns, and analysis of these problems can lead to changes in model construction that increase model accuracy. Examples of maps used to evaluate weighted and unweighted residuals are shown in Exercise 6.2 and in Chapter 15.

The runs test (Cooley, 1979; Draper and Smith, 1998, pp. 192–198) takes the order of the residuals into account, which is ignored in all the other summary statistics. The runs test produces a summary statistic that checks for the randomness of weighted residuals with respect to the order in which they are listed. A sequence of residuals of the same sign is called a run, and the number of runs is counted and the value assigned to the variable u. For example, the sequence of numbers $-5, -2, 4, 3, 6, -4, 2, -3, -9$ has the five runs $(-5, -2), (4, 3, 6), (-4), (2), (-3, -9)$, so that $u = 5$. By using the total number of positive residuals (n_1), and the total number of negative residuals (n_2), u can be defined as a random variable. If $n_1 > 10$ and $n_2 > 10$, u is normally distributed with mean, μ, and variance, σ^2, equal to

$$\mu = \left(\frac{2n_1 n_2}{n_1 + n_2}\right) + 1.0 \tag{6.14}$$

$$\sigma^2 = \frac{2n_1 n_2 (2n_1 n_2 - n_1 - n_2)}{(n_1 + n_2)^2 (n_1 + n_2 - 1)} \tag{6.15}$$

The actual number of runs in a data set is compared with the expected value using test statistics. The test statistic for too few runs is

$$z_f = (u - \mu + 0.5)/\sigma \tag{6.16}$$

The test statistic for too many runs is

$$z_m = (u - \mu - 0.5)/\sigma \tag{6.17}$$

Critical values for z_f and z_m are printed by UCODE_2005 and MODFLOW-2000. The critical values indicate the likelihood that the weighted residuals are in a random order. The critical values only apply when there are more than 10 positive residuals and more than 10 negative residuals.

For smaller numbers of positive and negative residuals, Table D.4 (Appendix D) can be used to assess the randomness of the ordered weighted residuals. This table is applicable for situations in which n_1 and n_2 are each greater than or equal to 3 and less than or equal to 10, and $10 \leq n_1 + n_2 \leq 20$. The table gives the lower-tail and upper-tail cumulative probabilities that a particular number of runs would occur, given the values of n_1 and n_2. Smaller probabilities indicate that it is less likely that the signs of the ordered weighted residuals are random.

In UCODE_2005 and MODFLOW-2000, the weighted residuals are analyzed using the order in which the observations are listed in the input file. The runs statistic can be made more meaningful by considering the ordering of the observations. For example,

```
STATISTICS FOR ALL RESIDUALS:
 AVERAGE WEIGHTED RESIDUAL: 0.691E+00
 # RESIDUALS >=0.: 50
 # RESISUALS <0.: 43
 NUMBER OF RUNS: 38 IN 93 OBSERVATIONS
INTERPRETING THE CALCUALTED RUNS STATISTIC VALUE OF -1.83
 NOTE: THE FOLLOWING APPLIES ONLY IF
          # RESIDUALS >= 0. IS GREATER THAN 10 AND
          # RESIDUALS < 0. IS GREATER THAN 10
 THE NEGATIVE VALUE MAY INDICATE TOO FEW RUNS:
    IF THE VALUE IS LESS THAN -1.28, THERE IS LESS THAN A
       10 PERCENT CHANCE THE VALUES ARE RANDOM,
    IF THE VALUE IS LESS THAN -1.645, THERE IS LESS THAN A
       5 PERCENT CHANCE THE VALUES ARE RANDOM,
    IF THE VALUE IS LESS THAN -1.96, THERE IS LESS THAN A
       2.5 PERCENT CHANCE THE VALUES ARE RANDOM.
```

FIGURE 6.4 Example runs test result printed by MODFLOW-2000 and UCODE_2005 (from the study described by Tiedeman et al., 1997).

in a groundwater model with pump-test data, listing the drawdowns at each location by increasing time produces a situation in which the runs statistic can be used to test whether the observed drawdowns are consistently greater than or less than the simulated values over time at each observation well. As the data are matched more randomly, the runs test will move away from indicating too few runs. In this situation it will rarely show too many runs. If spatial data are considered and observations are listed predominantly north to south, the runs statistic can provide a quick indication of whether spatial trends are diminishing as regression proceeds. Even if the runs statistic is used to evaluate trends in this manner, it is also necessary to conduct more thorough examinations using the graphical analyses of residuals described in this chapter.

The runs statistic information printed by MODFLOW-2000 is displayed in Figure 6.4. A two-tailed test is used, but the critical values from only one tail are printed. The information printed by UCODE_2005 is similar. The negative runs test statistic shown in Figure 6.4 indicates that, using the order in which they are listed in the input file, there are fewer runs than would be expected given 35 values consisting of 18 positive and 17 negative values. However, the -0.339 runs statistic is closer to zero than even -1.28, the critical value with the smallest absolute value. Thus, the hypothesis that the residuals are random is not rejected. This is one indication that the weighted residuals are sufficiently randomly distributed.

An example of using the runs test to evaluate weighted residuals along selected transects through a model area is shown in the discussion for Guideline 9 in Chapter 12.

6.4.5 Normal Probability Graphs and Correlation Coefficient R_N^2

The requirements for accurate simulated results are discussed in Chapter 3, Section 3.3. If the conditions listed in Section 3.3 are met, weighted residuals are expected to

6.4 ANALYZING MODEL FIT GRAPHICALLY AND RELATED STATISTICS

either (1) be random, normally distributed, and independent or (2) be random, normally distributed, and correlated in a way that is consistent with the fitting process of the regression. Possibility (1) is easiest to check and so is considered first. If the data do not satisfy the criteria, further testing is conducted to determine if the violations are consistent with the expected correlations produced by the fitting process.

The test for independent, normal weighted residuals is conducted using normal probability graphs of weighted residuals. If the weighted residuals are independent and normally distributed, they will fall on an approximately straight line in a normal probability graph (Cooley and Naff, 1990; Helsel and Hirsch, 2002, pp. 30–33). Normal probability graphs can be constructed by ordering the weighted residuals from smallest to largest and plotting them against the cumulative probability that would be expected for each value if they were independent and normally distributed. The expected cumulative probabilities depend on the number of weighted residuals considered and can be calculated in a number of ways (Looney and Gulledge, 1985a,b; Draper and Smith, 1998, p. 71). For the results presented in this work they are calculated as $(k - 0.5)/n$ (Hazen, 1914), where n equals the number of weighted residuals and k equals 1 for the smallest weighted residual, 2 for the next largest, and so on. For the largest weighted residual, k equals n. Calculating the cumulative probabilities in this way makes the normal probability graphs consistent with how the statistic R_N^2 is calculated, as discussed later in this section.

To obtain a graph on which random, normally distributed data are expected to lie on a straight line requires that the axis on which the probabilities are plotted be scaled for a normal probability distribution, as shown in Helsel and Hirsch (2002, Figures 2.7 and 2.9). This is called a normal probability axis. Many common plotting programs, such as Microsoft Excel, do not support normal probability axes. Fortunately, as shown by Helsel and Hirsch (2002, Figure 2.8), an alternative arithmetic scale can be used. The arithmetic scale requires that the probabilities be converted into what are called "standard normal statistics," "normal quantiles," or "normal score." The cumulative probability can be calculated from the standard normal statistics using, for example, the function NORMDIST in Excel. Common values printed on the axis of standard normal statistics and associated cumulative probabilities are as follows:

Standard Normal Statistic	Cumulative Probability
−4.0000	0.0000
−3.0000	0.0013
−2.0000	0.0228
−1.0000	0.1587
0.0000	0.5000
1.0000	0.8413
2.0000	0.9772
3.0000	0.9987
4.0000	1.0000

Based on the analysis above, given 101 ordered weighted residuals, the 51st largest value would have a cumulative probability of $(51 - 0.5)/101 = 0.5$; the standard normal statistic would be 0.0000. That is, the middle value would be expected, on average, to equal the mean of the standard normal distribution.

Helsel and Hirsch (2002, Figures 2.10–2.13) show and discuss normal probability graphs characterized by several common problems. In their graphs the term "normal quantile" is used instead of "standard normal statistic," and it is plotted on the horizontal axis instead of the vertical axis.

Regression problems commonly have small numbers of observations. To illustrate the variation that would be expected given a small sample size, Figure 6.5 shows normal probability plots generated with sample sizes of 10 and 40. As sample size increases, minor deviations from a straight line become more indicative of nonnormality.

The associated summary statistic, R_N^2, is the correlation coefficient between the weighted residuals ordered from smallest to largest and the normal order statistics (Brockwell and Davis, 1987, p. 304). R_N^2 is nearly equivalent to the PPCC statistic of Helsel and Hirsch (2002, Chap. 4.4). R_N^2 can be used to test for independent, normally distributed weighted residuals and was chosen instead of other statistics, such as chi-squared and Kolmogorov–Smirnov, because it is more powerful for commonly used sample sizes (Shapiro and Francia, 1972). The correlation coefficient is calculated as

$$R_N^2 = \frac{[(e_0 - m)^T \tau]^2}{[(e_0 - m)^T (e_0 - m)](\tau^T \tau)} \tag{6.18}$$

where all vectors are of length ND when R_N^2 is evaluated only for the ND observation weighted residuals, and of length $ND + NPR$ when R_N^2 is evaluated for the $ND + NPR$ observation and prior information weighted residuals; m is a vector

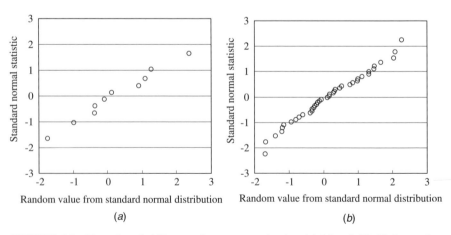

FIGURE 6.5 Normal probability graphs constructed using (a) 10 and (b) 40 data points generated from a normal probability distribution.

with all components equal to the average of the weighted residuals, e_0 is a vector of weighted residuals ordered from smallest to largest, and τ is a vector with the ith element equal to the standard normal statistic for a cumulative probability equal to $u_i = (i - 0.5)/ND$.

Values of R_N^2 close to 1.0 indicate that the weighted residuals are independent and normally distributed. If R_N^2 is too far below the ideal value of 1.0, the weighted residuals are not likely to be independent and normally distributed. To test whether R_N^2 is close enough to 1.0, it can be compared with critical values for R_N^2 at significance levels of 0.05 and 0.10. These critical values are shown in Table D.3 of Appendix D.

6.4.6 Acceptable Deviations from Random, Normally Distributed Weighted Residuals

Weighted residuals may appear to be nonrandom when evaluated using the methods described in Sections 6.4.1 to 6.4.5 because of (1) model inadequacy, (2) correlations induced by the fitting process of the regression (Cooley and Naff, 1990, p. 168; Draper and Smith, 1998, p. 206), or (3) too few residuals. Methods presented in this section can be used to test for the latter two reasons.

Otherwise unexplained deviations from expected attributes are likely due to model inadequacy. Problems could occur with the model construction, including the parameterization, the observation data, and/or the weights. If the model appears to be inadequate, then every attempt needs to be made to identify and resolve problems with the model, so that weighted residuals that are more random and normally distributed are achieved. Possible ways of dealing with an inadequate model are discussed in Guideline 9 in Chapter 12.

The correlations produced by the regression fitting process is most severe when there are few observations relative to the number of parameters. An extreme example occurs when only two data points are used to determine the slope and intercept of a simple linear model. In this situation, a perfect fit is achieved for both points, and the error is completely accommodated by the fitting process. As more points are added the situation becomes less dramatic, but the fit achieved by the regression always accommodates the error to some degree, and this can cause the weighted residuals to be correlated, rather than independent.

Too few residuals can cause normal probability graphs to appear nonnormal and can cause graphs of weighted residuals versus weighted simulated values to appear nonrandom, just by virtue of the small sample size. This problem was illustrated in Figure 6.5a for a normal probability graph.

Residuals that appear nonrandom and/or nonnormal can be tested by generating sets of values that have the expected correlations between the weighted residuals. The expected correlations can be calculated from the variance–covariance matrix of the weighted residuals, which equals (Bard, 1974, p. 194; similar to Cooley and Naff, 1990, p. 176)

$$V(\omega^{1/2}e) = (I - X(X^T \omega X)^{-1} X^T \omega)\sigma^2 \qquad (6.19)$$

The steps of the test are as follows (Cooley and Naff, 1990, p. 176).

1. Generate sets of independent, normally distributed random numbers, which do not have the regression-induced correlations (called d's by Cooley and Naff, 1990). Generate sets of correlated, normally distributed random numbers, which do have the regression-induced correlations (called g's by Cooley and Naff, 1990). Within each set, associate each generated number with one of the $ND + NPR$ observations or prior information values used in the regression using Eq. (6.19).
2. Compare graphs of the weighted residuals with graphs of the independent random numbers (d's) as follows:
   a. Evaluate graphs of weighted residuals and d's versus weighted or unweighted simulated values (as in Figure 6.1). If the graphs of the independent random numbers and of the weighted residuals have similar deviations from a random distribution about the zero line, the nonrandom distribution of the weighted residuals could result from the small number of observations.
   b. Evaluate normal probability graphs (as in Figure 6.5). If the graphs of the independent random numbers and of the weighted residuals have similar deviations from a straight line, the nonlinear shape of the weighted residuals graph could result from the small number of observations.
3. Compare graphs of the weighted residuals with graphs of the correlated random numbers (g's) as follows:
   a. Evaluate graphs of weighted residuals and g's versus weighted or unweighted simulated values (as in Figure 6.1). If the graphs of the correlated random numbers and of the weighted residuals have similar deviations from a random distribution about the zero line, the nonrandom distribution of the weighted residuals could result from the fitting process of the regression.
   b. Evaluate normal probability graphs (as in Figure 6.5). If the graphs of the correlated random numbers and of the weighted residuals have similar deviations from a straight line, the nonlinear shape of the weighted residuals graphs could result from the fitting process of the regression.

The d's and g's can be produced by MODFLOW-2000 and RESAN-2000 (Hill et al., 2000) or by UCODE_2005 and RESIDUAL_ANALYSIS (Poeter et al., 2005). Examples of graphs produced using data sets generated with RESAN-2000 are shown in Exercise 6.2e.

An alternative test is described by Cooley (2004) and Christensen and Cooley (2005). It involves generating hundreds or thousands of sets of correlated normal random numbers, calculating the mean and plus and minus two standard deviations for each normal probability plotting position, and plotting them with the weighted residuals on a normal probability graph. An example graph is presented by Christensen and Cooley (2005, p. 44). Data sets for these graphs can be produced by MODFLOW-2000's UNC Process (Christensen and Cooley, 2005) or UCODE_2005 and RESIDUAL_ANALYSIS_ADV (Poeter et al., 2005). Graphs can be produced using, for example, GWChart (Winston, 2000).

6.5 EXERCISES

These exercises consider the fit of the calibrated steady-state model of the flow system described in Chapter 2, Section 2.2. Transport predictions will not be credible if the model can not produce heads and flows that are reasonably similar to the observations. Predictions will also be suspect if the match to observations is so close that it appears observation error is being fit. To investigate model fit, Exercise 6.1 considers the overall statistical measures of model fit and Exercise 6.2 considers the graphical analyses and associated statistics.

Exercise 6.1: Statistical Measures of Overall Fit In this exercise, overall fit to the head, flow, and prior information data is evaluated. This evaluation uses statistics located in output files produced by the MODFLOW-2000 or UCODE_2005 regression run of Exercise 5.2c. For students who have not performed the simulations, this output file is available from the web site for this book; see Chapter 1, Section 1.1 for information about obtaining this file. The statistics also are included in tables accompanying the exercises.

(a) Examine objective-function values.

The values of the least-squares (Eq. (3.1)) and maximum-likelihood (Eq. (3.3)) objective functions for the final parameter values are shown in Figure 6.6.

Problem
- Use equation 3.3 in the text to verify the value of the maximum-likelihood objective function.
- Explain why the objective function values may not be the best indicators of model fit.

(b) *Demonstrate the circumstance in which the expected value of both the calculated error variance and the standard error is 1.0. (optional)*

In Section 6.3.2, it is claimed that if the fit achieved by regression is consistent with the data accuracy as reflected in the weighting, the expected value of both the calculated error variance and the standard error is 1.0. In this exercise, demonstrate this using generated random numbers instead of residuals. A diagonal weight matrix will be used, but the results are applicable to a full weight matrix as well. Proceed through the following steps:

1. Use a software package to generate $n = 100$ random numbers using any distribution (such as normal or uniform). These are equivalent to the residuals of Eq. (3.1) or (3.2).
2. Square each random number.

```
LEAST-SQUARES OBJ FUNC (DEP.VAR. ONLY) -    = 10.548
LEAST-SQUARES OBJ FUNC (W/PARAMETERS)-  -   = 10.556
CALCULATED ERROR VARIANCE------------       =  1.5080
STANDARD ERROR OF THE REGRESSION------      =  1.2280
CORRELATION COEFFICIENT------------         =  0.99979
      W/PARAMETERS-----------------         =  0.99989
ITERATIONS---------------------             =  5

MAX LIKE OBJ FUNC = -17.671
AIC STATISTIC--- =  -5.6713
BIC STATISTIC--- =  -2.2816
```

FIGURE 6.6 Selected statistics related to overall model fit, from the modified Gauss–Newton iterations of the regression run in Exercise 5.2c. This is a fragment from the global output file of MODFLOW-2000. "DEP.VAR.ONLY" means that only observations are included in the calculation. "W/PARAMETERS" means that prior information, if defined, is also included.

3. Divide each squared number by the variance of the distribution used. If weights are defined to be one divided by the variances, the resulting numbers are equivalent to squared, weighted residuals.
4. Sum the numbers from step 3 and divide by n.
5. Compare this value to 1.0. As n increases, the value should approach 1.0.
6. Repeat the analysis with two sets of n random numbers (total sample size is $2n$) generated with very different variances.

Problem: Discuss the results obtained.

(c) *Evaluate calculated error variance, standard error, and fitted error statistics.*

The values of the estimated error variance, s^2 (Eq. (6.1)), and its square root, the standard error of regression, s, are shown in Figure 6.6.

Problem
- How does s^2 compare to the expected value of 1.0? In the analysis, consider the confidence interval on the standard error of the regression. Use the χ^2 distribution in Table D.5 of Appendix D to obtain the critical values needed to calculate the confidence intervals. Here, $\chi^2_{(13-6),0.975} = 1.690$; $\chi^2_{(13-6),0.025} = 16.01$. Does 1.0 fall within the confidence interval?
- Using s and the standard deviation of measurement error used to calculate the weights for hydraulic-head observations (see Exercise 3.2d), calculate the fitted standard deviation for heads. Compare the fitted standard deviation to the total head loss across the flow system (i.e., the difference between the maximum and minimum head, derived from the contour map of heads in Figure 2.1), and use this to judge the model fit.

6.5 EXERCISES

(d) *Examine the AIC, AIC_c, and BIC statistics.*

The values of the AIC (Eq. (6.3)) and BIC (Eq. (6.4)) statistics are shown in Figure 6.6. As discussed in Section 6.2.4, these statistics can be useful when comparing different models.

Problem:
- Using Eqs. (6.3) and (6.4) and the values listed in the top part of Figure 6.6, verify the values of AIC and BIC shown in Figure 6.6. Calculate AIC_c. Should AIC or AIC_c be used?
- Suppose that parameters are added to the steady-state test-case model to better represent some feature of the true system. For each additional parameter added, how much does the model fit, as represented by the weighted least-squares objective function, need to improve to result in a reduced value of the AIC, AIC_c, and BIC statistics?

Exercise 6.2: Evaluate Graphs of Model Fit and Related Statistics In this exercise, the fit of the steady-state model calibrated in Exercise 5.2c to the head, flow, and prior observation data is evaluated using graphical methods and associated statistics. This evaluation uses residuals and statistics produced by the regression run of Exercise 5.2c. Students who have performed the simulations can create the graphs from model output files; see Chapter 1, Section 1.1 for the website where instructions are provided.

(a) *Graph of weighted residuals versus weighted simulated values and the minimum, maximum, and average weighted residuals.*

The graph of weighted residuals versus weighted simulated values is shown in Figure 6.7a. Ideally, the weighted residuals show no pattern relative to the simulated values.

Problem
- Comment on the graph in Figure 6.7a. Do the weighted residuals appear to be randomly distributed about zero? The very small residuals for the flows and prior information are discussed in subsequent exercises.
- Comment on the values of the maximum, minimum, and average weighted residuals shown in Figure 6.8.

(b) *Graphs of observations versus simulated values. Examine the correlation coefficient R.*

A graph of weighted observations versus weighted simulated values is shown in Figure 6.7b, and a graph of observed versus simulated values is shown in Figure 6.7c. The correlation coefficient between the weighted observed and simulated values, R (Eq. (6.11a)), equals 0.99979 for the head and flow

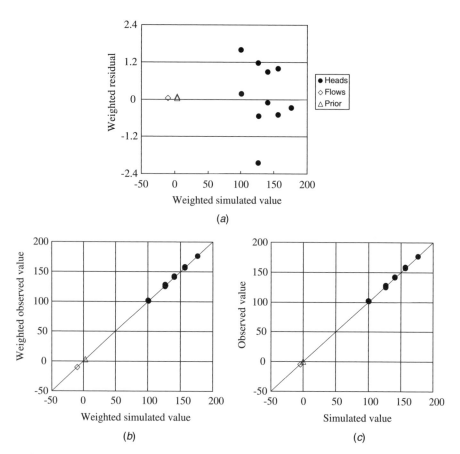

FIGURE 6.7 Plots for analyzing model fit for Exercise 5.2c. (*a*) Weighted residuals versus weighted simulated values (unweighted simulated values also could be used on the horizontal axis). The vertical gridlines are placed at increments of the standard error of the regression (1.2). (*b*) Plot of weighted observed values versus weighted simulated values. (*c*) Plot of observed versus simulated values.

observations, and 0.99989 for the observed values and prior values on K_RB and VK_CB. These values are shown in Figure 6.6. Ideally, the values plotted on the types of graphs shown in Figures 6.7*b* and 6.7*c* fall on a line with a slope of 1.0.

Problem
- Comment on the utility of the three different graphs shown in Figure 6.7. Which graph is likely to be more useful for diagnosing problems with the model fit to the observation data?

6.5 EXERCISES

```
SMALLEST AND LARGEST WEIGHTED RESIDUALS
                SMALLEST WEIGHTED              LARGEST WEIGHTED
                    RESIDUALS           |          RESIDUALS
NAME        WEIGHTED    PERCENT OF      |  NAME  WEIGHTED   PERCENT OF
            RESIDUAL    OBJ FUNC        |        RESIDUAL    OBJ FUNC
4.ss         -2.05        39.68         |  1.ss    1.59        24.01
6.ss         -0.552        2.89         |  2.ss    1.17        13.05
3.ss         -0.506        2.43         |  8.ss    0.993        9.34
9.ss         -0.275        0.72         | 10.ss    0.882        7.37
5.ss         -0.114        0.12         |  7.ss    0.178        0.30

STATISTICS FOR ALL RESIDUALS:
AVERAGE WEIGHTED RESIDUAL: 0.114E+00
```

FIGURE 6.8 Smallest, largest, and average weighted residuals from the regression run in Exercise 5.2c. This is a fragment from the global output file of MODFLOW-2000.

- Does the value of R indicate a good match between the trends in the weighted simulated and weighted observed values? Is R a useful diagnostic statistic in this situation? Why?

(c) *Graphs of weighted residuals against independent variables. Evaluate runs statistic.*

The weighted residuals from the regression of Exercise 5.2c are plotted on maps of the model layers in Figure 6.9.

Problem
- Do the weighted residuals shown in Figure 6.9 appear to be randomly distributed in space?

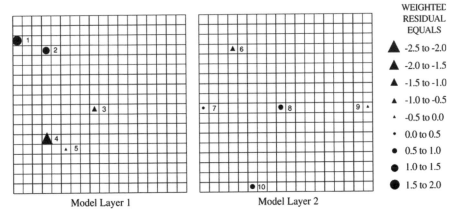

FIGURE 6.9 Weighted residuals for the steady-state regression plotted on maps of the two model layers.

```
# RESIDUALS >= 0. :  8
# RESIDUALS <  0. :  5
NUMBER OF RUNS : 5   IN 13 OBSERVATIONS

INTERPRETING THE CALCULATED RUNS STATISTIC VALUE OF
  -1.02
NOTE: THE FOLLOWING APPLIES ONLY IF
        # RESIDUALS >= 0 . IS GREATER THAN 10 AND
        # RESIDUALS <  0. IS GREATER THAN 10
THE NEGATIVE VALUE MAY INDICATE TOO FEW RUNS:
   IF THE VALUE IS LESS THAN -1.28, THERE IS LESS
      THAN A 10 PERCENT CHANCE THE VALUES ARE RANDOM,
   IF THE VALUE IS LESS THAN -1.645, THERE IS LESS
      THAN A 5 PERCENT CHANCE THE VALUES ARE RANDOM,
   IF THE VALUE IS LESS THAN -1.96, THERE IS LESS
      THAN A 2.5 PERCENT CHANCE THE VALUES ARE RANDOM.
```

FIGURE 6.10 Runs statistic and critical values from the regression run in Exercise 5.2c. This is a fragment from the global output file of MODFLOW-2000.

- Comment on the physical reasons for the three large weighted residuals in model layer 1. It may be helpful to consider the dimensionless scaled sensitivities of Table 4.1.

The runs statistic and critical values for this problem are shown in Figure 6.10. For the steady-state model regression, there are less than 10 positive residuals and less than 10 negative residuals, and thus the printed critical values for the runs statistic are not applicable. In most situations, there will be enough positive and negative residuals so that the critical values do apply. For cases where the critical values are applicable, understanding the runs statistic can be facilitated by locating the runs statistic and critical values on a normal probability distribution.

Problem
- Draw a normal probability distribution and locate the value of the test statistic and the critical values. Remember that this is a two-tailed test, so include the critical values printed in the file, and also the critical values of the other tail of the distribution.
- Given the runs test statistic value and the critical values, what do you conclude about the randomness of the weighted residuals with respect to their order in the MODFLOW-2000 or UCODE_2005 input files? When answering this question, ignore the problem that the steady-state regression has too few negative and positive residuals.

6.5 EXERCISES

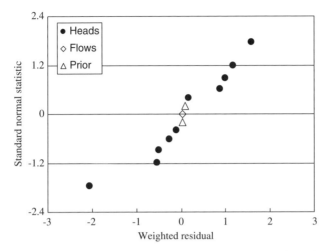

FIGURE 6.11 Normal probability graph of the weighted residuals from Exercise 5.2c.

(d) *Evaluate normal probability graphs and the correlation coefficient R_N^2.*

A normal probability graph of the weighted residuals from Exercise 5.2c is shown in Figure 6.11. R_N^2 (Eq. (6.18)) is shown in Figure 6.12. Use this plot and associated statistic to test the independence and normality of the weighted residuals.

Problem
- Do the weighted residuals appear to be normally distributed in Figure 6.11? Compare the results of this analysis with the calculated value of R_N^2 shown in Figure 6.12.
- Generate 10 sets of 13 normally distributed random numbers and calculate the R_N^2 statistic for each. Compare these values to the critical value for the 5-percent significance level. Compare how many of the 10 R_N^2 values are less than the critical value to how many are expected to be less than the critical value.

(e) *Determine acceptable deviations from random, independent, and normal weighted residuals.*

Graphs of independent and correlated random numbers versus weighted simulated values from Exercise 5.2c are shown in Figure 6.13, and normal probability graphs of independent and correlated random numbers are shown in Figure 6.14. These graphs are used to test expected correlation between the

```
CORRELATION BETWEEN ORDERED WEIGHTED RESIDUALS AND
NORMAL ORDER STATISTICS FOR OBSERVATIONS = 0.941

CORRELATION BETWEEN ORDERED WEIGHTED RESIDUALS AND
NORMAL ORDER STATISTICS FOR OBSERVATIONS AND
PRIOR INFORMATION = 0.926

-------------------------------------------
COMMENTS ON THE INTERPRETATION OF THE CORRELATION
BETWEEN WEIGHTED RESIDUALS AND NORMAL ORDER
STATISTICS:

Generally, IF the reported CORRELATION is LESS than the
critical value, at the selected significance level
(usually 5 or 10%), the hypothesis that the
weighted residuals are INDEPENDENT AND NORMALLY
DISTRIBUTED would be REJECTED. HOWEVER, in this case,
conditions are outside of the range of published
critical values as discussed below.

The sum of the number of observations and prior information
items is 13 which is less than 35, the minimum value
for which critical values are published. Therefore,
the critical values for the 5 and 10% significance levels
are less than 0.943 and 0.952, respectively.

CORRELATIONS GREATER than these critical values indicate
that, probably, the weighted residuals ARE INDEPENDENT
AND NORMALLY DISTRIBUTED.

Correlations LESS than these critical values MAY BE
ACCEPTABLE, and rejection of the hypothesis is not
necessarily warranted.

The Kolmogorov-Smirnov test can be used to further evaluate
the residuals.
-------------------------------------------
```

FIGURE 6.12 R_N^2 statistic and critical values from the regression run in Exercise 5.2c. This is a fragment from the global output file of MODFLOW-2000.

6.5 EXERCISES

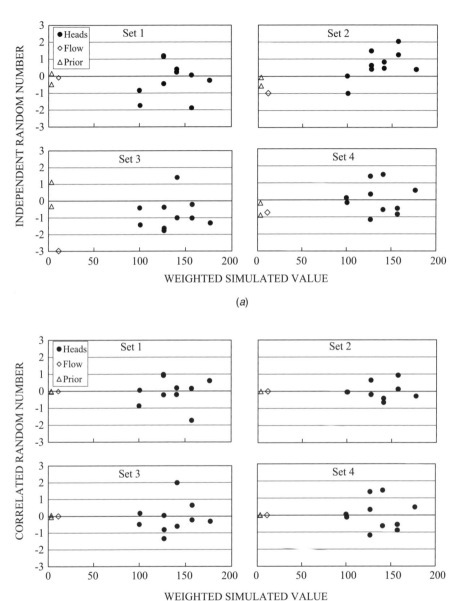

FIGURE 6.13 Graphs of four sets of normally distributed (*a*) independent and (*b*) correlated random numbers versus weighted simulated values from Exercise 5.2c, as needed in Exercise 6.2e.

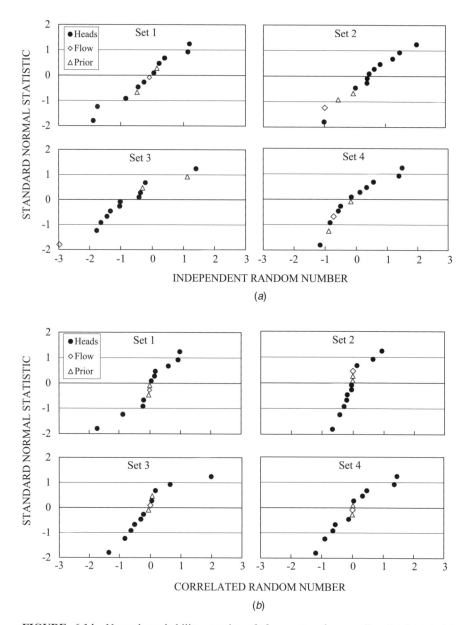

FIGURE 6.14 Normal probability graphs of four sets of normally distributed (*a*) independent and (*b*) correlated random numbers related to Exercise 5.2c, as needed in Exercise 6.2e.

6.5 EXERCISES

weighted residuals. Instructions for producing the data sets needed for these graphs are available from the web site for this book as described in Chapter 1, Section 1.1.

Problem
- Is the behavior of the weighted residuals more similar to that of the generated independent or correlated random numbers? To answer this question, compare Figure 6.7a with Figure 6.13, and compare Figure 6.11 with Figure 6.14. Explain your answer.
- What conclusion can be drawn about the reason for the nonrandomness of the flow and prior weighted residuals in Figure 6.7a, and for the deviation of the weighted residuals from a straight line in Figure 6.11? Use your knowledge of the model construction and the observations and prior information.

7

EVALUATING ESTIMATED PARAMETER VALUES AND PARAMETER UNCERTAINTY

Once parameter values are estimated, they need to be evaluated for a number of reasons. In this chapter we present methods for analyzing estimated parameters with brief explanations of how the analyses are used. Additional discussions and examples are provided in Guideline 10 of Chapter 12.

The methods described in this chapter start with reevaluation of composite scaled sensitivities. Next, five variations of the parameter variance-covariance matrix are introduced and statistics derived from the parameter variance-covariance matrix are defined. After comments about log-transformed parameters and when to use the five variations of the parameter variance-covariance matrix, five issues are discussed: (1) identification of individual observations that dominate the parameter estimates, (2) uniqueness and optimality of the estimates, (3) quantifying parameter uncertainty, (4) comparing parameter estimates against reasonable ranges, and (5) testing for model nonlinearity. The issues considered in this chapter span the first two components of the observation-parameter-prediction triad composed of entities that are directly connected by the model, as discussed in Chapters 1 and 10.

7.1 REEVALUATING COMPOSITE SCALED SENSITIVITIES

Composite scaled sensitivities (css) are a measure of the total information provided by the regression observations about a parameter value. These sensitivities were presented in Chapter 4, Section 4.3.4, which focused on using them to determine which

---

Effective Groundwater Model Calibration: With Analysis of Data, Sensitivities, Predictions, and Uncertainty. By Mary C. Hill and Claire R. Tiedeman
Published 2007 by John Wiley & Sons, Inc.

7.2 PARAMETER VARIANCE–COVARIANCE MATRIX

parameters to estimate by regression, and which to exclude because of insensitivity. At the optimal parameter values it is important to calculate *css* for all defined model parameters (those included in the regression as well as those excluded from the regression). These sensitivities will likely be different from those calculated for initial regression runs, because of model nonlinearity and because of scaling by b_j in Eq. (4.3). If any parameters that were initially excluded from the regression appear to have increased in stature, additional regression runs should be considered with these parameters included.

Although the *css* are useful measures of the information the data contain for a single parameter, they do not account for the many parameters being estimated simultaneously, and they do not measure the precision of the parameter estimates. Other statistics discussed in this chapter fill these roles.

7.2 USING STATISTICS FROM THE PARAMETER VARIANCE–COVARIANCE MATRIX

The variance–covariance matrix on the parameters contains important information about parameter uncertainty and correlation, and about the support that the observations offer to the estimated parameters, given the model as constructed. This section presents five alternate versions of the variance–covariance matrix, statistics derived from the variance–covariance matrix, and the circumstances in which these statistics are used for each of the five versions of the matrix. Finally, the section discusses alternate statistics that are commonly suggested and notes that we believe they are more complicated without providing much additional insight for the purposes of evaluating parameter uncertainty and correlation.

7.2.1 Five Versions of the Variance–Covariance Matrix

The parameter variance–covariance matrix is calculated using an equation of the form

$$V(b) = s^2(X^T \omega X)^{-1} \tag{7.1}$$

where $V(b)$ is an *NP* by *NP* matrix, s^2 is the calculated error variance (Eq. (6.1)), X is a matrix of sensitivities defined after Eq. (5.2b) and calculated for the parameters listed in the vector b, and ω is a weight matrix defined after Eq. (3.2) and in Appendix A.

It can be very useful to define X, b, and ω differently to investigate different aspects of the model, the data, and the predictions. Five versions are presented here and discussed further in Section 7.2.5.

1. *Variance–Covariance Matrix with Optimized Parameter Values.* Only optimized parameters are included. Sensitivities are calculated for the optimized

parameter values, $b = b'$, and X is a matrix of sensitivities for the parameters estimated and the observations and prior information used in the regression (Bard, 1974, p. 59; Draper and Smith, 1998, p. 223).

2. *Variance–Covariance Matrix with All Defined Parameters.* This is the same as option 1 except that any defined parameters for which values were set and not estimated in the regression are included. This means that there are additional columns in the sensitivity matrix, X. In addition, in some circumstances this variation of the variance–covariance matrix includes realistic weighting. This means that (1) any weights that were altered to obtain parameter estimates need to be returned to values representative of realistic levels of error in the observations and prior information, and (2) if available, prior information and associated realistic weighting needs to be included, and is most important for parameters not estimated by the regression.

For options 3 to 5, the useful statistics that are derived from the variance–covariance matrix do not depend on s^2.

3. *Variance–Covariance Matrix with Nonoptimal Parameter Values.* This is the same as option 1 or 2 except that any set of parameter values can be used.
4. *Variance–Covariance Matrix with Alternate Observation Sets.* This is the same as option 1 or 2 except that different observations are included. Existing observations may be omitted or information on new observations may be added. This requires changes in the weight matrix, ω, and, when adding observations, the sensitivity matrix, X.
5. *Variance–Covariance Matrix with Predictions.* This is the same as option 1 or 2 except that predictions are added. This requires changes in the weight matrix, ω, and the sensitivity matrix, X.

7.2.2 Parameter Variances, Covariances, Standard Deviations, Coefficients of Variation, and Correlation Coefficients

The precision (see definition in Chapter 1, Section 1.4.2) and correlation of parameter estimates can be analyzed by using the parameter variance–covariance matrix. The diagonal elements equal the parameter variances; the off-diagonal elements equal the parameter covariances. For a problem with three estimated parameters, the matrix would appear as

$$\begin{array}{ccc} \text{Var}(1) & \text{Cov}(1,2) & \text{Cov}(1,3) \\ \text{Cov}(2,1) & \text{Var}(2) & \text{Cov}(2,3) \\ \text{Cov}(3,1) & \text{Cov}(3,2) & \text{Var}(3) \end{array} \tag{7.2}$$

where Var(1) is the variance ($s_{b_1}^2$) of parameter 1, Cov(1, 2) is the covariance between parameters 1 and 2, and so on. The variance–covariance matrix is always symmetric, so that, for example, Cov(1, 2) = Cov(2, 1). Equation (7.1) is most useful if the model is nearly linear in the vicinity of b' (see Chapter 5, Section 5.1.2) and if the weight matrix is appropriately defined (see Chapter 3, Sections 3.3.3

7.2 PARAMETER VARIANCE–COVARIANCE MATRIX

and 3.4.2). A method of testing for model linearity is presented in Section 7.7. For nonlinear problems the variance–covariance matrix only approximates parameter uncertainty.

Variances and covariances commonly are not intuitively understood, but they can be used to calculate informative statistics. The first of these is the parameter standard deviation, which equals the square root of the parameter variances. That is,

$$s_{b_j} = (\mathrm{Var}(j))^{1/2} \tag{7.3}$$

where $\mathrm{Var}(j)$ is the jth diagonal of the variance–covariance matrix. Parameter standard deviations have the same units as do the parameter values and are more easily understood measures of parameter uncertainty. However, parameter standard deviations are perhaps most useful when processed further to calculate three other statistics: confidence intervals for parameter values (presented in Section 7.5.1), coefficients of variation, and the t-statistic. The coefficient of variation for each parameter equals the standard deviation divided by the parameter value:

$$c.v. = s_{b_j}/b_j \tag{7.4}$$

The coefficient of variation is a dimensionless number with which the relative accuracy of different parameter estimates can be compared. The t-statistic serves the same purpose and equals $1/c.v.$ or b_j/s_{b_j}. The coefficient of variation is used instead of the t-statistic in this book.

Correlation coefficients are calculated as the covariance between two parameters divided by the product of their standard deviations. Using the notation of Eq. (7.2), the correlation between the jth and kth parameter is

$$pcc(j,k) = \frac{\mathrm{Cov}(j,k)}{\mathrm{Var}(j)^{1/2}\mathrm{Var}(k)^{1/2}} \tag{7.5}$$

Characteristics of parameter correlation coefficients were discussed in Chapter 4, Sections 4.3.5 and 4.4.2. Briefly, unique values are nearly always assured if the absolute values of all pcc are all less than about 0.95. However, unique estimates can be obtained with larger absolute values. Suspected problems with uniqueness can be tested as discussed in Section 7.4.

7.2.3 Relation Between Sample and Regression Statistics

For students unfamiliar with means, variances, covariances, standard deviations, coefficients of variation, and correlation coefficients, it can be beneficial to compare how they are calculated for sample data with how they are calculated in regression. The two situations are similar in that both attempt to use data to estimate some quantity and express the precision of the estimate. Sample data used in a comparison are shown in Figure 7.1. The equations for calculating the sample

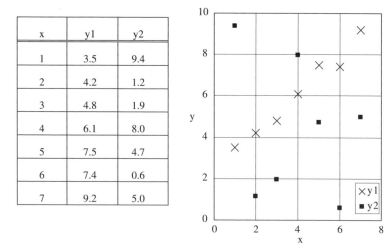

FIGURE 7.1 Values and graph of x and two sets of y variables used to investigate sample variances, covariances, and correlation coefficients. The values of y1 equal the x values plus and minus small deviations; the values of y2 were generated from a random normal distribution with a mean of 4.5 and a standard deviation of 3.0.

statistics are shown in Table 7.1, with the calculated values of the sample statistics for the two data sets shown in Figure 7.1. These equations differ from those used to calculate the analogous regression statistics, which are presented in Eq. 5.6 and Section 7.2.1.

The sample variance is a measure of the spread of the data. Table 7.1 shows that the sample variance for data set y1 is less than that for data set y2 (4.2 versus 11); this is expected given the wider range of values in data set y2 compared to data set y1. The sample covariance indicates whether x and y vary in a coordinated way, and the correlation coefficient is a scaled measure of the sample covariance. The covariance and the absolute value of the correlation coefficient for data set y1 are greater than the corresponding statistics for data set y2 (0.98 versus −0.26). This is because the y1 values vary in a systematic way with x, whereas the y2 values are more random with respect to the x values, as shown in Figure 7.1.

In regression, the parameter values are not estimated by direct sampling. Instead, they are estimated indirectly using observations of the state of the simulated system. This can be accomplished because the simulation model used in the regression is based on equations that relate the observations and the parameter values. Because of this indirect way of estimating parameter values, parameter variances and covariances are calculated in a different manner from the sample equations of Table 7.1, as indicated in the lower half of this table.

Interpretation of the variance and correlation of parameters estimated by regression is similar to that for the sample statistics but is not completely analogous. In regression, the variance indicates the range over which a parameter value could extend without affecting model fit too adversely, and the parameter correlation

7.2 PARAMETER VARIANCE–COVARIANCE MATRIX

TABLE 7.1 Equations for Sample Mean, Variance, Standard Deviation, Coefficient of Variation, Covariance, and Correlation Coefficients; the Values Calculated for the Data Sets Shown in Figure 7.1; and the Analogous Relations When Quantities Are Estimated by Regression Instead of Directly from Sample Data

	Sample Statistics		
Statistic	Equation (Davis, 2002)	y1	y2
Mean	$x' = (1/n) \sum_i x_i$ $y' = (1/n) \sum_i y_i$	$x' = 4$ $y' = 6.1$	$x' = 4$ $y' = 4.38$
Variance	$s_x^2 = 1/(n-1) \sum_i (x_i - x')^2$ $s_y^2 = 1/(n-1) \sum_i (y_i - y')^2$	$s_x^2 = 4.7$ $s_y^2 = 4.2$	$s_x^2 = 4.7$ $s_y^2 = 11$
Standard deviation	$s_x = (s_x^2)^{1/2}$ $s_y = (s_y^2)^{1/2}$	$s_x = 2.2$ $s_y = 2.0$	$s_x = 2.2$ $s_y = 3.3$
Standard deviation of the mean	$s_{x'} = s_x/n^{1/2}$ $s_{y'} = s_y/n^{1/2}$	$s_{x'} = 0.83$ $s_{y'} = 0.76$	$s_{x'} = 0.83$ $s_{y'} = 1.25$
Coefficient of variation	$c.v._x = s_x/x'$ $c.v._y = s_y/y'$	$c.v._x = 0.55$ $c.v._y = 0.33$	$c.v._x = 0.55$ $c.v._y = 0.75$
t-statistic	$t_x = x'/s_x$ $t_y = y'/s_y$	$t_x = 1.82$ $t_y = 3.03$	$t_x = 1.82$ $t_y = 1.33$
Covariance	$\mathrm{Cov} = 1/(n-1) \left[\sum_i (x_i - x')(y_i - y')\right]$	4.4	-1.9
Correlation coefficient	$r = \dfrac{\sum_i (x_i - x')(y_i - y')}{\left(\sum_i (x_i - x')^2\right)^{1/2} \left(\sum_j (y_j - y')^2\right)^{1/2}}$ $= \mathrm{Cov}/[s_x s_y]$	0.98	-0.26

When Parameters Are Estimated by Regression	
Statistic	Description
Mean (parameter estimate)	Symbol: b_j. Estimated using the observations, the model, and the modified Gauss–Newton normal equations (Eq. 5.6).
Parameter variance	Symbol: $s_{b_j}^2$; diagonals of Eq. (7.1) and (7.2). This equation uses the following quantities: (1) the sensitivities, as measures of the information provided for the parameter; (2) the weights, as measures of the error in the observations; and (3) the calculated variance of the regression, as a measure of model fit to the observations.
Parameter standard deviation	Symbol: s_{b_j}; Eq. (7.3). Equals the square root of the parameter variance. Analogous to the sample standard deviation of the mean instead of the standard deviation of the population.
Parameter coefficient of variation	Symbol: $c.v.$; Eq. (7.4). Equals (parameter standard deviation)/(parameter estimate)
Parameter t-statistic	Equals $1/c.v. = $ (parameter standard deviation)/(parameter estimate)
Parameter correlation coefficient	Symbol: $pcc(i, j)$; Eq. (7.5). Instead of measuring how closely x tracks y, it measures whether coordinated changes in two parameters would result in the same simulated values and, therefore, the same model fit to the observations and same objective function value.

coefficients indicate whether coordinated changes in the parameter values could produce the same simulated values and, therefore, the same model fit.

7.2.4 Statistics for Log-Transformed Parameters

For log-transformed parameters, the parameter estimates, coefficients of variation (Eq. (7.4)), and confidence intervals (discussed in Section 7.5.1) can be difficult to interpret. It is advantageous to present statistics related to native parameter values to encourage comparison with field data.

The native estimate is calculated as the exponential of the log-transformed estimate obtained by regression. The nature of the log-normal distribution means that the native value reported is the mode instead of the mean of the log-normal distribution. Using the mode as the measure of central tendency of the distribution has the advantage of producing a native parameter value that produces the regression results when used in the model input files, and this consideration overrides the need to use the mean of the log-normal distribution.

Confidence intervals on the native equivalent of log-transformed parameters are reported as the exponential of the confidence interval limits calculated for the log-transformed parameter. For log-transformed parameters, the linear confidence intervals for the true, unknown native parameters are symmetric when plotted on a log scale, but are not symmetric when plotted on an arithmetic scale. Despite this asymmetry of the intervals on an arithmetic plot, it is often easier for modelers to interpret and communicate to others the ranges for the native parameters than the ranges for the log-transformed parameters.

Standard deviations and coefficients of variation (the standard deviation divided by the estimate) for the native parameter estimates are obtained by converting the variance for the log-transformed parameter, $(s_{\log b})^2$, using the expression

$$s_b^2 = \exp[2.3(s_{\log b})^2 + 2.0 \times \log b][\exp(2.3(s_{\log b})^2) - 1.0] \qquad (7.6)$$

where the exponentials and logarithms are in base 10, b is the value of the native parameter, and $\log b$ is the estimated log-transformed parameter. The coefficient of variation of the native parameter is calculated by dividing the square root of its variance by the native parameter value.

7.2.5 When to Use the Five Versions of the Parameter Variance–Covariance Matrix

This section presents the circumstances for which the five variations of the variance–covariance matrix (Section 7.2.1) are used. Also discussed are the statistics calculated from the matrix that are most useful in each circumstance.

Matrix with Optimized Parameter Values This version of the variance–covariance matrix is routinely calculated if regression is used for model calibration. Useful

7.2 PARAMETER VARIANCE–COVARIANCE MATRIX

statistics include the parameter coefficients of variation and the parameter correlation coefficients.

Matrix with All Defined Parameters   In Eq. (7.1), the sensitivity and weight matrices usually contain entries only for the parameters estimated by regression. In many situations, there are additional defined model parameters that are excluded from the regression because of insensitivity and/or nonuniqueness detected using the sensitivity analysis discussed in Chapter 4, or for other reasons. It is important to periodically calculate sensitivities and the variance–covariance and correlation matrices for all defined model parameters, for two reasons.

First, it is important to determine whether updated parameter values or other modifications to the model have changed conclusions about insensitivity and nonuniqueness, and to evaluate observations and parameters from the perspective of predictions. Including all defined parameters can be accomplished easily using UCODE_2005 and MODFLOW-2000 by activating unestimated parameters.

Second, when evaluating the uncertainty of predictions or performing other related analyses, it is important to include all defined parameters to obtain realistic results. Parameters that may not have been important to observations may be important to predictions, and this can be determined only if all defined parameters are included in the analysis.

When activating all parameters to evaluate model predictions, it is important to include prior information (Chapter 3, Section 3.4.3) and associated weighting for the parameters that were not estimated by regression. This allows for a realistic degree of uncertainty in these parameters to be reflected in analyses of prediction uncertainty. If prior information on these parameters is not included, the contribution of parameter uncertainty will be unrealistically large. The prior value specified needs to equal the parameter value, so that the numerator of the s^2 term in Eq. (7.1) is not affected. The denominator will not be affected if one item of prior information is included for each added parameter. The weights on the prior information need to reflect the uncertainty in the independent information about the parameter values. Weighting strategies are discussed in more detail in Guideline 6 in Chapter 11.

Matrix with Nonoptimal Parameter Values   Equation (7.1) can be calculated for any set of parameter values, and some of the resulting statistics are very useful for diagnosing problems with the regression (Anderman et al., 1996; Poeter and Hill, 1997; Hill et al., 1998; Hill and Østerby, 2003). For example, parameter correlation coefficients calculated with the starting model parameter values are a very important aspect of the sensitivity analysis performed at the initial stages of the regression, as discussed in Chapter 4, Section 4.2.3.

Matrix with Alternate Observation Sets   The fourth version of Eq. (7.1) involves observation sets that are different from that used to calibrate the model. Two such alternative observation sets are used in this book. Both are used to calculate the observation-prediction (*opr*) statistic for evaluating the importance of observations to model predictions, discussed in Chapter 8, Section 8.3.2. The first set consists of the calibration observations with one or more of the observations used in model calibration omitted, which is used to evaluate existing observations in the context of model predictions. The second set consists of the calibration observations with one or more observations added, to evaluate potential new observations in the context of model predictions.

Usually these analyses with alternate observation sets are conducted using the variance–covariance matrix with all defined parameters represented and with weighting that reflects realistic errors in observations and prior information.

Matrix with Predictions A version of Eq. (7.1) can be used to determine if parameters that are highly correlated given the observations used in the regression are problematic to predictions of interest. This version is discussed in Chapter 8, Section 8.2.4 and is generally calculated using all defined model parameters.

7.2.6 Some Alternate Methods: Eigenvectors, Eigenvalues, and Singular Value Decomposition

Alternate methods available for evaluating parameter uncertainty and correlation include calculation of the eigenvectors and eigenvalues of the parameter variance–covariance matrix and singular value decomposition (SVD) of the weighted sensitivity matrix. Both produce eigenvectors and eigenvalues. Large eigenvalues identify important eigenvectors. Each eigenvector identifies a linear combination of parameters and parameters with larger coefficients dominate the eigenvector. Dominant parameters in important eigenvectors are important parameters.

Hill and Østerby (2003) compared the ability of parameter correlation coefficients and the SVD method to detect extreme parameter correlation. They found that both methods performed similarly for a simple hypothetical groundwater flow model similar to that used in the exercises of this book. Parameter correlation coefficients are emphasized in this work because they are easy to interpret, as discussed in Section 4.3.5. Sometimes parameter correlations are criticized because they only identify extreme correlation between pairs of parameters. However, as noted in Section 4.3.5, if more then two parameters are correlated all pairs will have correlation coefficients with absolute values close to 1.00, so the fact that *pcc* are calculated only for parameter pairs is rarely a meaningful limitation.

MODFLOW-2000 and UCODE_2005 can calculate the eigenvectors and eigenvalues of the parameter variance–covariance matrix, so modelers who prefer to use these measures can do so with these computer programs.

7.3 IDENTIFYING OBSERVATIONS IMPORTANT TO ESTIMATED PARAMETER VALUES

Different observations can play different roles in the regression. Even in regressions with hundreds of observations, one or two can profoundly affect parameter estimates. Important observations are not consistently associated with either very

7.3 OBSERVATIONS IMPORTANT TO ESTIMATED PARAMETER VALUES

large or very small weighted residuals, or large scaled sensitivities. As discussed in Chapter 4, dimensionless and composite scaled sensitivities can be used to identify observations important to individual parameters, but they cannot identify observations that reduce parameter correlation. Though parameter correlation coefficients can be used to address this concern, statistics that integrate the effects of sensitivity and correlation are needed.

The regression equations presented in Chapter 5 and the parameter variance–covariance matrix of this chapter can be used to create statistics that integrate effects measured separately by scaled sensitivities and parameter correlation coefficients. The statistics are similar to the scaled sensitivities and parameter correlation coefficients in that they take advantage of the model as a quantitative connection between the simulated equivalents to the observations and the model parameters. The scaled sensitivities and parameter correlation coefficients continue to be useful in part because they can be used to understand why different observations are important. As for the previous analyses, the statistics presented here depend on how the model is constructed.

Three statistics are presented: one is a leverage statistic; two are influence statistics.

Leverage and influence are two important measures of the role observations play in regression. Leverage statistics were mentioned in Chapter 4, Section 4.3.6, and depend only on the independent variables associated with an observation, such as its type, location, and time. Influence depends on the observed value as well. The concepts of leverage and influence are illustrated in Figure 7.2 for a simple linear regression problem. In Figure 7.2a, the outlier data point has high leverage because its x location is very different from that of all other observations. However, it does not have high influence, because its presence does not cause the regression results to significantly differ from the results that are obtained in its absence. In contrast, in Figure 7.2b, the outlier has both high leverage and high influence. It has the same

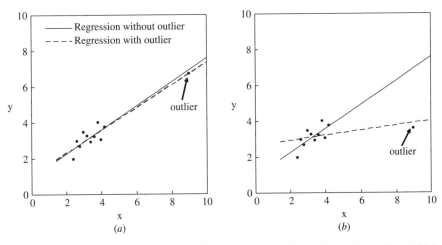

FIGURE 7.2 The effect on a simple linear regression of an observation with (a) high leverage and (b) high leverage and high influence. (From Helsel and Hirsch, 2002, Figure 9.19.)

x location as the outlier in Figure 7.2a, but it has a y value that causes the regression line to be significantly different from the line that is obtained in its absence. In general, whether or not the observation actually dominates the regression depends on how consistent its observed value is with the simulated equivalent calculated using all other observations. If it does dominate, it not only has high leverage but also has high influence.

In linear models, influential observations are a subset of the observations with substantial leverage; this relation applies approximately to nonlinear models.

Often it is useful to calculate leverage and influence statistics using all defined parameters. Dominant observations deserve extra attention to ensure that simulated equivalents are appropriate and simulated correctly, that the observation is correctly determined from field data, and that the observation errors are fully considered in the weighting.

7.3.1 Leverage Statistics

Leverage statistics identify observations that are sensitive in a way that causes the observed values to potentially have a profound effect on the regression results. In general, an observation is more likely to have high leverage if its location, time, circumstance, or type provides unusual information to the regression, as for the outlier points in Figure 7.2.

Leverage is calculated as (Helsel and Hirsch, 2002, p. 246):

$$h_{ii} = (\omega^{1/2} X)_i \, (X^T \omega X)^{-1} \, (X^T \omega^{1/2})_i \tag{7.7}$$

where $(\omega^{1/2} X)_i$ is a row vector of the weighted sensitivities associated with the ith observation, $(X^T \omega^{1/2})_i$ is a column vector equal to the transpose of $(\omega^{1/2} X)_i$, $(X^T \omega X)^{-1}$ is from Eq. (7.1), and h_{ii} is the leverage of the ith observation.

Values of h_{ii} range from 0.0 to less than 1.0. Values close to 1.0 identify observations with high leverage.

The leverage h_{ii} calculated in Eq. (7.7) is the ith diagonal of the "hat" matrix $\omega^{1/2} X (X^T \omega X)^{-1} X^T \omega^{1/2}$ (Belsley et al., 1980, p. 16; Draper and Smith, 1998, p. 205). The full matrix is used for advanced regression analyses that are not discussed in this book (see, e.g., Cook and Weisberg, 1982).

7.3.2 Influence Statistics

Whereas leverage statistics indicate the potential importance of an observation to the estimation of a parameter, the actual effect of the observation in the regression also depends on the observed values, as illustrated in Figure 7.2. The Cook's D and *DFBETAS* influence statistics incorporate this effect. The Cook's D statistics are calculated for each observation and measure the influence of each individual observation on the estimation of the set of parameters as a whole. *DFBETAS* are calculated for each parameter b_j and each observation y_i and measure an

7.3 OBSERVATIONS IMPORTANT TO ESTIMATED PARAMETER VALUES

observation's effect on a single parameter value. Both statistics were first applied to groundwater models by Yager (1998).

Cook's D Cook's D is a measure of how a set of parameter estimates would change with omission of an observation, relative to how well the parameters are estimated given the entire set of observations. Cook's D is defined as follows, where the first expression explicitly shows the components of the statistic and the second is used for calculation (Cook and Weisberg, 1982, p. 116; Draper and Smith, 1998, p. 211; Cook and Weisberg, 1999, pp. 357–360; Helsel and Hirsch, 2002, p. 248):

$$D_i = \frac{(b_{(i)} - b')^{\text{T}} \left[\sigma^2 (X^{\text{T}} \omega X)^{-1}\right]^{-1} (b_{(i)} - b')}{NP} = \frac{1}{NP} r_i^2 \frac{h_{ii}}{1 - h_{ii}} \quad (7.8)$$

where b' = the set of parameter values optimized using all observations;
$b_{(i)}$ = the linear estimate of the set of parameter values that would be estimated if the ith observation were omitted;
X = a matrix of sensitivities, as defined before Eq. (5.2b);
ω = the weight matrix of Eq. (3.2) and (5.1);
NP = the number of estimated parameters;
σ^2 = the variance of the regression;
r_i = the ith weighted residual divided by its standard error, calculated as $f_i / [\sigma (1 - h_{ii})^{1/2}]$;
f_i = the ith weighted residual of the regression with all observations;
h_{ii} = the leverage of the ith observation, calculated by Eq. (7.7).

The variance of the regression, σ^2, and its square root, σ, are estimated using s^2 (Eq. (6.1)) and s, the variance and standard error of the regression, respectively.

For Cook's D to be large, the misfit needs to be large relative to the expected accuracy of the observation (r_i is large) and/or the leverage term needs to be large.

Nonlinearity of influence measures that fill the same purpose as Cook's D were investigated by Ross (1987). The measures were based on likelihood distances. One performs like Cook's D and others were suggested by Cook and Weisberg (1982). The one that performed like Cook's D performed well in the presence of high parameter-effects curvature (the terminology of Bates and Watts, 1980), which is called nonintrinsic nonlinearity by Christensen and Cooley (2005). See Section 7.7 for a discussion of linearity measures. This means that Cook's D is more robust for many nonlinear models than, for example, the sensitivity methods discussed in Chapter 4. It is suspected that the fit-independence of the methods discussed in Chapter 4 is advantageous for initial models. However, more testing is needed and the advantage may depend on model nonlinearity and the misfit of the initial model. Cook's D can be calculated very quickly.

Two distinctly different critical values for Cook's D have been suggested. On the basis of comments by Cook (1977a), Helsel and Hirsch (2002, p. 248) suggest using $F_{\alpha=0.1}(NP+1, ND+NPR-NP)$, where F_α is the value of the F distribution (Table D.7 in Appendix D) with a significance level of 0.1 and with $NP+1$ and $ND+NPR-NP$ degrees of freedom. Cook (1977b) and Cook and Weisberg (1982, p. 116) note, however, that Cook's D is not distributed as F. Rawlings (1988) suggests using $4/(ND+NPR)$, which results in a much smaller critical value and more observations being identified as influential. The Rawlings (1988) critical value is used in this book.

The critical value for Cook's D lacks an associated significance level. That is, unlike confidence intervals, no probability level is suggested by the critical value of Cook's D. It simply identifies observations that are more influential than the other observations.

DFBETAS The *DFBETAS* statistic (pronounced d-f-beta-s) measures the influence of one observation, y_i, on one parameter, b_j. $DFBETAS_{ij}$ is calculated as follows, where the first expression again explicitly shows the components of the statistic and the second is used for calculation (Belsley et al., 1980):

$$DFBETAS_{ij} = \frac{(b'_j - b'_{j(i)})}{s(i)[(X^T \omega X)^{-1}_{jj}]^{1/2}} = \frac{c_{ji}}{\left(\sum_{k=1}^{ND} c_{jk}^2\right)^{1/2}} \frac{f_i}{s(i)(1-h_{ii})} \quad (7.9)$$

where $b'_j =$ the optimized value of the jth parameter using all observations;

$b'_{j(i)} =$ the optimized value of the jth parameter omitting only the ith observation;

$s(i) =$ an alternate to s as an estimate of σ, chosen to make the denominator statistically independent of the numerator under normal theory, and calculated as (Belsley et al., 1980, p. 14):

$$s(i) = [1/(ND+NPR-NP-1)][(ND+NPR-NP)s^2 - f_i^2/(1-h_{ii})]^{1/2};$$

$c_{ji} =$ an entry of the matrix product $C = (X^T \omega X)^{-1} \omega^{1/2} X^T$.

All other symbols are defined after Eq. (7.8).

A value of $DFBETAS_{ij}$ greater than the critical value of $2/(ND+NPR)^{1/2}$ (Belsley et al., 1980, p. 28) indicates that the ith observation is influential in the estimation of the jth parameter. The likelihood of a single observation being influential to a single parameter will generally decrease as the number of regression observations increases. The critical value for $DFBETAS_{ij}$ takes this into account, as it decreases with increasing $ND+NPR$. As a result, there are roughly the same proportion of influential observations identified regardless of the size of $ND+NPR$. *DFBETAS* can be calculated very quickly.

7.4 UNIQUENESS AND OPTIMALITY OF THE ESTIMATED PARAMETER VALUES

Two important questions are (1) given the constructed model, are the observations and prior information sufficient to have estimated the one and only set of parameter values that provide the best fit? and (2) does a set of parameter values exist that produces a better fit than that achieved? The first issue primarily requires investigation of uniqueness, the second primarily involves optimality.

Uniqueness of the estimated parameter values can be investigated by (a) evaluating parameter correlation coefficients and (b) repeating the regression using different starting values. These methods are the focus of Sections 4.3.5, 4.4.2, and 7.2.2, and Exercises 4.1c, 5.1a, and 7.1f. The method in (a) is considered to be a local method because parameter correlation coefficients apply locally in the objective function surface (see Figure 4.2). Optimality can not be investigated using local methods; they require more computationally demanding global methods such as that described in (b).

As mentioned previously, nonunique parameter estimates may be indicated if parameter correlation coefficients calculated at the optimal parameter estimates are greater than about 0.95 in absolute value or *pcc* accuracy is suspect because of inaccurate sensitivities and/or insensitive parameters. In these situations or to test optimality, additional regression runs can be useful, as shown in Exercise 7.1f.

If significantly different parameter estimates result from the regression runs with different starting values, and these estimates produce nearly identical values of the objective function (Eq. (3.1) or (3.2)), the parameter estimates are not unique. If smaller values are encountered, the orginal solution is not optimal. In the case of nonuniqueness, the identical objective function values generally are produced because coordinated changes in parameter values produce identical simulated equivalents. This indicates that the available observation data are insufficient to uniquely estimate each parameter value. To reduce correlation and improve the likelihood of obtaining a unique solution and to address nonoptimality, parameters can be redefined, observations can be added to the regression, or prior information on the correlated parameters can be added to the regression. Redesigning parameters and prior information are discussed in Chapters 11 and 12 in Guidelines 3, 5, and 10.

7.5 QUANTIFYING PARAMETER VALUE UNCERTAINTY

Two methods of quantifying prediction uncertainty are discussed—inferential statistics and Monte Carlo methods.

7.5.1 Inferential Statistics

Linear inferential statistical methods are used here to calculate confidence intervals on estimated parameter values. Nonlinear confidence intervals also are discussed briefly and are calculated in Exercise 7.1g, but details of calculating nonlinear intervals are presented in Chapter 8, which focuses on evaluating predictions. The most common use of nonlinear intervals is to assess prediction uncertainty.

Confidence intervals on parameter values are intervals that, with a specified likelihood, contain the true, unknown parameters, if the model is correct. Here we consider individual confidence intervals because they are most often used for parameters. Other types of intervals are discussed in Chapter 8 because they are often used to quantify the uncertainty of predictions. Other types of intervals on parameters can be calculated using the ideas and methods presented for intervals on predictions. Confidence intervals are discussed in many texts, such as Miller (1981), Seber and Wild (1989), Cooley and Naff (1990), Davis (2002), and Helsel and Hirsch (2002).

Linear Individual Confidence Intervals   An individual confidence interval on a quantity, such as a parameter estimate, has a specified probability of including the true value of the quantity, regardless of whether confidence intervals on other quantities, such as other parameter estimates, include their true values. Usually individual intervals are used to evaluate uncertainty in parameter estimates.

An individual linear confidence interval for the true, unknown jth parameter β_j is calculated as

$$b_j \pm t(n, 1.0 - \alpha/2)s_{b_j} \qquad (7.10)$$

where $t(n, 1.0 - \alpha/2)$ is the Student t-statistic (Appendix D, Table D.2) for n degrees of freedom and a significance level of α; n is the degrees of freedom, here equal to $ND + NPR - NP$; and s_{b_j} is the standard deviation of the jth parameter.

Because a confidence interval is a range that has a stated probability of containing the true value, it is stated in terms of the true, unknown value that is being estimated. Thus, Eq. (7.10) is said to be the confidence interval for the true value of the jth parameter, β_j, and the width of the confidence interval is a measure of the likely precision of the estimate. Narrower intervals indicate greater precision. If the model correctly represents the system, the interval also can be thought of as a measure of the likely accuracy of the estimate. Definitions of precision and accuracy relevant to parameter estimates are given in Chapter 1, Section 1.4.2.

Linear confidence intervals truly represent uncertainty at the given significance level only to the extent that the assumptions underlying the calculation of Eq. (7.10) are satisfied. These requirements are discussed in Chapter 3, Section 3.3. Normality of the parameter estimates is required because the Student t-statistic is used in Eq. (7.10). The Student t-distribution is similar to a normal distribution but accounts for small sample sizes. The normality assumption is tested using methods for assessing the normality of weighted residuals (Chapter 6, Sections 6.4.5 and 6.4.6), because the probability distribution of the true errors is unknown.

Linear confidence intervals require trivial amounts of execution time, and individual linear 95-percent confidence intervals are calculated and printed by UCODE_2005, MODFLOW-2000, and PEST. However, in many natural systems the assumptions discussed above are not met and calculated linear confidence intervals are not accurate. More accurate nonlinear intervals developed by Vecchia and Cooley (1987) can be calculated, as discussed next and in Chapter 8, Section 8.4.3, but require substantial execution time.

7.5 QUANTIFYING PARAMETER VALUE UNCERTAINTY

Nonlinear Individual Confidence Intervals Nonlinearity of simulated values with respect to parameters is discussed in Chapter 1, Section 1.4.1. The nonlinearity of a model with respect to its parameters can be evaluated using the methods described in Section 7.7.

For nonlinear models, linear confidence intervals on parameters calculated using Eq. (7.10) can be inaccurate. More accurate nonlinear confidence intervals can be calculated using inferential statistics, as described briefly here, or using Monte Carlo methods, as described in Section 7.5.2.

Vecchia and Cooley (1987) developed inferential methods to compute nonlinear confidence intervals on any function of the model parameters. For nonlinear intervals on a parameter, the function is specified to be the value of a single parameter. Calculating a nonlinear confidence interval involves finding the smallest and largest parameter values on a confidence region for the model parameters, as illustrated in Figure 7.3 for parameter b_1 of a hypothetical two-parameter model. Unlike a linear confidence interval, the nonlinear confidence interval generally is not symmetric about the optimal value of b_1: in Figure 7.3, the upper limit of the interval is much further from b'_1 than is the lower limit of the interval. The method for calculating nonlinear intervals is substantially more complicated and more computationally intensive than is the method for calculating linear intervals, as discussed in Chapter 8, Section 8.4.3.

MODFLOW-2000's UNC Process (Christensen and Cooley, 2005), UCODE_2005, and PEST support calculation of nonlinear confidence intervals for parameter values. Because they are expensive computationally, these intervals usually are calculated only for selected quantities of interest. If computation time is a limiting factor, it is likely that the intervals will be calculated for model predictions instead of for model parameters. Nonlinear intervals for parameters are presented here largely to introduce students to nonlinear intervals in as simple a context as possible.

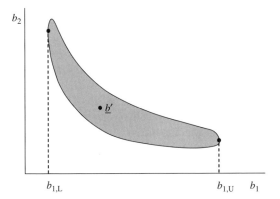

FIGURE 7.3 Confidence region (shaded area) and upper ($b_{1,U}$) and lower ($b_{1,L}$) limits of a nonlinear confidence interval on parameter b_1, for a hypothetical two-parameter model. (Adapted from Christensen and Cooley, 1999, Figure 9.)

7.5.2 Monte Carlo Methods

There are two approaches with which random sampling by Monte Carlo analysis can be used to evaluate uncertainty of estimated parameter values. The first approach involves investigating the variation in estimated parameter values that would result if the observations had a different realization of error. These methods are generally called bootstrap methods (Efron and Tibshirani, 1993; Chernick, 1999) and can produce measures of uncertainty that are consistent with confidence intervals. The calibrated model is used to generate sets of observations, which are then contaminated with noise. Prior information can be generated on the basis of the estimated parameter values. The generated observations and prior information are then used to estimate parameter values. This type of Monte Carlo analysis essentially addresses the question of how much the estimated parameter values would vary for different realizations of error in the observations and prior information.

The second way Monte Carlo methods can be used is to investigate how different model construction alternatives would affect the estimated parameter values (see Poeter and McKenna, 1995). These Monte Carlo runs could be combined with the first type, or confidence intervals could be calculated for the alternative model constructions using inferential methods. The Monte Carlo approach is discussed in Chapter 8, Section 8.5 in the context of predictions.

Beven and Binley (1992) and Binley and Beven (2003) present an interesting method of portraying Monte Carlo results called dottie plots. In their Monte Carlo analyses, many forward model runs are conducted using different parameter values, and for each a function of the sum of squared residuals is calculated such that larger values indicate a better fit. The dottie plots consist of x–y graphs with these statistics plotted against each parameter value. Optimal parameter values exist if there is a peak in the dottie plot.

7.6 CHECKING PARAMETER ESTIMATES AGAINST REASONABLE VALUES

When plotted on graphs with the related estimated values, linear confidence intervals can provide a vivid image of the approximate precision with which parameters are estimated using the data included as observations in the regression, given the constructed model. It often is useful to compare these intervals to ranges of reasonable parameter values. This comparison can be a powerful tool for diagnosing error in data interpretation and model construction. As discussed in Chapter 5, Section 5.5, avoiding limits that constrain the estimated parameters allows the regression to estimate unreasonable parameter values, and thus makes this comparison possible.

In Figure 7.4, the estimates and confidence intervals for three hypothetical parameters are plotted with reasonable ranges for each parameter. This figure shows three situations that might result from considering reasonable parameter ranges.

7.6 CHECKING PARAMETER ESTIMATES AGAINST REASONABLE VALUES

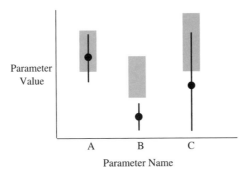

FIGURE 7.4 Graph illustrating the comparison of parameter estimates and confidence intervals with the reasonable range of parameter values. Closed circles are parameter estimates, black bars are confidence intervals, and grey bars represent the range of reasonable values for each parameter.

1. For parameter A, the parameter estimate and most of the confidence interval lie within the reasonable range of values. This suggests that the estimate is consistent with independent information about the parameter.
2. For parameter B, the estimate and entire confidence interval lie outside the range of reasonable values, meaning that the regression data together with the given model construction produce a parameter estimate that is inconsistent with independent information about the parameter. This indicates the existence of model bias and that the data, as represented by the observations and the reasonable range of parameter values, are sufficient to detect it. In this situation, the interpretation of the observations, prior information, model construction and reasonable range need to be carefully scrutinized.
3. For parameter C, the estimate is unreasonable, but the confidence interval partly lies in the reasonable range of values. This result indicates that there may or may not be model bias; the data are insufficient to make either conclusion. In this last situation, the modeler needs to consider both (a) the possibility of model error and (b) additional data that could provide information toward estimating the parameter value or the reasonable range more precisely.

Linear confidence intervals often are sufficient for this analysis, though the analyses also can be performed using nonlinear intervals.

The analysis described above does not evaluate one very important characteristic of reasonable parameter values. In some situations the reasonable ranges of two parameters may overlap, but it is known that the value of one should be greater or less than the value of the other. That is, in addition to the requirement that parameters lie in their respective reasonable ranges, the relative magnitudes of two or more different parameter values are important. This is a valuable test that needs to be considered when evaluating estimated parameter values (Poeter and McKenna, 1995; Poeter and Anderson, 2005). If the parameters cannot be estimated uniquely, a parameter that equals the ratio of the parameters could be estimated, and the ratio could be evaluated for its consistency with known relative properties. This manipulation of parameters can be accomplished with UCODE_2005 and PEST, and, in some circumstances with MODFLOW-2000.

7.7 TESTING LINEARITY

The application and utility of some of the methods presented in this chapter depend on the linearity of the model with respect to the parameter values. Although the modified Gauss–Newton optimization method and many of the statistical methods discussed are useful even for problems that are quite nonlinear, more stringent requirements on linearity are needed for linear confidence intervals to represent parameter uncertainty adequately.

Linearity can be tested using the modified Beale's measure (also called Linssen's measure) described by Cooley and Naff (1990, pp. 187–189). The original Beale's measure is described by Beale (1960). The modified Beale's measure indicates nonlinearity of the parameter confidence region and does not directly measure nonlinearity of confidence intervals. It is, however, a good measure of nonlinearity that is likely to affect confidence intervals on parameters.

The modified Beale's measure tests model linearity with respect to the regression observations. Use of this measure as an indicator of model linearity with respect to predictions becomes increasingly problematic as the predictive quantities or situations differ more from the calibration observations and situations. For more information, see Chapter 8, Section 8.7.

The modified Beale's measure is calculated by the following four steps.

Step 1. Sets of parameter values are generated that lie on the edge of the linear confidence region for the parameters. This is accomplished using the following equation:

$$\tilde{b} = b' \pm \frac{\sqrt{NP \times F_\alpha(NP, ND + NPR - NP)}}{s_{b_j}} [V(b')]_j, \quad j = 1, NP \quad (7.11)$$

where $\tilde{b}$ = a vector of generated parameter values;
b' = a vector of optimal parameter estimates;
$F_\alpha(NP, ND + NPR - NP)$ = the value from the F distribution with significance level α (equal to 0.05 for the calculations in this book) and with NP and $ND + NPR - NP$ degrees of freedom;
$[V(b')]_j$ = a vector equivalent to the jth column of $V(b')$;
s_{b_j} = the standard deviation of the jth parameter, defined in Eq. (7.3).

Because of the $\pm$, Eq. (7.11) yields two parameter vectors for each j, yielding a total of $2 \times NP$ generated parameter vectors. In the generated vectors, values of parameters with small variances generally are relatively near the optimal par-

7.7 TESTING LINEARITY

ameter estimates, and values of parameters with large variances generally are relatively far from optimal parameter estimates. The process by which Beale's measure generates parameter values is statistical, and sometimes one or more of the parameter sets generated do not yield valid or accurate solutions. Resolution of such problems is discussed later. For the jth pair of generated parameter vectors, the jth parameter value usually varies the most, but parameter correlation causes other parameter values to vary as well.

Step 2. Simulated equivalents of the calibration observations are computed by executing a forward model run for each generated set of parameter values. These simulated values are $\tilde{y}_{ik}$, where i refers to the ith observation and k refers to the kth generated parameter vector.

Step 3. Linearized estimates of the simulated values are calculated using the generated parameter sets as follows:

$$\tilde{y}^o_{ik} = y'_i + \sum_{j=1}^{NP} \left(b'_j - \tilde{b}^0_{jk}\right) \frac{\partial y'_i}{\partial b_j}\bigg|_{b'} \tag{7.12}$$

where $\tilde{y}^o_{ik}$ = the linearized simulated equivalent of the ith observation calculated using the kth parameter set;

y'_j = the simulated equivalent of the ith observation calculated using the original model and optimal parameter estimates;

b'_j = the jth optimal parameter estimate;

$\tilde{b}^0_{jk}$ = the jth parameter value from the kth generated parameter set.

Step 4. The modified Beale's measure, $\widehat{N}_b$, is calculated as a measure of the difference between the model-computed and the linearized estimates of the simulated values (Cooley and Naff, 1990, p. 188):

$$\widehat{N}_b = NP \times s^2 \frac{\sum_{k=1}^{2NP} \sum_{i=1}^{ND} \sum_{q=1}^{ND} \left(\tilde{y}_{ik} - \tilde{y}^o_{ik}\right) \omega_{iq} \left(\tilde{y}_{qk} - \tilde{y}^o_{qk}\right)}{\sum_{k=1}^{2NP} \sum_{i=1}^{ND} \sum_{q=1}^{ND} \left(\tilde{y}^o_{ik} - y'_i\right) \omega_{ij} \left(\tilde{y}^o_{qk} - y'_q\right)} \tag{7.13}$$

where $\tilde{y}_{ik}$ the simulated equivalent of the ith observation calculated using the original model and the kth parameter set.

The program BEALE-2000 (Hill et al., 2000) can be used to calculate the modified Beale's measure in conjunction with regression performed using MODFLOW-2000. The program MODEL_LINEARITY can be used to calculate the measure for regressions performed using UCODE_2005.

In step 2 above, problems can occur with obtaining a forward model solution using the parameter values generated in step 1. The most common problems are that physically impossible negative parameter values are generated (such as for hydraulic conductivity in groundwater models), or that solution with the specified solver convergence criteria is not possible within the number of allowed solver iterations. The problem of negative parameters can be solved by log-transforming the parameter(s) involved, possibly repeating the regression to account for any resulting change in optimized parameter values, and regenerating the parameter values for Beale's measure. For any parameter with prior information, the log-transformation may require alteration of the statistic used to calculate the prior information weights, as discussed in Guideline 6 in Chapter 11.

In some situations the solver will not converge for some of the data sets. As long as the final solution obtained is not too inaccurate, the resulting value of Beale's measure will still reflect model nonlinearity adequately. If the lack of convergence is accompanied by a very inaccurate solution, this is, of course, problematic. Sometimes better results can be obtained using a different solver.

To assess the degree of linearity of the model, the modified Beale's measure calculated by Eq. (7.13) is compared with two critical values (Cooley and Naff, 1990, p. 189), as follows. If $\widehat{N}_b < 0.09/F_\alpha(NP, ND + NPR - NP)$, the model is effectively linear, in that linear confidence intervals closely approximate the exact nonlinear confidence intervals of Vecchia and Cooley (1987). If $\widehat{N}_b > 1.0/F_\alpha(NP, ND + NPR - NP)$, the model is highly nonlinear. If the modified Beale's measure lies between these two critical values, then the model can be considered moderately nonlinear.

Alternative methods for evaluating model nonlinearity with respect to parameters described by Cooley (2004) and Christensen and Cooley (2005) are derived from Beale (1960) and Linssen (1975). Using the terminology used in Poeter et al. (2005), the statistics are called total model nonlinearity and intrinsic model nonlinearity. Intrinsic model nonlinearity is the nonlinearity that cannot be removed by a parameter transformation of any kind. For many circumstances it is the intrinsic nonlinearity that is problematic. The following equations apply if the weighting satisfies $\omega = V(\varepsilon)^{-1}$ (see Chapter 3, Section 3.4.2).

Total model nonlinearity can be calculated as follows. The sets of parameter values are calculated as

$$\tilde{b} = b' \pm \frac{\sqrt{NP}}{s_{b_j}}[V(b')]_j, \quad j = 1, NP \tag{7.14}$$

The total nonlinearity statistic is calculated as

$$\widehat{N} = \frac{1}{NP \times s^2} \frac{\sum_{k=1}^{2NP} \sum_{i=1}^{ND} \sum_{q=1}^{ND} (\tilde{y}_{ik} - \tilde{y}_{ik}^o) \omega_{iq} (\tilde{y}_{qk} - \tilde{y}_{qk}^o)}{2 \times NP} \tag{7.15}$$

Total model nonlinearity performs like the modified Beale's measure (Eq. (7.13)) but is scaled differently. Values of total model nonlinearity that are less than 0.09

indicate a linear model and values greater than 1.0 indicate a highly nonlinear model.

The intrinsic model nonlinearity uses the sets of parameters from Eq. (7.14). The intrinsic model linearity statistic is calculated as follows, using matrix notation instead of the summations of Eqs. (7.13) and (7.15):

$$\widehat{N}_{\min} = \frac{1}{NP \times s^2} \frac{\sum_{k=1}^{2 \times NP} \left((\tilde{y}_k - \tilde{y}_k^o) - X\psi\right)^T \omega_{iq} \left((\tilde{y}_k - \tilde{y}_k^o) - X\psi\right)}{2 \times NP} \quad (7.16)$$

where

$$\psi = (X^T \omega X)^{-1} X^T \omega (\tilde{y}_k - \tilde{y}_k^o) \quad (7.17)$$

As for total model nonlinearity, values of intrinsic model nonlinearity that are less than 0.09 indicate a linear model and values greater than 1.0 indicate a highly nonlinear model.

The total and instrinsic model linearity are calculated using UCODE_2005 and associated code MODEL_LINEARITY_ADV, or with the UNC Process of MODFLOW-2000. Calculation of the statistics is illustrated in Exercise 7.3.

7.8 EXERCISES

Exercise 7.1: Parameter Statistics This exercise uses parameter statistics to evaluate the optimal parameter estimates from the steady-state model regression run of Exercise 5.2c. These statistics are used to reevaluate the importance of the observations to the parameter estimates, and to evaluate parameter uncertainty and correlation.

  (*a*) *Evaluate composite scaled sensitivities.*

For nonlinear models, the composite scaled sensitivities calculated for the final estimated parameters are likely to be different from those for the starting parameter values. In the regression of Exercise 5.2c, the initial *css* considered in Exercise 4.1b (shown in Figure 4.3 and 7.5*a*) suggested that prior information (actually, regularization) was needed for parameters K_RB and VK_CB. It is important to examine the final *css*, shown in Figure 7.5*b*, to assess whether their relative values for the model parameters are similar. If the final *css* for the parameters with regularization have become larger (relative to the *css* for other parameters), then it is important to try to estimate these parameters without the regularization imposed.

Problem
- Discuss the differences between the initial (Figure 7.5*a*) and final (Figure 7.5*b*) *css* values in terms of model nonlinearity and scaling.

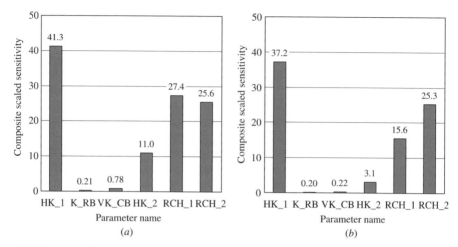

FIGURE 7.5 Composite scaled sensitivities from the (*a*) starting and (*b*) final steady-state model.

- How does nonlinearity and the scaling used affect the utility of the *css*?
- Do the *css* in Figure 7.5*b* suggest that the regression should be attempted without prior information specified for parameters K_RB or VK_CB?
- Use the *css* to explain why the weighted residuals for the prior information are so small (Figure 6.7*a*).

(*b*) *Evaluate leverage statistics.*

Problem: Compare the leverage statistics of Exercise 4.1e and Table 7.2 and comment on any differences. Refer to system dynamics and the added prior information. To help explain the leverage of hd07.ss, consider Table 7.4, which shows the parameter correlation coefficients calculated using the final parameter estimates from Exercise 5.2e and all the regression data except observation hd07.ss.

(*c*) *Evaluate importance using influence statistics.*

In this exercise, the importance of individual observations and prior information to the estimation of the model parameters is assessed using the Cook's *D* and *DFBETAS* measures. Table 7.2 shows the Cook's *D* values for the steady-state regression. The critical value of Cook's *D* is 0.308. Table 7.3 shows the *DFBETAS* statistics for the steady-state regression; the critical value is 0.555.

Problem

- Which observations have values of Cook's *D* and of *DFBETAS* that exceed the critical values? Why would these observations be most influential to the

TABLE 7.2 Leverage Statistics[a] and Cook's D Values[b] for the Steady-State Regression

Observation Name	Leverage Statistic	Cook's D
hd01.ss	0.00	0.0013
hd02.ss	0.14	0.0283
hd03.ss	0.18	0.0079
hd04.ss	0.14	0.0860
hd05.ss	0.22	0.0005
hd06.ss	0.19	0.0093
hd07.ss	**0.96**	**0.5934**
hd08.ss	0.18	0.0300
hd09.ss	0.84	0.2879
hd10.ss	0.19	0.0247
flow01.ss	**1.00**	**589.1**
K_RB prior	**1.00**	**35.94**
VK_CB prior	**1.00**	**72.16**

[a] The values in bold type are larger than 0.90.
[b] The values in bold type are larger than the critical value of 0.308.

TABLE 7.3 *DFBETAS* Values[a] for the Steady-State Regression

	DFBETAS					
Observation Name	HK_1	K_RB	VK_CB	HK_2	RCH_1	RCH_2
hd01.ss	0.030	−0.089	0.000	−0.006	0.018	0.005
hd02.ss	−0.074	−0.013	0.018	−0.001	0.069	−0.076
hd03.ss	0.020	−0.006	0.003	−0.005	0.001	0.000
hd04.ss	0.162	0.029	−0.040	0.001	−0.149	0.167
hd05.ss	0.0012	0.0003	0.002	0.016	−0.021	0.022
hd06.ss	0.075	−0.001	−0.007	−0.114	0.066	−0.070
hd07.ss	**−0.586**	−0.001	−0.012	**1.74**	**−1.46**	**1.605**
hd08.ss	−0.030	0.009	−0.001	−0.019	0.020	−0.024
hd09.ss	−0.098	0.011	−0.009	−0.032	0.223	−0.245
hd10.ss	−0.046	0.003	−0.013	−0.037	0.084	−0.089
flow01.ss	**−59.4**	−0.088	−0.018	**1.14**	**−32.8**	**−21.1**
K_RB prior	**−0.611**	**14.7**	0.007	**0.929**	**−0.607**	**0.685**
VK_CB prior	**−1.60**	0.010	**20.1**	**4.05**	**−3.10**	**3.41**

[a] The values in bold type are larger than the critical value of 0.555.

estimation of the model parameters? As for leverage, to help explain the influence of observation hd07.ss, consider Table 7.4.

- Compare the *DFBETAS* values in Table 7.3 to the dimensionless scaled sensitivities shown in Table 7.5. Explain why observation–parameter combinations with the largest *DFBETAS* values can have very small dimensionless scaled sensitivities. What is the implication for using dimensionless scaled

TABLE 7.4 Parameter Correlation Coefficient Matrix Calculated by MODFLOW-2000 for the Final Parameter Estimates, Using All Hydraulic-Head and Flow Observations and Prior Information Except Observation hd07.ss[a]

	HK_1	K_RB	VK_CB	HK_2	RCH_1	RCH_2
HK_1	1.00	−0.025	−0.035	−0.79	0.87	−0.72
K_RB		1.00	0.0006	0.015	−0.011	0.011
VK_CB			1.00	0.034	−0.029	0.029
HK_2		Symmetric		1.00	**−0.98**	**0.99**
RCH_1					1.00	**−0.97**
RCH_2						1.00

[a]Bold values have absolute value greater than 0.95.

TABLE 7.5 Dimensionless Scaled Sensitivities Calculated for the Final Parameter Values Estimated for the Steady-State Regression

	Dimensionless Scaled Sensitivities					
Observation Name	HK_1	K_RB	VK_CB	HK_2	RCH_1	RCH_2
hd01.ss	1.18E−05	−0.210	−8.23E−09	1.29E−06	0.116	0.094
hd02.ss	−25.5	−0.210	−0.020	−1.16	13.2	13.7
hd03.ss	−52.7	−0.210	−0.041	−4.13	22.1	35.1
hd04.ss	−25.5	−0.210	−0.020	−1.16	13.2	13.7
hd05.ss	−38.5	−0.210	−0.028	−2.30	18.6	22.4
hd06.ss	−25.6	−0.210	−0.181	−1.02	13.2	13.9
hd07.ss	−0.699	−0.210	−0.677	0.657	0.490	0.440
hd08.ss	−52.7	−0.210	0.003	−4.19	21.9	35.2
hd09.ss	−68.9	−0.210	0.184	−7.67	22.0	54.5
hd10.ss	−38.5	−0.210	−0.096	−2.25	18.5	22.6
flow01.ss	−5.65E−04	−2.38E−05	4.10E−07	−6.20E−05	−5.54	−4.50

sensitivities and composite scaled sensitivities to determine which observations are most important to estimating the parameters?

(*d*) *Evaluate the uniqueness of the parameter estimates using correlation coefficients.*

Parameter correlation coefficients were introduced in Exercise 4.1a, to assess likely parameter uniqueness using the starting parameter values, and in Exercise 5.1a, to demonstrate the relation between these coefficients and objective-function surfaces and to illustrate the necessity of flow observations in preventing complete correlation between groundwater flow model parameters. Here, the correlation coefficients are used to evaluate uniqueness of the parameter estimates from Exercise 5.2c. The correlation coefficients calculated by MODFLOW-2000 are shown in Table 7.6a, and those calculated by UCODE_2005 are shown in Table 7.6b.

7.8 EXERCISES

TABLE 7.6 Parameter Correlation Coefficient Matrix for Final Parameter Values Using the Hydraulic-Head Observations, the Streamflow Observation, and Prior Information Calculated for the Steady-State Problem by MODFLOW-2000 and UCODE_2005[a]

	HK_1	K_RB	VK_CB	HK_2	RCH_1	RCH_2
(a) MODFLOW-2000						
HK_1	1.00	−0.042	−0.080	−0.36	0.72	0.025
K_RB		1.00	0.0005	0.063	−0.041	0.047
VK_CB			1.00	0.20	−0.15	0.17
HK_2		Symmetric		1.00	−0.85	0.91
RCH_1					1.00	−0.65
RCH_2						1.00
(b) UCODE_2005						
HK_1	1.00	−0.039	−0.077	−0.36	0.72	0.024
K_RB		1.00	0.0005	0.059	−0.038	0.043
VK_CB			1.00	0.19	−0.15	0.16
HK_2		Symmetric		1.00	−0.85	0.91
RCH_1					1.00	−0.65
RCH_2						1.00

Problem
- Which parameter pairs are most highly correlated? What physical arguments can be used to explain why these parameters are correlated?
- Do the parameter correlations calculated by MODFLOW-2000 at the final parameter values (Table 7.6a) differ from those calculated at the starting parameter values (Table 4.2a)? Is this expected?
- Are there any significant differences between the correlations calculated by MODFLOW-2000 (Table 7.6a) and by UCODE_2005 (Table 7.6b)? What would produce the differences?

(e) *Detecting nonunique parameter estimates.*

The next part of this Exercise repeats some of the types of regression runs performed in Exercise 5.1b for the two-parameter combined model, to demonstrate further the effects of parameter correlation and methods for detecting it. Recall that the data available for estimating the six parameters of the steady-state model consist of 10 hydraulic heads (five in each model layer) and the gain in streamflow. When the streamflow gain observation is omitted, no prior information on parameters is specified, and only the 10 head observations are used, then all parameter correlation coefficients equal 1.0. This result is a direct consequence of Darcy's Law, as discussed in the answer to Exercise 5.1a (available on the web site described in Chapter 1, Section 1.1). When the absolute values of any correlations are 1.00 or very close to 1.00, it may be that no single set of parameter values will produce the smallest value of the sum of squared, weighted residuals, and the nonlinear regression may have trouble converging, or the solution may be nonunique in that different solutions would result from using different initial parameter values.

Instructions for these simulations are available from the web site for this book described in Chapter 1, Section 1.1.

(1) Perform a regression run in which the flow observation is omitted from the calibration data set, and there is no prior information on parameters. Statistics from the modified Gauss–Newton iterations of this run are shown in Figure 7.6.

Problem
- What happened in this regression run? Discuss the calculated value of the maximum change (column 3 in the top of the figure).

SELECTED STATISTICS FROM MODIFIED GAUSS-NEWTON ITERATIONS

ITER.	MAX. PARAMETER PARNAM	CALC. CHANGE MAX. CHANGE	MAX. CHANGE ALLOWED	DAMPING PARAMETER
1	K_RB	-6.09916	2.00000	0.32791
2	HK_2	-0.902458	2.00000	1.0000
3	K_RB	0.849936	2.00000	1.0000
4	HK_2	-4.83286	2.00000	0.41383
5	VK_CB	39.8648	2.00000	0.50170E-01
6	VK_CB	-329.378	2.00000	0.30360E-02
7	VK_CB	13.8278	2.00000	0.36159E-01
8	VK_CB	12.3276	2.00000	0.16224
9	VK_CB	1.96435	2.00000	1.0000
10	HK_2	-1.84155	2.00000	1.0000

SUMS OF SQUARED WEIGHTED RESIDUALS FOR EACH ITERATION

ITER.	SUMS OF SQUARED WEIGHTED RESIDUALS		
	OBSERVATIONS	PRIOR INFO.	TOTAL
1	1751.1	0.0000	1751.1
2	7941.6	0.0000	7941.6
3	695.77	0.0000	695.77
4	148.52	0.0000	148.52
5	63.538	0.0000	63.538
6	58.476	0.0000	58.476
7	62.818	0.0000	62.818
8	59.196	0.0000	59.196
9	46.333	0.0000	46.333
10	16.519	0.0000	16.519

PARAMETER ESTIMATION DID NOT CONVERGE IN THE ALLOTTED NUMBER OF ITERATIONS

FIGURE 7.6 Selected statistics from the modified Gauss–Newton iterations of the regression run with only hydraulic-head observations in Exercise 7.1e. This is a fragment from the global output file of MODFLOW-2000.

7.8 EXERCISES

- Explain the parameter correlations resulting from this run. These correlations are shown in Table 4.3 in Exercise 4.1c; in that exercise a regression run was not performed but correlations for the six-parameter steady-state model were calculated using only the head observations.

(2) Test model nonuniqueness by starting the regression from the different sets of initial parameter values listed in Table 7.7, and comparing the resulting estimates, as suggested in Section 7.5.1. Include the flow observation and prior information. The results are shown in Table 7.8.

Problem: How much do the estimated parameter values differ from those produced using the original initial values? Are any differences large when compared to the associated parameter standard deviations? What are the strengths and weaknesses of this test?

(*f*) *Evaluate the precision of the estimates using standard deviations, linear confidence intervals, and coefficients of variation.*

Table 7.9 and Figure 7.7 show the starting and estimated (optimal) parameter values for the steady-state regression of Exercise 5.2c, and the approximate linear, individual, 95-percent confidence intervals on the estimated parameter values.

Problem
- Which estimated parameters have the largest individual, linear, 95-percent confidence intervals as a percentage of the estimated value? Do these same parameters have the largest coefficients of variation? Explain.
- What conclusions can be drawn about the relative uncertainty among the six parameters?
- Theoretically, 95-percent confidence intervals should include the true value 95 percent of the time. Use the last column in Table 7.9 to note how many of these

TABLE 7.7 New Sets of Starting Parameter Values for Exercise 7.1e

	HK_1	K_RB	VK_CB	HK_2	RCH_1	RCH_2
Original	3×10^{-4}	1.2×10^{-3}	1×10^{-7}	4×10^{-5}	63.072	31.536
Set 1	1.5×10^{-4}	0.6×10^{-3}	0.5×10^{-7}	2×10^{-5}	31.536	15.768
Set 2	6×10^{-4}	2.4×10^{-3}	2×10^{-7}	8×10^{-5}	126.144	63.072

TABLE 7.8 Estimated Parameter Values for Exercise 7.1e

	HK_1	K_RB	VK_CB	HK_2	RCH_1	RCH_2
Original	4.62×10^{-4}	1.17×10^{-3}	9.90×10^{-8}	1.54×10^{-5}	47.45	38.53
Set 1	4.62×10^{-4}	1.17×10^{-3}	9.90×10^{-8}	1.54×10^{-5}	47.45	38.53
Set 2	4.62×10^{-4}	1.17×10^{-3}	9.90×10^{-8}	1.54×10^{-5}	47.43	38.54

TABLE 7.9 Estimated Values of the Steady-State Flow System Parameters; Coefficients of Variation; Individual, Linear, 95-percent Confidence Intervals; and True Parameter Values

Parameter Name	Estimated Value	Coefficient of Variation	Individual, Linear, 95-percent Confidence Interval	True Value	Coefficient Interval Includes True Value?
HK_1	4.62×10^{-4}	0.14	3.11×10^{-4}; 6.13×10^{-4}	4.0×10^{-4}	
K_RB	1.17×10^{-3}	0.38	1.26×10^{-4}; 2.21×10^{-3}	1.0×10^{-3}	
VK_CB	9.90×10^{-8}	0.37	1.19×10^{-8}; 1.86×10^{-7}	2.0×10^{-7}	
HK_2	1.54×10^{-5}	1.77	-4.91×10^{-5}; 7.99×10^{-5}	4.4×10^{-5}	
RCH_1	47.45	0.27	16.6; 78.2	31.536	
RCH_2	38.53	0.31	10.5; 66.6	47.304	

7.8 EXERCISES

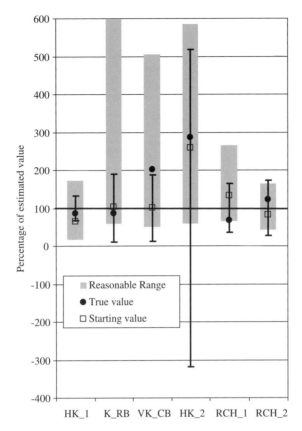

FIGURE 7.7 Starting and true parameter values, limits of approximate, individual, linear, 95-percent confidence intervals (black bars), and limits of reasonable ranges of parameter values, expressed as percentage of the estimated values, for the steady-state regression run. Note that linear confidence intervals can have a negative lower limit, even if physically implausible, when parameters are not log-transformed in the regression. Here, none of the parameters are log-transformed.

linear 95-percent confidence intervals include the true value. If the percent is significantly smaller than 95 percent, explain why. In your answer, consider the prior information imposed, whether it constitutes regularization, and its effect on measures of uncertainty such as confidence intervals.

(g) *Compare estimated parameter values with reasonable ranges.*

Figure 7.7 shows the estimated parameter values and individual linear confidence intervals in relation to the reasonable ranges of parameter values.

Problem: Are the estimated parameter values reasonable on the basis of the specified reasonable ranges? Are parameter confidence intervals needed to answer this question for this problem?

(h) *Evaluate the precision of the estimates using nonlinear confidence intervals.*

Figure 7.8 shows the nonlinear 95-percent confidence intervals calculated on parameter values for the steady-state regression of Exercise 5.2c, together with the approximate, individual, linear 95-percent confidence intervals and reasonable ranges of parameter values from Figure 7.7. The nonlinear confidence intervals were computed using the UNC Process (Christensen and Cooley, 2005) of MODFLOW-2000. UCODE_ 2005 produced the same results. The web site for this book (see Chapter 1, Section 1.1) provides instructions for calculating the nonlinear intervals.

Problem
- Compare the individual linear and nonlinear intervals, in terms of their size and symmetry.
- How many of the nonlinear 95-percent confidence intervals include the true parameter value? How does this analysis compare to that performed in Exercise 7.1c for the linear intervals?

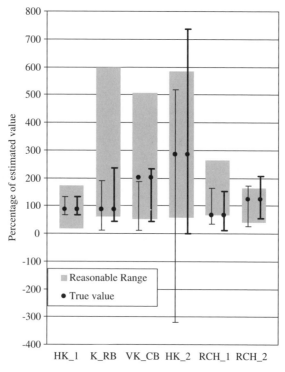

FIGURE 7.8 Limits of individual, linear, 95-percent confidence intervals (thin error bars); individual, nonlinear 95-percent confidence intervals (thick error bars); and reasonable ranges of parameter values, for the steady-state regression run. All values are expressed as percentage of the estimated parameter values.

7.8 EXERCISES

- Using the nonlinear intervals, assess the relative uncertainty among the six parameter values. How do the conclusions about relative parameter uncertainty compare to the conclusions drawn in Exercise 7.1f using the linear intervals?

Exercise 7.2: Consider All the Different Correlation Coefficients Presented
Three different statistics referred to as correlation coefficients have been presented: R (Eq. (6.11)), the correlation between weighted simulated and observed values; R_N^2 (Eq. (6.18)), the correlation coefficient between weighted residuals ordered from smallest to largest and the order statistics from a $N(0, 1)$ probability distribution; and pcc, parameter correlation coefficients (Eq. (7.5)), which measure whether coordinated changes in parameter values would produce the same simulated values and, therefore, the same value of the objective function. For all of these correlation coefficients, values range between -1.0 and 1.0, and values close to these extremes indicate high correlation. For R and R_N^2, values close to 1.0 are good: for R, this means the simulated values are in some ways similar to the observed values; for R_N^2, this means the weighted residuals are normally distributed. For pcc, values close to -1.0 or 1.0 are bad: it means that the available data are insufficient to uniquely estimate the parameter values being estimated.

Problem: Consider the equations for these three statistics. Note how they are similar and different, and use the equations to explain why extreme values of R_N^2 and of R are good, whereas extreme values of the parameter correlation coefficients are problematic.

Exercise 7.3: Test for Linearity

(a) *Use the modified Beale's measure.*

In this exercise, the linearity of the steady-state model is tested using the modified Beale's measure. First, calculate the measure using the weights on the prior values for K_RB and VK_CB that were used in the regression. Then, recalculate the measure using weights that more realistically reflect likely uncertainty in these hydraulic-conductivity parameters.

Instructions for model and postprocessor simulations needed to calculate the modified Beale's measure are available from the web site for this book described in Chapter 1, Section 1.1. For students not performing the simulations, the information shown in Figures 7.9 and 7.10 can be used to complete the exercise.

Problem
- Does the modified Beale's measure indicate that the model is effectively linear so that linear confidence intervals accurately display the uncertainty in the parameters?
- Would using nonlinear instead of linear confidence intervals change the conclusions reached in Exercise 7.1g?

```
USING FSTAT = 3.8700, BEALES MEASURE = 35.564
IF BEALES MEASURE IS GREATER THAN 0.26, THE MODEL IS
NONLINEAR.
IF BEALES MEASURE IS LESS THAN 0.23E-01, THE MODEL IS
EFFECTIVELY LINEAR, AND LINEAR CONFIDENCE INTERVALS ARE
FAIRLY ACCURATE IF THE RESIDUALS ARE NORMALLY DISTRIBUTED.
```

FIGURE 7.9 Part of BEALE-2000 output file showing Beale's measure calculated with the prior weights used in the regression for Exercise 5.2c.

```
USING FSTAT = 3.8700, BEALES MEASURE = 61.107
IF BEALES MEASURE IS GREATER THAN 0.26, THE MODEL IS
NONLINEAR.
IF BEALES MEASURE IS LESS THAN 0.23E-01, THE MODEL IS
EFFECTIVELY LINEAR, AND LINEAR CONFIDENCE INTERVALS ARE
FAIRLY ACCURATE IF THE RESIDUALS ARE NORMALLY DISTRIBUTED.
```

FIGURE 7.10 Part of BEALE-2000 output file showing Beale's measure calculated with a more realistic coefficient of variation of 1.0 used to compute the weights on prior values for both K_RB and VK_CB.

- How does the modified Beale's measure change when the weights on the prior values for K_RB and VK_CB are changed? Which calculated measure more realistically reflects the nonlinearity of the steady-state model?

(b) *Use total and intrinsic model nonlinearity.*

In this exercise, the linearity of the steady-state model is tested using total and intrinsic model nonlinearity measures mentioned at the end of Section 7.7. As in Exercise 7.3a, first calculate these measures using the weights on the prior values for K_RB and VK_CB that were used in the regression. Then, recalculate the measures using weights that more realistically reflect the uncertainty in these two parameters.

Instructions for the simulations needed to calculate total and intrinsic nonlinearity are available from the web site for this book described in Chapter 1, Section 1.1. For students not performing the simulations, the results are as follows. With the weights used in the regression, total model nonlinearity is 223.7 and intrinsic model nonlinearity is 0.142, and with a more realistic coefficient of variation of 1.0, total model nonlinearity is 359.0 and intrinsic model nonlinearity is 0.138. See Section 7.7 for critical values against which to compare the total model nonlinearity measures.

7.8 EXERCISES

Problem
- Does the total model nonlinearity statistic indicate that the model is effectively linear so that linear confidence intervals accurately display the uncertainty in the parameters? Is this result consistent with the analysis of the modified Beale's measure?
- Does the steady-state model have a large degree of intrinsic model nonlinearity?
- How do the statistics change when the weights on the prior values for K_RB and VK_CB are changed? Which calculated measure more realistically reflects the nonlinearity of the steady-state model?

8

EVALUATING MODEL PREDICTIONS, DATA NEEDS, AND PREDICTION UNCERTAINTY

This chapter presents methods for evaluating model predictions and focuses on three broad topics: (1) defining the predictions of interest and calculating the predictions, their sensitivities, and their standard deviations; (2) using simulated predictions to assess future data needs, which involves statistics that indicate which parameters and which existing or potential observations are important to the predictions; and (3) quantifying prediction uncertainty using linear and nonlinear inferential statistics and Monte Carlo methods.

8.1 SIMULATING PREDICTIONS AND PREDICTION SENSITIVITIES AND STANDARD DEVIATIONS

Model predictions typically are made to investigate the simulated system at a past or future time, under stress conditions that may differ from those used to calibrate the model and/or at spatial locations where no observations exist. For example, in a groundwater transport model, the predictions might be future solute concentrations resulting from evolution of a contaminant plume under the steady-state flow conditions for which the model was calibrated. Or, in a groundwater flow model, the predictions might be simulated hydraulic heads under future pumping conditions that are substantially different from those for which the model was calibrated. Or, both future transport and changes in pumpage might be of interest. Simulating predictions involves imposing the appropriate stresses and conditions and then

---

Effective Groundwater Model Calibration: With Analysis of Data, Sensitivities, Predictions, and Uncertainty. By Mary C. Hill and Claire R. Tiedeman
Published 2007 by John Wiley & Sons, Inc.

8.2 COLLECTION OF DATA THAT CHARACTERIZE SYSTEM PROPERTIES

calculating the predicted quantity. After the predictions have been simulated, their sensitivities can be calculated by the same methods used to calculate sensitivities for the simulated equivalents of the observations.

In MODFLOW-2000, UCODE_2005, and PEST, any type of quantity that can be treated as an observation also can be treated as a prediction. The quantities that can be used as observations are listed in Table 2.1.

After the predictions have been defined and simulated, and their sensitivities have been calculated, the prediction standard deviations can be calculated as

$$s_{z'_\ell} = \left[\sum_{i=1}^{NP} \sum_{j=1}^{NP} \frac{\partial z'_\ell}{\partial b_j} V(b) \frac{\partial z'_\ell}{\partial b_i} \right]^{1/2} \quad (8.1a)$$

where z'_ℓ = the ℓth prediction;
$s_{z'_\ell}$ = the standard deviation of the predictions;
$\partial z'_\ell / \partial b_j$ = the sensitivity of the ℓth prediction with respect to b_j, the jth parameter;
$V(b)$ = the parameter variance–covariance matrix (Eq. (7.1)), often calculated for all parameters, as described in Sections 7.2.1 and 7.2.5.

Expressing the sensitivities of prediction z'_ℓ as vector $x_{z'_\ell}$ and expanding $V(b)$ using Eq. (7.1) yields

$$s_{z'_\ell} = \left[s^2 (x_{z'_\ell} (X^T \omega X)^{-1} x_{z'_\ell}^T) \right]^{1/2} \quad (8.1b)$$

Prediction sensitivities and standard deviations are not often used directly, but rather are scaled or used to derive measures for evaluating prediction uncertainty and assessing data needs, as discussed later in this chapter. They can be used to measure and communicate substantial insight about the system information, parameters, and observations that are most important to the calculated predictions and, to the extent that the model is accurate, the actual predictions. Prediction standard deviations also can be used to quantify the uncertainty of the predictions. This chapter describes methods for accomplishing these tasks.

Use of measures derived from prediction sensitivities and from standard deviations computed using Eq. (8.1) assumes that the model is linear. When calculating these measures for nonlinear models, it is important to conduct the analyses with likely sets of parameter values that differ from the optimal parameter estimates. This tests the robustness of the conclusions drawn from these measures when applying them to nonlinear models.

8.2 USING PREDICTIONS TO GUIDE COLLECTION OF DATA THAT DIRECTLY CHARACTERIZE SYSTEM PROPERTIES

The expense of data collection and the inaccessibility of many natural systems typically limits the amount of information that can be obtained about the properties and state of a simulated system. It is, therefore, important to design data collection strategies that provide as much information as possible about aspects of the system that are important to the predictions. Here, we consider collection of data that directly characterize the properties of the simulated system. For a groundwater system, these data might include information about stratigraphy, hydrogeologic unit geometries, hydraulic conductivities, and areal recharge values. In Section 8.3, we consider collection of data related to quantities that can be used as observations in model calibration, such as hydraulic heads, streamflow gains and losses, and concentrations.

One way to use a calibrated model to identify system properties that are most important to the predictions is by identifying model parameters that are most important to the predictions. The issue of parameter importance to predictions spans the last two components of the observation–parameter–prediction triad composed of entities that are directly connected by the model, as discussed in Chapters 1 and 10. Using predictions to design and evaluate strategies for collecting data related to the model parameters requires statistics that measure the importance of the parameters to the predictions. In nonlinear regression, predictions may be distinctly different kinds of quantities from the observations. For example, in groundwater models, the observations might be heads and flows, while the predictions might be solute concentrations. Any analysis needs to accommodate this.

The methods discussed in this book for identifying parameter importance to predictions include (1) prediction scaled sensitivities (pss); (2) combined use of prediction, composite, and dimensionless scaled sensitivities (pss, css, and dss), to identify parameters important to predictions that are not well supported by the observations; (3) parameter correlation coefficients (pcc) that include both observations and predictions, to evaluate whether parameters that are highly correlated in the calibrated model are individually important to the predictions; and (4) the parameter–prediction (ppr) statistic, which includes the effects of parameter uncertainty and correlation, in addition to prediction sensitivities. The pss, css, dss, and pcc statistics can be used to help reveal why different parameters are important; the ppr statistic does the best job of identifying parameters for which additional data is most advantageous. Foglia et al. (in press) compare these statistics to the results of cross validation.

The broad field of model sensitivity analysis offers many additional measures for evaluating the importance of model inputs to model predictions (e.g., Saltelli et al., 2000, 2004). These range from simple measures like pss, to computationally intensive measures that account for model nonlinearity. In this book, we focus on a set of methods that are conceptually intuitive and fairly simple to calculate.

8.2.1 Prediction Scaled Sensitivities (pss)

Prediction sensitivities ($\partial z'_\ell / \partial b_j$) indicate the importance of the parameter values to the predictions, but need to be scaled when used for comparing the relative importance of different parameters. These are called prediction scaled sensitivities regardless of the exact scaling used. When calculating and presenting these measures, it is important to state clearly the scaling used.

8.2 COLLECTION OF DATA THAT CHARACTERIZE SYSTEM PROPERTIES

Generally, prediction scaled sensitivities are calculated in one of four ways:

$$\begin{array}{lll} & \text{Sensitivity} & \text{Scaling} \\ pss_{\ell j} = (\partial z'_\ell / \partial b_j) & (b_j/100)(100/r'_\ell) & (8.2\text{a}) \\ pss_{\ell j} = (\partial z'_\ell / \partial b_j) & (s_{bj}/100)(100/r'_\ell) & (8.2\text{b}) \\ pss_{\ell j} = (\partial z'_\ell / \partial b_j) & (b_j/100)(100/z'_\ell) & (8.2\text{c}) \\ pss_{\ell j} = (\partial z'_\ell / \partial b_j) & (s_{bj}/100)(100/z'_\ell) & (8.2\text{d}) \end{array}$$

where $pss_{\ell j}$ is the scaled sensitivity of prediction z'_ℓ to parameter b_j; r'_ℓ is a reference value defined by the modeler, as described below; s_{bj} is the standard deviation of parameter b_j calculated in Eq. (7.3); and z'_ℓ is the simulated value of the prediction. As noted, the first term is the sensitivity; the following terms are the applied scaling. UCODE_2005 produces data-exchange files with these prediction scaled sensitivities (Poeter et al., 2005, Table 20).

The multiplication by $b_j/100$ in the scaling of Eq. (8.2a) is equivalent to the scaling for the one-percent scaled sensitivity of Eq. (4.7). When the scaling by $100/r'_\ell$ or $100/z'_\ell$ also is included, the resulting statistic is the change in the predicted value, expressed as a percentage of r'_ℓ or z'_ℓ, caused by a one-percent change in the parameter value. This scaling can be produced rather awkwardly with MODFLOW-2000 by setting the statistic for the weighting of the predictions to r'_ℓ or z'_ℓ, and specifying the STAT-FLAG as 1 (Hill et al., 2000, p. 53). Prediction scaled sensitivities with this scaling will then be listed in the table of dimensionless scaled sensitivities printed by the programs.

The multiplication by $s_{bj}/100$ in Eq. (8.2b) produces scaled sensitivities that equal the change in the predicted value, expressed as a percentage of r'_ℓ or z'_ℓ, caused by changing the parameter value by an amount equal to one-percent of the parameter standard deviation. This scaling expresses prediction sensitivity in the context of parameter uncertainty and has two advantages and one disadvantage. The first advantage is that, unlike parameter values, s_{bj} almost never equals zero. The second advantage is that it is valid for parameters that are affected by the datum of the model, such as groundwater model parameters representing the head at constant-head boundaries. Its disadvantage is that it is not fit-independent because the value of the objective function is a term of the variance–covariance matrix (Eq. (7.1)). For all *pss* calculated using results from a single regression run, this disadvantage will affect all parameters proportionately, so the relative importance of parameters in a single run can be evaluated. This scaling can be accomplished awkwardly with MODFLOW-2000 by printing unscaled sensitivities and then applying the scaling using spreadsheet software.

Possible alternatives for r'_ℓ in Eq. (8.2) include a regulatory limit or another quantity relevant to a given modeling situation. Clearly, using 0.0 in Eq. (8.2) is not mathematically valid.

In some circumstances the prediction is the difference between two simulations, as discussed in Section 8.4.5. For example, in groundwater models, a common prediction is the drawdown or the change in flow to a stream caused by pumpage.

UCODE_2005 and MODFLOW-2000 are designed to calculate sensitivities related to differences, and thus *pss* also can be computed for these differences. For UCODE_2005, this is accomplished using the Predictions mode and derived predictions; for MODFLOW-2000, this is achieved using the computer program YCINT-2000 (Hill et al., 2000, pp. 87–91).

8.2.2 Prediction Scaled Sensitivities Used in Conjunction with Composite Scaled Sensitivities

In Figure 8.1, selected *pss* values calculated using Eq. (8.2c) for the model discussed in Chapter 15, Section 15.2.1 are compared to the *css* of Eq. (4.6). In the example, the predictions are the Cartesian components of advective travel simulated by particle tracking using the ADV Package of Anderman and Hill (2001). The model grid is oriented with the north compass direction, so the predictions are the particle travel distance in the north or south, east or west, and vertical directions. Figure 8.1*b* shows results for the north–south component of travel. The figure shows the mean and range of the *pss* values for five transported particles. Here, the *pss* values are defined to equal the percent change in the advective transport caused by a one-percent change in parameter value. The simulated value equals the simulated length of advective transport in the north or south coordinate direction.

In Figure 8.1*b*, the *pss* show that HK2, HK3, and RCH2 (the hydraulic conductivity of two rock types and the recharge potential of one area) are the most important parameters to the determination of advective transport. In Figure 8.1*a*, the *css* show that the observations used in the regression provide more information for parameters HK2 and HK3 than for RCH2. This suggests that of these three parameters, it is probably most important to collect additional information about RCH2 for improving the transport predictions. This parameter was estimated by the regression, as shown by its black bar, but collecting additional information about its characteristics or additional observation data that support it could help improve its representation in the model and its estimated value, and thereby probably also improve the predictions.

This analysis can be taken one step further by evaluating the *dss* shown in Figure 8.1*c*. Although the *css* for parameter HK4 is large, the *dss* show that the support primarily comes from just four observations, suggesting that these observations should be closely investigated. This type of analysis can be used to understand and communicate model strengths and weaknesses and to justify and plan additional model development and data collection efforts. It can also be used to better understand more sophisticated statistics such as the parameter–prediction (*ppr*) statistic described in Section 8.2.5.

8.2.3 Parameter Correlation Coefficients without and with Predictions

To determine whether parameters that are highly correlated for the calibrated model are individually important to predictions of interest, two different sets of parameter correlation coefficients are compared: those calculated using one of the first two

8.2 COLLECTION OF DATA THAT CHARACTERIZE SYSTEM PROPERTIES

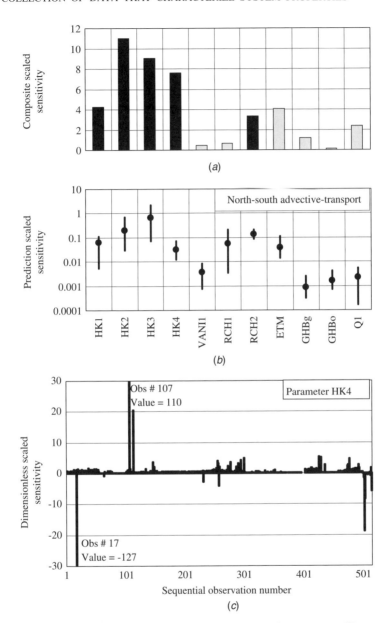

FIGURE 8.1 (*a*) Composite scaled sensitivities for selected parameters, (*b*) one-percent prediction scaled sensitivities for the north–south component of predicted advective transport, and (*c*) dimensionless scaled sensitivities showing the support provided by the observations for parameter HK4. In (*a*), the composite scaled sensitivities for parameters estimated in the regression are shown using black bars; those not estimated in the regression are shown using gray bars. In (*b*), the prediction scaled sensitivities are defined as the percent change in the prediction given a one-percent change in the parameter value. These are selected results from simulations of the model discussed in Chapter 15, Section 15.2.1.

versions of the parameter variance–covariance matrix (Eq. (7.1)) described in Chapter 7, Sections 7.2.1 and 7.2.5, and those calculated with predictions as well (the fifth version of the parameter variance–covariance matrix described in Chapter 7, Sections 7.2.1 and 7.2.5).

The *pcc* calculated with the predictions as well as the observations used in the regression is produced by augmenting the terms of Eq. (7.1) to include information related to the predictions. This produces an alternate parameter variance–covariance matrix that can be represented as

$$V_\ell(b') = s^2 (X_\ell^T \omega_\ell X_\ell)^{-1} \tag{8.3}$$

where the sensitivity matrix, X_ℓ, and the weight matrix, ω_ℓ, are augmented to include the predictions. Predictions can be included individually or in groups, as appropriate for the particular problem to be addressed.

These augmentations can be implemented easily when using MODFLOW-2000 or UCODE_2005 by adding the predictions to the list of observations, and executing the sensitivity analysis mode of either computer program. The value specified for the prediction as the "observed value" does not affect the calculated parameter correlation coefficients with predictions because the s^2 term in Eq. (8.3) cancels out in the calculation of Eq. (7.5). However, the specified weight does affect the calculation. One implication of this is that in MODFLOW-2000 and UCODE_2005 it is generally not desirable to specify the statistic used to calculate the prediction weight as a coefficient of variation because in that situation the "observed value" is used to calculate the weight. Additional comments about determining "weights" for predictions are provided at the end of this section.

The resulting *pcc* with predictions are compared with *pcc* calculated only with the observations used in the regression. If adding the predictions causes some highly correlated parameter pairs to become much less correlated, this indicates that the predictions are likely to depend on individual parameter values that the regression could not estimate uniquely. This identifies a weakness in the calibrated model.

The utility of *pcc* with predictions is illustrated by a groundwater modeling example. Consider a groundwater flow model calibrated by estimating parameter values using observations of hydraulic-head and streamflow gain or loss. The calibrated model is used to predict (a) hydraulic head at a location where no measurement can be obtained and (b) advective transport from the site of a contaminant spill. Parameter correlation coefficients are first calculated using the calibrated model, all calibration observations, and all defined parameters. Two sets of *pcc* with predictions are then obtained, by first including the predicted hydraulic-head location and then the predicted advective transport. Using an analysis with calculations similar to these, Anderman et al. (1996) show that prediction of the hydraulic head using the calibrated model did not require uncorrelated parameter estimates and thus this prediction could be used with some confidence. Prediction of advective transport did require uncorrelated estimates, and thus the transport prediction is highly suspect.

8.2 COLLECTION OF DATA THAT CHARACTERIZE SYSTEM PROPERTIES

The weights for the predictions can be established using one of the two following approaches:

1. The weight can be established using a statistic (standard deviation or variance; see Chapter 3, Section 3.4.4 and Guideline 6 in Chapter 11) that reflects an acceptable range of uncertainty in the prediction. Compared to approach 2, this approach is more consistent with the scaling of the CTB statistic of Sun and Yeh (1990a) and Sun (1994).
2. The weight determined using approach 1 can be increased (by decreasing the value of the statistic used to calculate the weight) so that the results clearly indicate whether unique parameter values are important to predictions.

By approach 1, predictions for which a larger amount of uncertainty is acceptable have smaller weights. Predictions that are desired to be more certain have larger weights, which increases the absolute values of the *pcc* for parameters to which these predictions are sensitive. Approach 2 allows weights for certain predictions to be subjectively increased. This option ensures that if individual parameter values are important to predictions, this will be revealed by the *pcc*.

8.2.4 Composite and Prediction Scaled Sensitivities Used with Parameter Correlation Coefficients

Composite and prediction scaled sensitivities and parameter correlation coefficients (*css*, *pss*, and *pcc*) can be used together to assess whether an improved estimate of a parameter is needed. A classification system for this analysis is shown in Figure 8.2. The upper portion of this figure classifies the precision of a parameter estimate in combination with the importance of the parameter to the predictions. The lower portion of the figure classifies the uniqueness of the estimates for a parameter pair in combination with the importance to the predictions of having unique estimates of the two parameters.

If the analysis of a parameter indicates a classification in box IV of Figure 8.2a or 8.2b, this means that improved estimation of this parameter and/or improved representation of the system features with which it is associated are likely to improve prediction accuracy. If the analysis indicates a classification in boxes I, II, or III, the term "acceptable" in these boxes means that a parameter is estimated well, is unimportant to the predictions, or both. Improved estimation of the parameter and improved representation of the system features with which it is associated are likely to be less beneficial to improving prediction accuracy than for parameters that are classified in box IV.

The *pcc* are used in Figure 8.2b as measures of both the uniqueness of the parameter estimate and the importance of unique parameter values to the predictions. To measure the uniqueness of the parameter estimate, *pcc* are calculated with only the observations and prior information used in the calibration. To measure whether unique parameter estimates are important to the predictions, *pcc* are calculated with

	Precision of the parameter estimate	
(a)	**Imprecise:** *composite scaled sensitivity* < 1.0 or small relative to other parameters[1]; large coefficient of variation or confidence interval	**Precise:** *composite scaled sensitivity* > 1.0 and large relative to other parameters[1]; small coefficient of variation or confidence interval
Importance of the parameter to predictions of interest — **Not important:** Small *prediction scaled sensitivity*	I. Acceptable	II. Acceptable
Importance of the parameter to predictions of interest — **Important:** Large *prediction scaled sensitivity*	IV. Improve estimation of this parameter and representation of associated system features	III. Acceptable

	Uniqueness of the estimates for a parameter pair	
(b)	**Nonunique:** parameter correlation coefficient[2] is close to |1.00|	[2]**Unique:** parameter correlation coefficient[2] is less than about |0.95|
Importance of unique parameter estimates to predictions of interest — [3]**Not important:** *parameter correlation coefficient with predictions*[3] is close to |1.00|	I. Acceptable	II. Acceptable
Importance of unique parameter estimates to predictions of interest — [3]**Important:** *parameter correlation coefficient with predictions*[3] is less than about |0.95|	IV. Improve estimation of one or both of the parameters and improve the representation of associated system features	III. Acceptable

[1] A composite scaled sensitivity (*css*) is small relative to other parameters if it is less than about one percent of the largest *css*. This is a rule of thumb but is sufficient for identifying potentially problematic parameters.
[2] Calculated using unestimated as well as estimated parameters, and including only the observations and prior information used in the calibration.
[3] Calculated as in footnote 2, but also including the predictions.

FIGURE 8.2 Classification of the need for improved estimation of a parameter and, perhaps, associated system features. The classification is based on statistics that indicate (*a*) the precision and importance to predictions of a single parameter and (*b*) the uniqueness and importance to predictions of a pair of parameters. See text for additional explanation.

the predictions as well as the calibration observations and prior information. These two different types of *pcc* are discussed in Section 8.2.3.

The classification system illustrated in Figure 8.2*b* only addresses uniqueness caused by lack of parameter correlation. Methods for detecting nonuniqueness caused by multiple minima are discussed in Chapter 7, Section 7.4.

This method of using *css*, *pss*, and *pcc* can be revealing but is awkward. Fortunately, the *ppr* statistic described next incorporates the effects of parameter correlation as well as observation and prediction sensitivity.

8.2.5 Parameter–Prediction (*ppr*) Statistic

Unlike prediction scaled sensitivities (*pss*), the parameter–prediction (*ppr*) statistic assesses the importance of parameters to predictions in a way that accounts for

8.2 COLLECTION OF DATA THAT CHARACTERIZE SYSTEM PROPERTIES

parameter correlations. The *ppr* statistic also takes advantage of the connection between parameter uncertainty, parameter correlation, and prediction uncertainty provided by Eq. (8.1) for the prediction standard deviation. The drawback of the *ppr* statistic is that the equation and procedure for obtaining the statistic is more complicated than for *pss*, which can cause the results to be less clear. Evaluating the *dss*, *css*, *pss*, and *pcc* statistics as described above can help explain the *ppr* results. The *ppr* statistic was developed by Tiedeman et al. (2003) and was called the value of improved information (*voii*) statistic in that work. In this book, the statistic name has been changed to better reflect its purpose.

The equation for the *ppr* statistic is derived using Eq. (8.1), which calculates the prediction standard deviation using a calibrated model with existing independent information about parameter values included as prior information. In this calculation, it is important that the parameter variance–covariance matrix for all parameters (defined in Chapter 7, Sections 7.2.1 and 7.2.5) be used. Then, the standard deviation is recomputed under the assumption of increased certainty in one or more parameter values. The difference in prediction standard deviation is used to calculate the *ppr* statistic.

The parameter variance–covariance matrix in Eq. (8.1) is calculated as

$$V(b) = s^2(X^T \omega X)^{-1} \tag{8.4}$$

where X and ω include sensitivities and weights for prior information as well as for observations used in the regression. To explain the method for calculating the *ppr* statistic, it is convenient to express X and ω as

$$X = \begin{bmatrix} X_{Y,PRI} \\ I \end{bmatrix} \tag{8.5}$$

$$\omega = \begin{bmatrix} \omega_{Y,PRI} & 0 \\ 0 & \omega_{ppr} \end{bmatrix} \tag{8.6}$$

where $X_{Y,PRI}$ = the *NP* by *ND* + *NPR* matrix of sensitivities of the *ND* calibration observations and the *NPR* prior equations with respect to the *NP* model parameters, with elements equal to $\partial y'_i/\partial b_j$ (Eq. 4.1);

NP = the total number of defined model parameters and may be greater than the number of estimated model parameters;

I = the *NP* by *NP* identity matrix (all elements equal 1.0);

ω = the weight matrix expressed here as in Appendix B;

$\omega_{Y,PRI}$ = the *ND* by *ND* + *NPR* matrix of weights on observations and prior equations;

ω_{ppr} = the *NP* by *NP* matrix used to calculate *ppr* statistics, defined after Eq. 8.7.

In calculating the variance–covariance matrix for all parameters, there is usually no prior information on parameters for which the calibration observations supply abundant information. For parameters supported better by independent information than by the calibration observations (commonly these parameters are not estimated by the regression), it is important that prior information and associated weighting be specified, as discussed in Chapter 7, Section 7.2.5. By specifying prior weights in this manner, the parameter variance–covariance matrix calculated using Eq. (8.4) reflects actual levels of uncertainty, and the prediction uncertainty calculated using Eq. (8.1) reflects these realistic parameter uncertainties.

The prediction uncertainty produced with improved information on one parameter is calculated using a modified form of Eq. (8.1):

$$s_{z'_\ell(j)} = \left[\sum_{i=1}^{NP'} \sum_{j=1}^{NP'} \frac{\partial z'_\ell}{\partial b_j} V(b)_{(j)} \frac{\partial z'_\ell}{\partial b_j} \right]^{1/2} \tag{8.7}$$

where $s_{z'_\ell(j)}$ = the standard deviation of the ℓth predicted value, z'_ℓ, calculated with improved information on the jth parameter, b_j;

NP = the total number of defined parameters;

$V(b)_{(j)}$ = the symmetric, square NP by NP parameter variance–covariance matrix for all parameters, calculated with improved information on the jth parameter, expressed as $V(b)_{(j)} = s^2(X^T \omega_{(j)} X)^{-1}$;

s^2 = identical to s^2 in Eq. (8.1), because the model has not been recalibrated and s^2 is still considered the best estimate of the true error variance σ^2;

$\omega_{(j)}$ = the weight matrix in which the jth parameter has improved information, expressed as

$$\omega_{(j)} = \begin{bmatrix} \omega_{Y,PRI} & 0 \\ 0 & \omega_{ppr(j)} \end{bmatrix};$$

$\omega_{ppr(j)}$ = a NP by NP matrix, in which all entries are zero except for the diagonal entry related to the jth parameter.

The matrix $\omega_{ppr(j)}$ is central to calculating the *ppr* statistic. Improved information on the jth parameter is implemented in this matrix by specifying a positive value on its jth diagonal. Conceptually, this positive value represents the increased certainty in the prior value that might result from collection of additional field data; that is, from improved information about the parameter.

The consequence of including $\omega_{ppr(j)}$ is that the variance of the jth parameter, which has improved information, will be smaller in $V(b)_{(j)}$ than in $V(b)$. Parameters that do not have improved information, but that are correlated with the jth parameter, also tend to have smaller variances in $V(b)_{(j)}$ compared to those in $V(b)$. Primarily because of the reductions in parameter variances, the prediction standard deviation calculated with improved information ($s_{z'_\ell(j)}$ of Eq. (8.7)) generally is smaller than

8.2 COLLECTION OF DATA THAT CHARACTERIZE SYSTEM PROPERTIES

the prediction standard deviation calculated without improved information ($s_{z'_\ell}$ of Eq. (8.1)).

The scaled difference between $s_{z'_\ell(j)}$ and $s_{z'_\ell}$ measures the value of the improved information on the jth parameter with respect to prediction z'_ℓ and is calculated as

$$ppr_{\ell(j)} = 100 \times \left(\frac{s_{z_\ell} - s_{z'_\ell(j)}}{s_{z_\ell}}\right) = 100 \times \left(1 - \frac{s_{z'_\ell(j)}}{s_{z_\ell}}\right) \tag{8.8}$$

where $ppr_{\ell(j)}$ is the parameter–prediction statistic and equals the percent reduction in the standard deviation of prediction z'_ℓ that results from improved information on the jth parameter. To rank the importance of individual parameters to prediction z'_ℓ, $ppr_{\ell(j)}$ is calculated NP times, each time with improved information on one parameter. The parameter associated with the largest value of $ppr_{\ell(j)}$ ranks as most important to prediction z'_ℓ.

To implement improved information on each model parameter in a consistent manner, Tiedeman et al. (2003) suggest increasing the positive value on the diagonal of the $\omega_{ppr(j)}$ matrix until the standard deviation on the parameter estimate is decreased by a specified percent. This requires an iterative procedure. Tiedeman et al. (2003) specified a 10 percent decrease, which represents the situation in which improved, but not perfect, information is collected about a parameter. For diagonal $\omega_{ppr(j)}$ matrices, implementing this specified percent decrease is accomplished by increasing the value on the diagonal associated with the parameter in question; for full weight matrices it is less clear how to proceed and this issue has not been investigated.

The method presented above is easily extended to the case of evaluating improved information on more than one parameter. When evaluating multiple parameters, the effect of parameter correlations can strongly influence which parameters are important to a prediction. This effect can produce situations in which the set of parameters with the highest individual $ppr_{\ell(j)}$ values is not identical to the set of parameters that are most important when improved information on multiple parameters is considered. Tiedeman et al. (2003) present the method for the general case of improved information on any number of parameters.

The computer program OPR-PPR (Tonkin et al., in press) can be used to calculate the ppr statistic. OPR-PPR easily can calculate the ppr statistic for models developed and calibrated using UCODE_2005 or MODFLOW-2000 because it is designed to use their output files directly. OPR-PPR also can be used with other models if appropriate files are produced.

Other methods for evaluating the importance of model parameters to model predictions include those developed for hydrologic models by Walker (1982), Melching et al. (1990), Indelman et al. (1996), Høybye (1998), Levy et al. (1998), and Levy and Ludy (2000). These are similar to the ppr statistic in that they incorporate parameter uncertainty and prediction sensitivity, but unlike the ppr statistic, most of these methods do not include the effects of parameter correlations.

The calculation of the *ppr* statistic assumes that the model is linear with respect to the parameter values (see Chapter 1, Section 1.4.1). However, Tiedeman et al. (2003) found that the method is fairly robust for a mildly nonlinear groundwater model, for an application that is summarized in Chapter 15. Methods that account for model nonlinearity are presented by Sulieman et al. (2001) for models calibrated by regression, and by Saltelli et al. (2000) for the general case in which the model may not have been calibrated by regression. These methods are more complex than that for calculating the *ppr* statistic and also are substantially more computationally intensive.

The basic concept embodied by the *ppr* statistic is the use of the first-order second-moment equation for prediction uncertainty (Eq. (8.1)) as a basis for assessing parameter importance. This concept has been used by other researchers as the basis for designing sampling networks for collecting data about the properties or parameters of groundwater systems. McLaughlin and Wood (1988) were among the first to investigate aquifer property sampling strategies in this context. McKinney and Loucks (1992), Sun and Yeh (1992), and Wagner (1995, 1999) incorporated this type of analysis into an optimization framework and developed methods for designing aquifer property sampling networks that minimize prediction uncertainty. The *ppr* statistic differs in that it is used as a tool for ranking the importance of all model parameters to any individual prediction.

8.3 USING PREDICTIONS TO GUIDE COLLECTION OF OBSERVATION DATA

Evaluating the importance of observations to predictions spans the entire observation–parameter–prediction triad discussed in Chapters 1 and 10. We present two methods for evaluating the importance of observations to predictions, both of which use sensitivity statistics and are computationally fast. The first, using scaled sensitivities and parameter correlation coefficients (*pss*, *css*, and *dss*), is more awkward than the second, which uses the observation–prediction (*opr*) statistic.

The statistics available in classical regression methods to address evaluation of observation importance to predictions include jackknife and bootstrap methods, both of which require many regressions and therefore often require prohibitive amounts of computer execution time for models of environmental systems. Foglia et al. (in press) demonstrate that *opr* statistics perform comparably to leave-one-out cross-validation in the evaluation of a groundwater model.

8.3.1 Use of Prediction, Composite, and Dimensionless Scaled Sensitivities and Parameter Correlation Coefficients

Using *pss*, *css*, and *dss* together as illustrated in Figure 8.1 is one method for spanning the observation–parameter–prediction sequence to identify existing and potential observations important to the predictions. This figure was presented and discussed in Section 8.2.2 in the context of identifying parameters that are important to the predictions but are not well-supported by the existing observations. A similar approach is also suggested by Merry et al. (2003).

The *pss*, *css*, and *dss* also can be evaluated with the primary objective of identifying important existing and potential observations. Identification of the most important existing observations involves first identifying the parameters most important to the predictions using the methods illustrated in Figure 8.1, then identifying observations with large *dss* for these parameters. In Figure 8.1, this analysis revealed the potentially problematic situation of only four observations providing information for parameter HK_1. This same type of analysis can be used to identify observations most important to the predictions; in this case, observation types and locations with large *dss* are likely to be important, for example, to continue monitoring in the future (see Section 8.3.4). This type of analysis also can be used to identify important potential new observation types and locations, by calculating the *dss* for potential observations instead of for existing observations.

The importance of potential new observations to the predictions also can be evaluated with respect to parameter correlations because *pcc* does not depend on the value of the observation. If the analysis of *pcc* without and with predictions (Section 8.2.3) shows that the predictions are likely to depend on parameter values that the regression could not estimate uniquely, then potential new observations could improve this situation if they enable unique estimation of the parameters. This can be evaluated prior to actually collecting the observations, by calculating the *pcc* with both the existing and the potential observations. In this calculation, the simulated conditions might be different for the existing and potential observations, and both sets of conditions need to be properly represented. The *pcc* calculated with the existing and potential observations are then compared to those calculated with only the existing observations. If adding the potential observations reduces the absolute values of *pcc* that are very large when only the existing observations are included, then the potential observations probably are important to predictions that depend on the individual parameter values.

A drawback of using the *pss*, *css*, *dss*, and *pcc* together in this manner is that this procedure is awkward. It can result in many graphs from which it can be difficult to extract the key results. However, these methods can be quite useful in providing insight about values of observation–prediction (*opr*) statistics.

8.3.2 Observation–Prediction (*opr*) Statistic

The *opr* statistic integrates the information contained in the fit-independent statistics dimensionless and composite scaled sensitivities (*css* and *dss* of Chapter 4), parameter correlation coefficients (*pcc* of Chapter 7), and prediction scaled sensitivities (*pss* of Chapter 8). As indicated in Sections 8.3.1 and 8.3.3, it often is useful to investigate those statistics to better understand *opr* results.

The methodology for the *opr* statistic assumes that the model is linear with respect to the model parameters. Tests of this assumption are presented in Chapter 7, Section 7.7 and Section 8.7.

The observation–prediction (*opr*) statistic assesses the effect on the prediction standard deviation of either removing one or more existing observations or adding one or more new observations. This evaluation can address issues related to monitoring the state of the simulated system, as discussed in Section 8.3.4. Calculating the *opr* statistic does not involve recalibrating the model with these observations added or removed. Thus, the leverage of the observations, rather than their influence, is determined (leverage and influence are defined in Chapter 7, Section 7.3).

The *opr* statistic requires trivial computational effort. In contrast, identifying existing observations that are influential with respect to the predictions requires jackknifing or similar methods, which repeat the nonlinear regression with one or more observations omitted (Efron, 1982; Good, 2001). In addition, influence cannot be determined for potential observations because assessing influence requires the observed value in addition to other information associated with the potential observation, such as its type, location, and time.

A modified version of Eq. (8.1) is used to evaluate the effect on prediction uncertainty of omitting or adding one observation (Hill et al., 2000; Tiedeman et al., 2004):

$$s_{z'_\ell(\pm i)} = \left[\sum_{i=1}^{NP'} \sum_{j=1}^{NP'} \frac{\partial z'_\ell}{\partial b_j} V(b)_{(\pm i)} \frac{\partial z'_\ell}{\partial b_i} \right]^{1/2} \quad (8.9)$$

where $s_{z'_\ell(\pm i)}$ = the standard deviation of the ℓth predicted value, z'_ℓ, calculated with the ith observation either added (+) or removed (−);

NP = the number of defined parameters, which may exceed the number of estimated parameters;

$V(b)_{(\pm i)}$ = the symmetric, square NP by NP parameter variance–covariance matrix for all parameters, with the ith observation either added or removed and is calculated as

$$V(b)_{(\pm i)} = s^2 (X^T_{(\pm i)} \omega_{(\pm i)} X_{(\pm i)})^{-1} \quad (8.10)$$

where s^2 = identical to s^2 in Eq. (8.1), because the model has not been recalibrated and s^2 is still considered the best estimate of the true error variance σ^2;

$X_{(\pm i)}$ = a sensitivity matrix formed either by adding (+) or removing (−) the sensitivities of the simulated equivalent of the ith observation;

$\omega_{(\pm i)}$ = formed by modifying matrix ω (defined after Eq. (8.6)), either by adding (+) or by removing (−) the weight associated with the ith observation.

As for the *ppr* statistic, it is important that the parameter variance–covariance matrix for all parameters (Chapter 7, Sections 7.2.1 and 7.2.5) be calculated in Eq. (8.10), and that prior information and associated weighting be specified for parameters that are supported better by independent information than by the calibration observations.

8.3 USING PREDICTIONS TO GUIDE COLLECTION OF OBSERVATION DATA

In practice, the ith observation is removed by setting its weight equal to zero, and by leaving the sensitivity matrix X unchanged. An observation is added by calculating the related sensitivities and assigning weights on the basis of an analysis of errors that would be expected for the potential observed values. The observed value itself does not affect the *opr* statistic because s^2 from the regression is used in Eq. (8.10).

The percent change in prediction uncertainty that results from removing or adding the ith observation is used as the measure of its importance to prediction z'_ℓ:

$$opr_{\ell(\pm i)} = 100 \times \left| \frac{s_{z'_\ell} - s_{z'_\ell(\pm i)}}{s_{z'_\ell}} \right| = 100 \times \left| 1 - \frac{s_{z'_\ell(\pm i)}}{s_{z'_\ell}} \right| \qquad (8.11)$$

where $opr_{\ell(\pm i)}$ is the observation–prediction statistic, and the vertical lines indicate absolute value.

This method can easily be extended to evaluate adding or omitting any combination of existing or potential observations (Tiedeman et al., 2004).

The computer program OPR-PPR (Tonkin et al., in press) can be used to calculate the *opr* statistic. As discussed in Section 8.2.5, this program can calculate the statistic for UCODE_2005 and MODFLOW-2000 models, as well as for any other model that produces the needed output files.

Some of the strengths and weaknesses of the *opr* statistic, and, indeed of all statistics calculated using a model, are that they reflect model simplifications and approximations. Generally, the model is the best available representation of the system in question, and as such it is important to consider model-calculated statistics. Close evaluation of results that do not make sense can help improve model results, as discussed in the next section.

8.3.3 Insights About the *opr* Statistic from Other Fit-Independent Statistics

The reasons that certain observations rank as important to the model predictions by the *opr* statistic can determine what action is advised on the basis of the *opr* results. For example, large values of *opr* might be caused by aspects of model construction that are unrealistic. The appropriate response is to fix the model, which is likely to have the advantageous consequence of allowing other observations that are more accurately simulated to have greater influence on simulated results. In other circumstances the *opr* analysis may reveal plausible improvements in data collection strategies. Several of the fit-independent statistics discussed in previous chapters can help reveal why particular observations have large *opr* statistics.

The contribution to the *opr* statistic of the prediction sensitivities in Eq. (8.9) can be investigated using the *pss* of Eq. (8.2). The contribution of the variance–covariance matrix can be investigated by first noting that Eq. (8.11) is designed so that the s^2 term of Eq. (7.1) cancels out, causing the *opr* statistic to be fit-independent. Thus, only the term $(X^T \omega X)^{-1}$ from Eq. (7.1) remains. The contribution of this term to the *opr* statistic can be investigated by considering dimensionless and composite scaled sensitivities (*dss* of Eq. (4.3) or (4.5) and *css* of Eq. (4.6)) and the parameter correlation coefficients (pcc of Eq. (7.5)). Generally, if (1) observation y_i provides substantial information about parameter b_j or about parameter b_k correlated with b_j (dss_{ij} is large), and (2) b_j is important to prediction z'_ℓ ($pss_{\ell j}$ is large), then it is likely that $opr_{\ell(\pm j)}$ will be large.

It is important to be aware that, in some situations, values of dss can be large because of simplifications made during model construction rather than because of actual hydrogeologic conditions. For example, consider a groundwater system in which a pumping well draws water from a localized zone of high hydraulic conductivity, but in a regional model of this system, the cell containing the well has a much lower hydraulic conductivity. Because the pumping well is located in a zone of low conductivity, hydraulic head in this cell will have a relatively large sensitivity to the hydraulic conductivity of the cell. In this case, the importance of an existing or potential hydraulic-head observation in this cell may not actually be as important to a prediction as the opr statistic may indicate.

Additional insight into why certain observations can rank as important using opr is provided in Exercises 8.1d and 8.1f. Also, Tiedeman et al. (2004) apply the opr statistic to a groundwater flow model with advective-transport predictions. This application is summarized in Chapter 15.

8.3.4 Implications for Monitoring Network Design

Minsker (2003) summarizes methods and applications for groundwater monitoring network design. The prediction standard deviation of Eq. (8.1) has been used by many authors for monitoring network design. For example, Sun and Yeh (1990b) and Wagner (1995, 1999) used Eq. (8.1) together with optimization methods to determine an optimal set of groundwater observations for minimizing prediction uncertainty. Reeves et al. (2000) used it to identify new data locations most beneficial to groundwater remediation designs. Valstar and Minnema (2003) used a Bayesian method that considers prediction uncertainty. A strength of these methods and the opr statistic of Eq. (8.11) is that they can be used to evaluate and rank individual and user-defined groups of observations by their importance to predictions.

8.4 QUANTIFYING PREDICTION UNCERTAINTY USING INFERENTIAL STATISTICS

Prediction uncertainty can be evaluated and quantified using inferential statistics and/or Monte Carlo analysis. The Monte Carlo method is discussed in Section 8.5. In both techniques, the magnitude of prediction uncertainty is related to the uncertainty in the model parameters and the sensitivity of the predicted quantities to the model parameters.

The inferential methods discussed here produce intervals on predictions. Larger intervals indicate greater uncertainty. The methods are sometimes called first order, second moment (FOSM) methods: first order because they are linear, second moment because they use standard deviations, which are second moment statistics.

8.4 INFERENTIAL STATISTICS

Elementary texts that discuss inferential methods include Ott (1993, pp. 201–204) and Davis (2002, pp. 200–204). More advanced references include Seber and Wild (1989), Cooley and Naff (1990), Hill (1994), Helsel and Hirsch (2002), Glasgow et al. (2003), and Stauffer et al. (2004).

8.4.1 Definitions

The intervals discussed in this book can be individual or simultaneous intervals, and they can be confidence or prediction intervals. Thus, four types of intervals are possible: individual confidence intervals, individual prediction intervals, simultaneous confidence intervals, and simultaneous prediction intervals. The four terms are described in the following sections.

Individual Intervals An individual confidence or prediction interval is said to have a $(1 - \alpha)$ probability of including the true value of one predicted quantity. α is the significance level; $\alpha = 0.05$ produces 95-percent confidence intervals.

Simultaneous Intervals Simultaneous intervals have the specified probability of containing their respective true predicted values simultaneously. Because they simultaneously account for uncertainty in more than one quantity, simultaneous intervals are always of equal size or larger than equivalent individual intervals. To understand this, consider 95-percent intervals on a set of predictions. If calculated using Monte Carlo methods, individual intervals would need to be set so that each interval contains the predictions produced by 950 of 1000 randomly generated sets of parameter values. Simultaneous intervals, on the other hand, would need to be set so that all intervals contain the predictions produced by 950 of 1000 randomly generated sets of parameter values. As more intervals are considered, the intervals tend to become larger.

The size of linear simultaneous intervals increases until the number of intervals equals the number of parameters included in the uncertainty analysis. Additional intervals do not increase the size of linear simultaneous intervals. Nonlinear simultaneous intervals generally are similar, but there may be exceptions.

Confidence Intervals Confidence intervals on predictions are intervals that, with a specified likelihood, contain the true, unknown predictions, if the model is correct. Confidence intervals reflect the uncertainty with which the parameters are estimated, as represented by the variance–covariance matrix on the parameters, projected using prediction sensitivities (Eq. (8.1)).

Prediction Intervals Prediction intervals account for the same uncertainty in the parameter values reflected in confidence intervals, but also account for random error incurred when the predicted quantity is measured. A prediction interval is needed if the interval is to be compared with a measurement of the prediction.

Prediction intervals are most often calculated for predictions and rarely for parameters. The use of the term "prediction" to describe both a type of interval and the

8.4.2 Linear Confidence and Prediction Intervals on Predictions

All linear confidence intervals have the form

$$z'_\ell \pm [critical\ value]s_{z'_\ell} \quad (8.12)$$

where z'_ℓ is the ℓth simulated value; $s_{z'_\ell}$ is the standard deviation of the prediction, calculated as shown in Eq. (8.1); and *critical value* is a critical value from a statistical distribution. Critical values for four types of intervals are defined in Table 8.1.

All linear prediction intervals have the form

$$z'_\ell \pm [critical\ value](s^2_{z'_\ell} + s^2_a)^{1/2} \quad (8.13)$$

where s_a is the product of (1) the standard error of the regression s (defined after Eq. (6.1) in Chapter 6, Section 6.3.2) and (2) the standard deviation of the error associated with a measured equivalent of the prediction (Hill, 1994, p. 32; Miller, 1981). Thus, to calculate prediction intervals the modeler needs to estimate the likely uncertainty in a measurement of the predicted value. Strategies for estimating this uncertainty are similar to those for observations discussed in Chapter 3, Sections 3.3.3, 3.4.2 and in Guideline 6 in Chapter 11.

The calculation of linear confidence and prediction intervals can (and often should) include more parameters than were included in the regressions performed for model calibration. Thus, when calculating $s_{z'_\ell}$ by Eq. (8.1), the parameter variance–covariance matrix of Eq. (7.1) often will be the parameter variance–covariance matrix with all parameters and with realistic weighting, as defined in Chapter 7, Sections 7.2.1 and 7.2.5 and discussed in Sections 8.2.5 and 8.3.2.

TABLE 8.1 Critical Values Required in Eqs. (8.12) and (8.13) to Calculate the Linear Confidence and Prediction Intervals Used in This Book

Type of Interval	Critical Value	Table in Appendix D
Individual	$t_s(n, 1.0 - \alpha/2)$	D.2, Student-t distribution
Bonferroni simultaneous	$t_B(n, 1.0 - \alpha/2k)$	D.6, Bonferroni t statistic
Scheffé $d = k$ simultaneous	$[d \times F_\alpha(d, n)]^{1/2}$ $= [k \times F_\alpha(k, n)]^{1/2}$	D.7, F distribution
Scheffé $d = NP$ simultaneous	$[d \times F_\alpha(d, n)]^{1/2}$ $= [NP \times F_\alpha(NP, n)]^{1/2}$	D.7, F distribution

Note: α is the significance level and is commonly 0.05 or 0.10 (5 or 10 percent), which results in 95- or 90-percent intervals, respectively; n is the degrees of freedom, here equal to $ND + NPR - NP$; k is the number of simultaneous intervals or NP, whichever is smaller; NP is the number of parameters for which sensitivities are used in Eq. (8.1). NP commonly equals either the number of estimated parameters or the number of defined parameters.

8.4 INFERENTIAL STATISTICS

Linear, individual and simultaneous, confidence and prediction intervals for predictions are listed in output files produced by MODFLOW-2000 and computer program YCINT-2000 (Hill et al., 2000), or by UCODE_2005 and computer program LINEAR_UNCERTAINTY (Poeter et al., 2005). Calculation of linear intervals requires only the sensitivities calculated for the optimized parameter values and, therefore, takes very little computer execution time.

Individual confidence intervals are exact when constructed using the critical value of Table 8.1 and Eq. (8.12) if the model is linear and satisfies the requirements of Chapter 3, Section 3.3, as tested for using the methods of Chapters 6, 7, and 8. "Exact" means that intervals have the stated probability of including the true value.

Exact critical values for linear simultaneous intervals are difficult to calculate, but can be approximated using the Bonferroni, Scheffé $d = k$, and Scheffé $d = NP$ critical values of Table 8.1, as discussed by Miller (1981).

The Bonferroni, Scheffé $d = k$, and Scheffé $d = NP$ approximate critical values tend to be large. For example, an interval calculated for a 5 percent significance level (a 95 percent interval) may be large enough to satisfy a smaller significance level such as 3 percent (resulting in a 97 percent interval). The linear simultaneous intervals tend to indicate that the uncertainty is greater than it really is.

If k is less than NP, either the Bonferroni or Scheffé $d = k$ critical values could be used. Both tend to be too large; using the smaller critical value reduces the error. If k is larger than NP, Scheffé $d = NP$ critical values are needed.

In some cases, k is not finite. For example, if a prediction of interest is the largest simulated value over a defined area, the predicted quantity cannot be exactly specified before performing a model simulation. An infinite number of simultaneous predictions then need to be considered. In this circumstance, the Scheffé $d = NP$ critical value is needed. These intervals are denoted Scheffé $d = NP$ intervals in Table 8.1 and throughout this book; in other publications the term Scheffé interval almost always refers to these $d = NP$ intervals.

For all linear intervals, as the model becomes nonlinear and violates the requirements of Chapter 3, Section 3.3 the calculated intervals become less accurate. This means that the actual significance level can be substantially different than intended, and is a serious concern for nonlinear models (Donaldson and Schnabel, 1987). For some non-ideal situations linear intervals may be accurate enough to be useful, as discussed in Section 8.4.3. Hopefully, evolving experience will provide additional guidance on when the computationally expensive nonlinear intervals are needed.

8.4.3 Nonlinear Confidence and Prediction Intervals

For nonlinear models, nonlinear intervals are sometimes much more accurate than linear intervals. Nonlinear intervals can be calculated using the methods of Vecchia and Cooley (1987). These methods compute individual or simultaneous intervals on any function of the model parameters $g(\boldsymbol{b})$. The intervals can be individual or simultaneous Scheffé $d = NP$ confidence intervals or individual prediction intervals. To obtain a nonlinear interval on a parameter, the function $g(\boldsymbol{b})$ is specified to represent the parameter; this situation was discussed in Chapter 7, Section 7.5.1. Here, we consider that $g(\boldsymbol{b})$ represents a prediction.

Calculating nonlinear intervals involves determining the minimum and maximum values of $g(b)$ over a confidence region on the parameter set listed in vector b. The confidence region is defined in NP-dimensional parameter space and has a specified probability of containing the true set of parameter values.

Vecchia and Cooley (1987) present methods for calculating intervals using exact confidence regions and using approximate likelihood confidence regions. The method that uses the likelihood confidence region is presented here for three reasons: (1) determining the exact confidence region is mathematically more difficult than determining the likelihood region; (2) it has been shown that the probability that b lies in the likelihood region is very close to the true probability determined using the exact confidence region (Donaldson and Schnabel, 1987); and (3) this method is used in MODFLOW-2000's UNC Process (Christensen and Cooley, 2005), UCODE_2005 (Poeter et al., 2005), and PEST (Doherty, 2005).

The method for computing nonlinear confidence intervals involves first defining the $(1 - \alpha)100$-percent likelihood parameter confidence region. This region is defined as the set of parameter values for which the objective-function values, $S(b)$, satisfy the following condition (modified from Vecchia and Cooley, 1987, Eq. (10) and Christensen and Cooley, 2005, Eqs. (8) and (19)):

$$S(b) \leq S(b') + s^2 \times \text{critical value} + a \qquad (8.14)$$

where $S(b')$ = the objective function for the optimal parameter values b';

s^2 = the calculated error variance defined in Eq. (6.1);

critical value = a critical value from a statistical distribution; the critical values required for different types of intervals are defined in Table 8.2;

$a = 0.0$ for confidence intervals and for prediction intervals reflects the accuracy of a measured observed equivalent of the prediction.

TABLE 8.2 Critical Values for Eq. (8.14) Required to Calculate the Nonlinear Confidence and Prediction Intervals Used in This Book

Type of Interval	Critical Value	Table in Appendix D
Individual confidence	$c_c[t_s(n, 1.0 - \alpha/2)]^2$	D.2, Student-t distribution
Scheffé $d = NP$ simultaneous confidence	$c_r[NP \times F_\alpha (NP, n)]$	D.7, F distribution
Individual prediction	$c_p[t_s(n, 1.0 - \alpha/2)]^2$	D.2, Student-t distribution

Note: α is the significance level and is commonly 0.05 or 0.10 (5 or 10 percent), which results, respectively, in 95- or 90-percent parameter confidence region and intervals. n is the degrees of freedom, here equal to $ND + NPR - NP$. NP is the dimension of the parameter space, which commonly equals the number of estimated parameters. ND is the number of observations and NPR is the number of prior information equations. c_c, c_r, and c_p are correction factors defined by Christensen and Cooley (2005), as discussed by Poeter et al. (2005, Table 38). The correction factors are set to 1.0 for the results presented in this book.

8.4 INFERENTIAL STATISTICS

The quantity on the right-hand side of Eq. (8.14) defines the bounding surface of the parameter confidence region. The term a is not discussed further in this book. For additional information see Christensen and Cooley (2005, pp. 11–12).

Some characteristics of nonlinear intervals can be investigated using Eq. (8.14) and the critical value from Table 8.2 with $ND + NPR - NP$ substituted for n. All of the critical values generally increase as NP increases or $ND + NPR - NP$ decreases. The product $NP \times F_\alpha(NP, ND + NPR - NP)$ generally increases with NP. Thus, the size of the parameter confidence region and nonlinear intervals are larger for poorer model fits (larger values of s^2), more parameters (larger NP), and fewer observations (smaller ND).

After defining the $(1 - \alpha)100$-percent parameter confidence region, the method finds the minimum and maximum values of the prediction $g(b)$ on the boundaries of this region. These extreme values are the lower and upper limits of the $(1 - \alpha)100$-percent nonlinear confidence interval on $g(b)$. Figure 8.3 illustrates the confidence region and the limits of a nonlinear confidence interval on a simple prediction $g(b)$ made with a hypothetical two-parameter model. Unlike a linear interval, the nonlinear confidence interval is not symmetric about the value calculated using the optimized parameter values, $g(b')$: the upper limit of the interval ($g(b) = c_4$) is much further from $g(b')$ than is the lower limit of the interval ($g(b) = c_2$).

Nonlinear confidence intervals can be larger or smaller than corresponding linear confidence intervals and, as shown in Figure 8.3, can be asymmetric about the estimated value. These characteristics are illustrated in Figure 8.4, which shows linear and nonlinear intervals calculated by Christensen and Cooley (1999) for

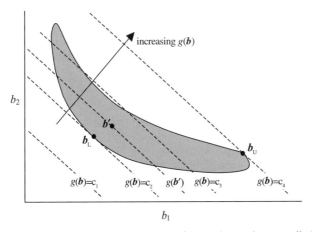

FIGURE 8.3 The geometry of a nonlinear confidence interval on prediction $g(b')$. The parameter confidence region (shaded area), contours of constant $g(b)$ (dashed lines), and locations of the minimum ($g(b) = c_2$, with $b = b_L$) and maximum ($g(b) = c_4$, with $b = b_U$) values of the prediction on the confidence region are shown. The lower and upper limits of the nonlinear confidence interval on prediction $g(b)$ are thus c_2 and c_4, respectively. (Adapted from Christensen and Cooley, 1999, Figure 9.)

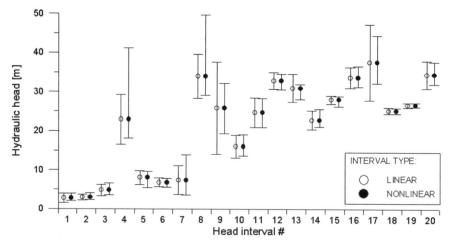

FIGURE 8.4 Linear and nonlinear Scheffé simultaneous confidence intervals on hydraulic heads predicted by a steady-state groundwater flow model of an aquifer in Denmark. (From Christensen and Cooley, 1999, Figure 5a.)

hydraulic-head predictions in a groundwater flow model of an aquifer in Denmark. For these predictions, the differences between the linear and nonlinear intervals tend to increase as the size of the intervals increase. For relatively small intervals, the linear and nonlinear intervals are roughly the same size. For some of the larger intervals, such as 9, 13, 17, and 20, the linear confidence interval is larger than the nonlinear confidence interval. For others, such as 4, 7, and 8, the nonlinear confidence interval is larger. Some of the nonlinear intervals, such as 4, 7, 8, and 13, are highly asymmetric.

A significant difference in the evaluation of linear and nonlinear confidence intervals involves the assumptions that apply to the calculation of the different intervals. Recall that three important assumptions apply for linear confidence intervals to be accurate: (1) the model is correct, (2) the model is linear, and (3) the true errors are normally distributed. For nonlinear confidence intervals, only the first assumption is needed. To the extent that model nonlinearity and deviations from normality of the weighted residuals are problematic, nonlinear intervals are likely to be more accurate than associated linear intervals.

Calculating nonlinear intervals is computationally intensive because of the difficulty of determining the extreme values of $g(b)$ over the confidence region. Calculation of each limit of each nonlinear confidence interval involves a computational effort approximately equivalent to a full nonlinear regression simulation. Furthermore, it is good practice to calculate each limit using a few different starting parameter values, as the results can depend on these values.

It has been stressed in this book that it is important to include defined parameters that were not estimated in the regression in evaluations of prediction uncertainty. Though including such parameters in the calculation of nonlinear intervals has not been investigated, it is likely that their inclusion will cause difficulties in

8.4 INFERENTIAL STATISTICS

determining the interval limits. If that proves to be the case, an advantage of the linear intervals would be their ability to include the effects of these parameters. This difference can be significant if the added parameters are important to predictions. The importance of parameters to predictions can be evaluated using the methods described earlier in this chapter, in Section 8.2.

The substantial effort required to compute nonlinear confidence and prediction intervals suggests that a practical approach is first to calculate linear intervals, and then to calculate nonlinear intervals for selected predictions. Additional nonlinear intervals may be needed depending on the discrepancies between the linear and nonlinear intervals and the requirements of the uncertainty evaluation.

8.4.4 Using the Theis Example to Understand Linear and Nonlinear Confidence Intervals

The nonlinear and linearized objective-function surfaces shown in Figure 8.5 (modified from Figure 5.3) can be used to better understand linear and nonlinear confidence intervals on predictions. Each contour of the nonlinear objective-function surface (Figure 8.5a) can be related to a significance level for inferential statistics. Contours that are closer to the minimum of the nonlinear surface relate to larger significance levels.

Consider a situation like that in Figure 8.5b, in which the objective-function surface has been linearized around a point close to or equal to the minimum of the objective function. If the designated significance level is large enough, then the contour of this linearized surface will be close to the associated contour on

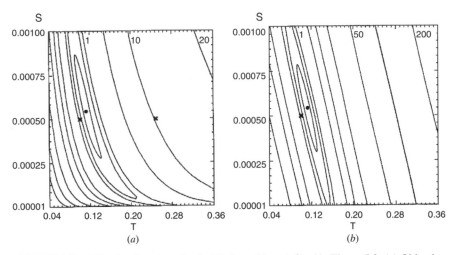

FIGURE 8.5 Objective functions for the Theis problem defined in Figure 5.3. (a) Objective function for the nonlinear model, with the minima (•) and the linearization (×) points near to and far from the minima. (b) Objective function for model linearized about the S and T values at ×. The point • is the minimum for the nonlinear model.

the nonlinear surface. This is illustrated by comparing the contours close to the large dot in Figure 8.5a and 8.5b. In this case, the inferential statistics calculated using linear theory are likely to be accurate if the other required assumptions hold (that the model is correct and the weighted residuals are normally distributed).

As the significance level declines, a contour more distant from the optimum parameter values is needed, a broader range of parameter values is included in the interval of interest, and more nonlinear parts of the objective-function surface become important. This is illustrated by comparing contours distant from the large dot in Figure 8.5a and 8.5b. In this circumstance, the stated significance level for the critical value used to calculate linear confidence intervals (α in Table 8.1) becomes less reliable. Thus, while a 90-percent confidence interval (10-percent significance level) might be well-estimated using linear theory in a certain situation, a 99-percent confidence interval (1-percent significance level) might not.

8.4.5 Differences and Their Standard Deviations, Confidence Intervals, and Prediction Intervals

In many management situations, the prediction is the change in a simulated quantity under certain conditions. For example, in a groundwater model, the relevant predictions might be drawdowns or changes in flow to a stream caused by changes in pumping rates. Such changes are called differences here and are calculated by subtracting values produced by a base simulation from values produced by a predictive simulation. Thus, the difference, u'_ℓ, is calculated as

$$u'_\ell = z'_{p_\ell} - z'_{q_\ell} \tag{8.15}$$

where p represents the predictive conditions and q represents the base conditions.

Figure 8.6 illustrates the concept of differences. In the simple groundwater flow system shown in Figure 8.6a, hydraulic heads and flow to a river were simulated for steady-state calibration conditions and for steady-state conditions representing two pumping scenarios. If the management criterion is that hydraulic heads cannot decline by more than 2 meters, simulated drawdown is of interest. This difference is calculated by subtracting simulated hydraulic heads for pumping scenario 1 or 2 from simulated hydraulic heads for the calibration conditions. If the management criterion is that the streamflow gain along a reach must not decrease by more than 20 percent of an observed flow, the simulated change in streamflow is of interest. This difference is calculated by subtracting streamflow gain simulated for pumping scenarios 1 and 2 from the streamflow gain simulated for the calibration conditions.

To compute the standard deviation of a difference, z'_ℓ of Eq. (8.1a) is replaced with the difference, u'_ℓ, yielding

$$s_{u'_\ell} = \left[\sum_{i=1}^{NP} \sum_{j=1}^{NP} \left(\frac{\partial z'_{p_\ell}}{\partial b_j} - \frac{\partial z'_{q_\ell}}{\partial b_j} \right) V(b')_{ij} \left(\frac{\partial z'_{p_\ell}}{\partial b_i} - \frac{\partial z'_{q_\ell}}{\partial b_i} \right) \right]^{1/2} \tag{8.16}$$

8.4 INFERENTIAL STATISTICS

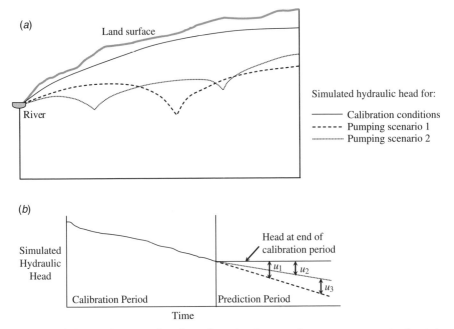

FIGURE 8.6 (a) Cross section through a simple groundwater system at steady state, showing simulated hydraulic head for calibration conditions and two predictive scenarios with pumping. (b) Hydrograph of a transient groundwater system showing simulated hydraulic head during a calibration period and during two predictive scenarios with pumping. Quantities u_1, u_2, and u_3 are differences that may be of interest.

where $s_{u'_\ell}$ is the standard deviation of the difference u'_ℓ. To compute confidence or prediction intervals on differences, z'_ℓ of Eqs. (8.12) and (8.13) is replaced with u'_ℓ, and $s_{z'_\ell}$ of these equations is replaced with $s_{u'_\ell}$. In Eq. (8.13), s_a^2 is calculated as

$$s_a^2 = s_{a_p}^2 + s_{a_q}^2 \qquad (8.17)$$

The calculation of confidence and prediction intervals on differences involves differences in the sensitivities, as shown in Eq. (8.16). If the sensitivities to each of the parameters are the same for the two subtracted values, $s_{u'_\ell}$ equals zero and the limits of calculated confidence intervals on differences each equal the simulated difference u'_ℓ. As a result, the width of these confidence intervals equals zero and prediction intervals only reflect s_a^2.

An unrealistic but informative example illustrates this circumstance. If all conditions, including stresses, are the same in the two simulations for which differences are calculated, then all differences equal zero and the confidence interval limits on differences all equal zero. This result indicates the certainty that if simulated conditions do not change, the simulated values do not change. A more realistic situation is that the differences between two simulations are small, in which case the confidence intervals on the differences also will tend to be small.

For some parameters, sensitivities might be identical for the two predictions being subtracted, so the difference in sensitivity will equal zero. For example, consider a groundwater flow simulation with areal recharge and pumpage in which all model layers are confined and all boundary conditions are linear. Predictions are changes in head or flow to a stream. In such a simulation, a specified increase in areal recharge produces the same increase in hydraulic head or flow at any location in the system regardless of the simulated pumpage. Thus, the sensitivities related to areal recharge are independent of the pumpage that causes a difference in heads or flows, and uncertainty in the recharge rate would not affect the uncertainty of the predicted changes in head and flow.

Differences need not be between calibration conditions and alternative conditions. For example, in Figure 8.6a, the relevant predictions might be the differences between hydraulic heads simulated under pumping scenarios 1 and 2. Figure 8.6b shows the hydrograph for a simulated well in such a model. Differences that might be useful are (1) the decline in hydraulic head since the end of the calibration period (differences u_1 and u_2 of Figure 8.6b) or (2) the additional decline in hydraulic head that would occur under pumping scenario 1 compared to scenario 2 (difference u_3 of Figure 8.6b).

Differences could also be spatial. For example, in a groundwater system the predicted head loss across a confining layer might be of interest.

With MODFLOW-2000, YCINT-2000 (Hill et al., 2000, pp. 87–91) can be used to calculate differences and their linear intervals. In UCODE_2005, differences can be defined using derived predictions (Poeter et al., 2005); their linear intervals can be calculated using LINEAR_UNCERTAINTY and nonlinear intervals can be calculated as described by Poeter et al. (2005, Chapter 17).

8.4.6 Using Confidence Intervals to Serve the Purposes of Traditional Sensitivity Analysis

Confidence intervals on simulated values can be employed to replace the traditional procedure used to perform sensitivity analyses. According to Anderson and Woessner (1992, p. 246), "the purpose of a sensitivity analysis is to quantify the uncertainty in the calibrated model caused by uncertainty in the estimated parameter values"; and in the procedure traditionally followed to fulfill this purpose, "calibrated values for hydraulic conductivity, storage parameters, and recharge and boundary conditions are systematically changed within the previously established plausible range." The results of several traditional sensitivity analyses are shown in Anderson and Woessner (1992, pp. 247–254). The major weaknesses of the traditional procedure are as follows:

1. The "plausible range" usually is determined subjectively prior to model calibration (Anderson and Woessner, 1992, p. 231). Thus, this range does not reflect the possibly substantial information provided by model calibration on the parameter values. One effect of this is that many sets of parameter values used in a traditional sensitivity analysis may result in a much poorer

match to the observations than was achieved in model calibration. Yet if the model reasonably represents the system, the poor fit produced by these parameter values suggests that the values are unlikely. Thus, by using the previously defined plausible ranges, traditional sensitivity analysis tends to produce unrealistically large measures of uncertainty for results of calibrated models.

2. Coordinated changes in two or more parameter values are rarely considered, though they are often important, and some attempts to consider coordinated parameter changes can actually be detrimental. For example, Anderson and Woessner (1992, p. 248) suggest that, in traditional sensitivity analysis of groundwater models, hydraulic conductivity and recharge values be changed in opposite directions because such parameters are often positively correlated. In some cases this can be useful, but in others it can exacerbate the problem noted in weakness 1, producing an even more severely exaggerated impression of model uncertainty.

Because of these weaknesses, confidence intervals often can be used to fulfill the purpose of sensitivity analyses more effectively than the traditional approach.

8.5 QUANTIFYING PREDICTION UNCERTAINTY USING MONTE CARLO ANALYSIS

As discussed in Chapter 7, Section 7.5.2, in Monte Carlo analysis, uncertain aspects of the model input data are changed and for each change or set of changes a model run is conducted and changes in selected simulated results are evaluated. The changed model input data are often parameter values but can be any other model attributes. Monte Carlo analysis often is used to evaluate prediction uncertainty and can be used to test the significance level of confidence intervals (e.g., Hill, 1989). The present work seeks only to introduce the reader to Monte Carlo analysis. More detailed description of Monte Carlo methods can be found in a number of texts, including Skinner (1999), Vose (2000), and Bedford and Cook (2001).

8.5.1 Elements of a Monte Carlo Analysis

The elements that define a Monte Carlo analysis include:

1. What model inputs are changed.
2. How changed model inputs are generated.
3. What constitutes a model run.
4. How many Monte Carlo runs are conducted.
5. What simulated values or quantities calculated using the simulated values are saved.
6. How the saved results are analyzed.

Each of these elements is discussed in more detail below.

1. Commonly, parameter values are changed during Monte Carlo analysis, but other entities such as aspects of the conceptual model also could be changed. For example, in groundwater models hydrogeologic interpretations might be changed. In this situation, the modified model inputs could be the configuration of hydraulic-conductivity zones that represent different hydrogeologic units, or the variation of hydraulic conductivity within such zones using, perhaps, pilot points.

A very different type of analysis involves changing the errors on the observations. These analyses generally are called bootstrap methods and are discussed briefly in Chapter 7, Section 7.5.2.

2. If parameter values are changed, commonly they are assumed to be normally or log-normally distributed with their means equal to the parameter estimates from model calibration and their variation characterized by the parameter variance–covariance matrix. Parameter values that honor these means and variation can be generated randomly, or more frugal methods can be used that require fewer Monte Carlo runs to produce an equivalently accurate evaluation. One such method is Latin hypercube sampling (Gwo et al., 1996; Zhang and Pinder, 2003). Another method, Markov chain Monte Carlo, has received substantial attention recently, as noted by the contributions posted at the web site http://www.statslab.cam.ac.uk/~mcmc/. For other types of changed quantities, analogous options can be used. UCODE_2005 supports a very simple method in which parameter values are sampled at equal intervals within a stated range.

If other aspects of a model are changed, often it is useful to consider a discrete number of changes that produce deterministically derived alternative models. For example, in groundwater modeling, a limited number of alternative interpretations of the hydrogeologic framework could be tested.

3. The model run may be a forward simulation or can be a more complicated run such as an inverse simulation. Poeter and McKenna (1995) provide an example of using inverse simulations. This application is summarized in Guideline 8 (Chapter 11, Section G8.3) and Guideline 14 (Chapter 14, Section G14.2).

4. The number of Monte Carlo runs required depends on many factors, including the number of model inputs changed, how they are changed, and whether they are continuous or discrete variables. The desired results of the Monte Carlo analysis also are important. For example, using Monte Carlo to determine an estimate of the mean of a predicted value to a given accuracy generally takes far fewer Monte Carlo runs than determining 90-percent confidence intervals, which in turn takes fewer runs than determining 95-percent confidence intervals. In addition, accounting for model nonlinearity can require an increased number of Monte Carlo runs. The number of runs required often is determined by calculating the desired result after some number of Monte Carlo runs, and then examining whether this result changes after conducting an additional set of runs. When the result becomes stable, sufficient runs have been conducted. The number of runs can be very large in some circumstances and can commonly exceed 1000. Even if the

model run takes only 1 minute of execution time, a 1000-run Monte Carlo analysis requires almost 17 hours of computer time. This is problematic in fields such as groundwater, in which even with modern computers model runs can take 30 minutes or more. For a 30-minute run, conducting 1000 runs requires 21 days of execution time. Availability of parallel processors often can be used to great advantage as long as each processor has enough random access memory to run the model. Additional guidance on determining the required number of Monte Carlo runs, with application to groundwater models, is presented by Ballio and Guadagnini (2004).

5. Important results to save from Monte Carlo runs can include any model output, or any quantity calculated from the output. The obvious items to save are values of the defined model predictions. Less obvious items might be measures that reflect the numerical accuracy of the solution, to ensure that any solutions that did not satisfy the convergence criteria are identified; and statistics that measure model fit to observations such as the objective function or standard error. Often it also is useful to save information about what was changed for each solution, such as parameter values, so that unusual results can easily be evaluated. It is important to carefully choose what results to save from each Monte Carlo run, to avoid (a) repeating the analysis to obtain model results that were not saved and (b) saving too much output, which can be unwieldy and difficult to process.

6. There are many ways to analyze and display Monte Carlo results, including calculation of histograms and confidence limits and presentation on maps or cross sections. In many circumstances the main criterion motivating the presentation of results is to convey the essence of the results to resource managers.

Many authors have suggested that (a) for any type of change considered, simulated values need to be compared to observations, and (b) including simulated values that produce a poor match to the observations can result in an overstatement of model uncertainty (Beven and Binley, 1992; Brooks et al., 1994; Evers and Lerner, 1998; Binley and Beven, 2003; and Morse et al., 2003). To account for this, a measure of model fit needs to be saved for each run, and runs with poor matches need to be omitted from the analyses and display of the Monte Carlo results.

8.5.2 Relation Between Monte Carlo Analysis and Linear and Nonlinear Confidence Intervals

Some Monte Carlo analyses produce results that approximate those produced by inferential statistics. Differences occur because of approximations made in the inferential methods or inadequate sampling by the Monte Carlo analysis. For example, nonlinear intervals are routinely estimated using an approximate likelihood-function approach (e.g., Cooley, 2004; Christensen and Cooley, 2005).

Considering the approximate equivalence of Monte Carlo and inferential results helps to understand both methods. For example, if the following conditions are met, the results of a Monte Carlo analysis approximate nonlinear confidence intervals (described in Section 8.4.3).

a. Only parameter values are changed for the Monte Carlo runs, and the parameters changed are the same as those that are active for the nonlinear confidence interval calculations.
b. The same observations, prior information, and weighting are used for the Monte Carlo analysis and to calculate the nonlinear intervals. In the Monte Carlo runs the least-squares objective function of Eq. (3.1) is used to compare the simulated values to the observations, and runs are omitted if the result exceeds a specified value (denoted the "fit criterion" here), as suggested by Beven and Binley (1992) and Binley and Beven (2003). This objective-function fit criterion is the quantity on the right-hand side of Eq. (8.14).
c. No local minima exist in the objective-function surface and predictions are continuous and monotonic with respect to the parameter values in the range of interest.
d. The ranges of parameter values used for the Monte Carlo analysis are sufficiently large to cover the entire parameter confidence region, as defined by the objective function being less than the fit criterion.
e. The Monte Carlo interval is constructed as the maximum and minimum predicted values that occur for the objective-function fit criterion.

Advantages of the Monte Carlo method are that it is possible to consider (1) changes in aspects of the system other than parameter values, and (2) highly nonlinear models that violate (c) above. In the first situation, inferential statistical methods currently have no equivalent capability. In the second situation, inferential methods would be unable to produce meaningful limits on the predictions.

8.5.3 Using the Theis Example to Understand Monte Carlo Methods

The Theis example of Figure 8.5 can be used to understand Monte Carlo analysis, in a manner similar to that for understanding linear and nonlinear confidence intervals (Section 8.4.4). In this example, suppose a Monte Carlo analysis is conducted that involves changing parameter values, and that an objective-function value of 1.0 is chosen as the fit criterion defined in Section 8.5.2. Figure 8.5a illustrates the importance of choosing an appropriate range of parameter values sampled. For example, to ensure that the analysis samples the entire parameter space within the objective-function contour of 1.0, transmissivity values greater than 0.28 ft^2/s and storage coefficients larger than 0.001 need to be included. Objective-function surfaces such as that in Figure 8.5 are almost never available to guide selection of the range of parameter values, and thus in Monte Carlo analysis often it is difficult to verify that the ranges are sufficiently large.

After the ranges of parameter values are chosen, it also is important to carefully select the sampled parameter values within these ranges. If prediction uncertainty is being evaluated, and the prediction consistently increases or decreases with the parameter values, as for the example in Figure 8.3, then the extreme predictions will occur for an objective-function value equal to the chosen fit criterion. In this case, it is most important to sample parameter sets that produce an objective-function value near the fit criterion. However, for some types of nonlinearity, the extreme predictions may occur for smaller values of the objective function. In this case, Monte Carlo methods are needed to identify extreme values, and it is important to thoroughly sample sets of parameters that produce objective-function values smaller than the fit criterion. Clearly, it can be difficult to determine the extreme predictions with Monte Carlo methods and, when applicable, the nonlinear confidence interval calculation is much more efficient for evaluating prediction uncertainty.

8.6 QUANTIFYING PREDICTION UNCERTAINTY USING ALTERNATIVE MODELS

Recent work has proposed a number of methods for quantifying prediction uncertainty in a way that accounts for alternative models. The methods involve calculation of confidence intervals for each alternative model, weighting the intervals to reflect the validity of the related model, and producing composite intervals. Methods have been presented that calculate the weighting based on the AIC_c statistic (Eq. (6.3)) (Burnham and Anderson, 2004) and Bayes factors (Kass and Raftery, 1995; Neuman, 2003; Meyer et al., 2004). Poeter and Anderson (2005) compared these two methods using the Multi-Model Analysis (MMA) computer code (Poeter and Hill, in press) and found the intervals produced using the weights based on the AIC_c statistic to be more useful. Additional testing in a variety of circumstances is needed, however, to determine the applicability of that conclusion.

8.7 TESTING MODEL NONLINEARITY WITH RESPECT TO THE PREDICTIONS

The test for linearity described in Chapter 7, Section 7.7 uses the modified Beale's measure to evaluate the linearity of the model with respect to observed quantities. If the predictions are similar in type, location, and time to the observations, and if the predictive conditions are similar to the calibration conditions, then that test is sufficient. In many circumstances, however, we are in the unenviable position of trying to make predictions that are in some way very different from the observations. The prediction conditions may be very different, or the predicted quantity could be very different. In this circumstance, a model that is linear on the basis of the analysis presented in Chapter 7, Section 7.7 may be nonlinear with respect to the predictions. In this situation, the judgment of linearity could lead to the incorrect conclusion that linear intervals on predictions are accurate measures of prediction uncertainty.

Methods for testing model nonlinearity with respect to predictions are just being developed. As an introduction to these methods, consider first the following statistic, which is a direct extension of the modified Beale's measure.

1. Use the same sets of parameters produced from step 1 of Chapter 7, Section 7.7, using Eq. (7.11). These parameter values are on the edge of the linearized parameter confidence region.

2. Compute the predictions by executing a forward model run for each generated set of parameter values. These simulated values are $\tilde{z}_{k\ell}$, where k refers to the kth generated parameter vector, ℓ refers to the ℓth prediction, and the tilde ($\sim$) is used to designate values associated with the generated parameter values.
3. Calculate linearized estimates of the predictions using the generated parameter sets as follows:

$$\tilde{z}_{\ell k}^o = z'_\ell + \sum_{j=1}^{NP} (b'_j - \tilde{b}_{kj}) \left. \frac{\partial z'_\ell}{\partial b_j} \right|_{b'} \qquad (8.18)$$

where $\tilde{z}_{\ell i}^o =$ the linearized estimate of the ith prediction;
$z'_\ell =$ the ℓth simulated prediction calculated using the optimal parameter estimates;
$b'_j =$ the jth optimal parameter estimate;
$\tilde{b}_{kj} =$ the jth parameter value from the kth generated parameter set.
4. Calculate the proposed modified Beale's measure for predictions, $\hat{N}_b^z$, which is a measure of the difference between the model-computed and the linearized estimates of the predictions:

$$\hat{N}_b^z = \sum_{k=1}^{2NP} \frac{|\tilde{z}_{\ell k} - \tilde{z}_{\ell k}^o|}{|\tilde{z}_{\ell k}^o - z'_\ell|} \qquad (8.19)$$

Equation (8.19) produces a unique measure of linearity for each prediction. For a truly linear model, the numerator of Eq. (8.19) equals zero. Thus, values of $\hat{N}_b^z$ that are close to zero generally indicate that the model is close to being linear. As model nonlinearity increases, the magnitude of the numerator increases. Testing of Eq. (8.19) would be needed to develop critical values of $\hat{N}_b^z$ that can be used as objective criteria against which to evaluate model linearity.

A deficiency of Eq. (8.19) is that it measures only the nonlinearity of the predictions with respect to the parameters. If Eq. (8.19) is close to zero, but the measures discussed in Chapter 7, Section 7.7 indicate nonlinearity of the observation with respect to the parameters, linear intervals on predictions may be in error. The combined nonlinearity can be measured using the combined intrinsic model nonlinearity measures of Cooley (2004). The method is described by Cooley (2004) and Christensen and Cooley (2005, pp. 20–24) and is available in MODFLOW-2000's UNC Process with BEALE2-2K (Christensen and Cooley, 2005) and in UCODE_2005 with MODEL_LINEARITY_ADV (Poeter et al., 2005). While the complete equations are not repeated here, the following analysis applicable to individual confidence intervals on predictions provides an introduction.

Two issues of concern in Christensen and Cooley (2005) are (1) the validity of linear confidence intervals and (2) the validity of correction factors that account for unrepresented heterogeneity (the c_c, c_r, and c_p of Table 8.2). The equations are presented below along with many of the steps required to go from the more

8.7 TESTING MODEL NONLINEARITY

general equations presented in Christensen and Cooley (2005) to the equations presented here, which apply if the weight matrix used in the regression has been defined in Chapter 3, Section 3.4.2 (i.e., $\boldsymbol{\omega} = V(\boldsymbol{\varepsilon})^{-1}$).

If intrinsic model nonlinearity is small, the first issue is addressed by what we call the combined intrinsic model nonlinearity measure for confidence intervals. Using Christensen and Cooley (2005, Eq. (48)), the notation of this book, and rearranging terms, the measure is calculated for each prediction as

$$\widehat{M}_{\min} = \frac{1}{2s^2} \sum_{p=1}^{2} (\tilde{y}_p - \tilde{y}_p^o - X\boldsymbol{\varphi}_p)^T \boldsymbol{\omega} (\tilde{y}_p - \tilde{y}_p^o - X\boldsymbol{\varphi}_p) \tag{8.20}$$

X is defined for Eq. (5.2b) and is calculated using the optimal parameter values, b'. Terms covered by a tilde ($\sim$) and with a p subscript are calculated for parameter values generated as (Christensen and Cooley, 2005, p. 21; Cooley, 2004, pp. 86–87)

$$\tilde{b} = b' \pm \frac{1}{s_{\tilde{z}_\ell}} [V(b')] \frac{\partial z_\ell}{\partial b} \tag{8.21}$$

where addition is used for $p = 1$, substraction is used for $p = 2$

$\tilde{b}$ = a vector of generated parameter values;
b' = a vector of optimal parameter estimates;
$s_{\tilde{z}_\ell}$ = the standard deviation of the ℓth prediction, defined in Eq. (8.1);
$[V(b')]$ = the parameter variance–covariance matrix of Eq. (7.1);
$\partial z_\ell / \partial b$ = the sensitivity of the ℓth prediction with respect to the optimal parameter values; this is a vector with NP elements.

Superscript o indicates values calculated using a model linearized about the optimized parameter values—that is, $\tilde{y}_p^o = y'(b') + X(\tilde{b} - b')$. The remaining term, $\boldsymbol{\varphi}_p$, accounts for prediction nonlinearity and is calculated as

$$\boldsymbol{\varphi}_p = \boldsymbol{\varphi}_\ell^0 + \frac{1}{s_{\tilde{z}_\ell}^2} [V(b')] \frac{\partial z_\ell}{\partial b} (\tilde{z}_p - \tilde{z}_p^0) \tag{8.22}$$

where

$$\boldsymbol{\varphi}_\ell^0 = \left[\left(\frac{1}{s_{\tilde{z}_\ell}} [V(b')] \frac{\partial z_\ell}{\partial b} \left(\frac{\partial z_\ell}{\partial b} \right)^T (X^T \boldsymbol{\omega} X)^{-1} X^T \boldsymbol{\omega}^{1/2} \right) \right. \\ \left. - (X^T \boldsymbol{\omega} X)^{-1} X^T \boldsymbol{\omega}^{1/2} \right] \boldsymbol{\omega}^{1/2} (\tilde{y}_p - \tilde{y}_p^0) \tag{8.23}$$

To obtain Eqs. (8.22) and (8.23) from the results presented by Christensen and Cooley (2005, p. 21, Eqs. (50) and (51)), their term $Q^T Q$ needs to be expanded using the definition of Q on their page 8. The definition of $s_{\tilde{z}_\ell}^2$ using the square of

192 MODEL PREDICTIONS, DATA NEEDS, AND PREDICTION UNCERTAINTY

Eq. (8.1b) also is used. This proof is valid for all types of intervals.

$$Q^T Q = \frac{\partial z_\ell}{\partial b}(X^T \omega X)^{-1} X^T \omega^{1/2} \omega^{1/2} X (X^T \omega X)^{-1} \left(\frac{\partial z_\ell}{\partial b}\right)^T$$

$$= \frac{\partial z_\ell}{\partial b}(X^T \omega X)^{-1} X^T \omega X (X^T \omega X)^{-1} \left(\frac{\partial z_\ell}{\partial b}\right)^T$$

$$= \frac{\partial z_\ell}{\partial b}(X^T \omega X)^{-1} \left(\frac{\partial z_\ell}{\partial b}\right)^T$$

$$= \frac{1}{s^2}\frac{\partial z_\ell}{\partial b} s^2 (X^T \omega X)^{-1} \left(\frac{\partial z_\ell}{\partial b}\right)^T$$

$$= \frac{s^2_{z_\ell}}{s^2} \tag{8.24}$$

It is assumed that $V(b) = s^2(X^T \omega X)^{-1}$, which is valid for linear systems with the correct model (see Appendix C), and approximate otherwise.

One value of $\widehat{M}_{\min}$ is calculated for each prediction. If intrinsic nonlinearity is small, values less than 0.01 indicate that a standard linear individual confidence interval, calculated using the equations in Section 8.4, should not be affected significantly by nonlinearity.

If intrinsic model nonlinearity is small, the second issue is addressed by what we call the combined intrinsic model nonlinearity for correction factors. The measure is calculated as the largest of two values: $\widehat{M}_{\min} + 2\widehat{B}_U$ or $|\widehat{M}_{\min} - 2\widehat{B}_L|$. $\widehat{M}_{\min}$ would equal $\widehat{B}_U$ if y were linear, so only the last term in the parentheses of Eq. (8.20) is nonzero. Thus,

$$\widehat{B}_U = \frac{1}{2s^2}\sum_{p=1}^{2}(X\varphi_p)^T \omega_{ij}(X\varphi_p)$$

$$= \frac{1}{2s^2}\sum_{p=1}^{2}\left(X\frac{1}{s^2_{z_\ell}}[V(b')]\frac{\partial z_\ell}{\partial b}(\tilde{z}_p - \tilde{z}_p^0)\right)^T \omega \left(X\frac{1}{s^2_{z_\ell}}[V(b')]\frac{\partial z_\ell}{\partial b}(\tilde{z}_p - \tilde{z}_p^0)\right)$$

$$= \frac{1}{2s^2 s^2_{z_\ell}}\sum_{p=1}^{2}\left(\frac{1}{s^2_{z_\ell}}\right)(\tilde{z}_p - \tilde{z}_p^0)^T \frac{\partial z_\ell^T}{\partial b}[V(b')] X^T \omega X [V(b')] \frac{\partial z_\ell}{\partial b}(\tilde{z}_p - \tilde{z}_p^0)$$

$$= \frac{1}{2s^2 s^2_{z_\ell}}\sum_{p=1}^{2}\left(\frac{1}{s^2_{z_\ell}}\right)(\tilde{z}_p - \tilde{z}_p^0)^T \frac{\partial z_\ell^T}{\partial b}$$

$$\times s^2 (X^T \omega X)^{-1} X^T \omega X [V(b')] \frac{\partial z_\ell}{\partial b}(\tilde{z}_p - \tilde{z}_p^0)$$

$$= \frac{s^2}{2s^2 s^2_{z_\ell}}\sum_{p=1}^{2}\left(\frac{1}{s^2_{z_\ell}}\right)(\tilde{z}_p - \tilde{z}_p^0)^T \frac{\partial z_\ell^T}{\partial b}[V(b')] \frac{\partial z_\ell}{\partial b}(\tilde{z}_p - \tilde{z}_p^0)$$

$$= \frac{s^2}{2s^2 s_{\tilde{z}_\ell}^2} \sum_{p=1}^{2} \left(\frac{1}{s_{\tilde{z}_\ell}^2}\right) (\tilde{z}_p - \tilde{z}_p^0)^T s_{\tilde{z}_\ell}^2 (\tilde{z}_p - \tilde{z}_p^0)$$

$$= \frac{1}{2s_{\tilde{z}_\ell}^2} \sum_{p=1}^{2} ((\tilde{z}_p - \tilde{z}_p^0)^T (\tilde{z}_p - \tilde{z}_p^0)) \quad (8.25)$$

This can be compared to Eq. (53) of Christensen and Cooley (2005, p. 21). $\hat{M}_{\min}$ would equal $\hat{B}_L$ if z were linear. In this circumstance, the last term in the parentheses of Eq. (8.20) is zero. Thus,

$$\hat{B}_L = \frac{1}{2s^2} \sum_{p=1}^{2} (\tilde{y}_p - \tilde{y}_p^o - X\varphi_p^0)^T \omega_{ij} (\tilde{y}_p - \tilde{y}_p^o - X\varphi_p^0) \quad (8.26)$$

This can be compared with Eq. (54) of Christensen and Cooley (2005, p. 21).

One value of $\hat{B}_U$ and $\hat{B}_L$ is calculated for each parameter. Two values are considered: $\widehat{M}_{\min} + 2\hat{B}_U$ and $|\widehat{M}_{\min} - 2\hat{B}_L|$. If the largest of these values is less than 0.09, the correction factors used to determine the confidence intervals are valid. For individual confidence intervals on predictions, the correction factor is 1.0, so that satisfying the linearity requirements means that the individual confidence intervals of Section 8.4 are not adversely affected by nonlinearity from the perspective of correction factors.

8.8 EXERCISES

As discussed in Chapter 2, Section 2.2, water-supply wells are being completed and a landfill has been proposed for the groundwater flow system considered in the exercises of this book. The wells are located in the center of the area, at row 9, column 10; one well would pump from model layer 1, one would pump from layer 2, and each is expected to pump, on average, 1.1 m^3/s. The proposed site for the landfill is near the center of row 2, column 16 (Figure 2.1a,b).

The landfill developers claim that if the landfill liner leaks, effluent from the landfill would flow toward the river, not toward the supply wells. Also, the landfill developers (who are knowledgeable about regression) argue that it is inappropriate to use this model to evaluate potential advective travel from the landfill because it is calibrated using hydraulic-head and streamflow-gain observations (no transport observations), and they claim that the need for prior information indicates clearly that the data used to calibrate the model are insufficient. They claim, therefore, that there is no reason to believe the model predictions related to transport. In Exercises 8.1 and 8.2, we evaluate these claims.

The county wants to use the steady-state model to evaluate the potential transport and the likely utility of additional data in addressing the complaints of the developer. It would be possible to collect hydraulic-head and streamflow-interaction observations under pumping conditions and to use these data to further calibrate the model, but obtaining these model results would require substantial delay of the landfill permitting process. County officials would like to know whether the additional information is likely to be important enough to warrant such a delay. As an initial evaluation, all parties agree to use the steady-state model to evaluate two potential observations using simulated steady-state pumping conditions: (1) the streamflow gain or loss and (2) hydraulic head in one location far from the river (row 9, column 18).

While a thorough analysis of the potential transport requires an advective–dispersive transport model, a preliminary analysis can be conducted by simulating advective transport alone, using the Advective-Transport Observation Package (ADV) (Anderman and Hill, 2001) of MODFLOW-2000. If the advective-travel path goes to the well, advective–dispersive modeling is not necessary. If the advective path does not go to the well, it is still possible that landfill effluent will reach the well by dispersive transport, and an analysis using a model that includes dispersive processes will be needed.

The analysis of advective transport from the potential landfill site will address the following questions:

Question 1. When the supply wells are pumped, where does an advective path from the landfill travel? Does it go to the well or the river? If it goes to the well, how long does it take to get there?

Question 2. What parameter values are most important to the predicted advective transport and how does this compare to the information provided for their estimation by the observation data?

Question 3. Of the existing head and flow observations used for calibration of the steady-state model, which are most important to the advective-transport predictions?

Question 4. Is collection of the streamflow and hydraulic-head data under pumping conditions likely to contribute additional information that is important to the simulation of advective transport, and if so, which of these potential observations would contribute the most information?

Question 5. What is the uncertainty with which the predictions are simulated using the steady-state calibrated model?

Question 1 is addressed in Exercise 8.1a with a forward model run that includes calculation of advective travel from the landfill site using the ADV Package.

Question 2 is partly addressed in Exercise 8.1b using two comparisons. First, prediction scaled sensitivities (pss) are compared with composite scaled sensitivities (css) of Exercise 5.2c. Second, parameter correlation coefficients (pcc) calculated with observations and predictions are compared to those calculated with only observations. These calculated pcc each have prior information omitted, to address the developer's concern about the use of prior information. Question 2 is further addressed in Exercise 8.1c using the parameter–prediction (ppr) statistic.

8.8 EXERCISES

Question 3 is addressed in Exercise 8.1d using the observation–prediction (*opr*) statistic for evaluating existing observations.

Question 4 is first addressed in Exercise 8.1e by considering dimensionless scaled sensitivities (*dss*) for the possible new observations and *pcc* calculated including the new observations. It is further addressed in Exercise 8.1f by using the observation–prediction (*opr*) statistic.

Question 5 is addressed in Exercise 8.2 using confidence intervals.

Exercise 8.1: Predict Advective Transport and Perform Sensitivity Analysis This exercise addresses Questions 1–4. Parts of this exercise involve simulations using either MODFLOW-2000 or UCODE_2005, and in Exercises 8.1c,d and 8.1f calculations are performed using the computer program OPR-PPR (Tonkin et al., in press). For students performing the simulations and calculations, instructions are available from the web site for this book listed in Chapter 1, Section 1.1.

(*a*) *Predict advective transport.*

This exercise addresses Question 1 using a forward MODFLOW-2000 run with the ADV Package and with steady-state pumping imposed to predict the advective-transport path originating at the proposed landfill location.

The ADV Package uses particle-tracking methods comparable to those of Pollock (1994) to determine advective-transport paths. To compute a particle path, total particle movement is decomposed into displacements in the three spatial grid dimensions, resulting in three advective-transport predictions at every location of interest along a path. An additional system property, effective porosity, is needed to simulate advective transport. Effective porosity does not affect the path trajectory but does affect the particle travel time.

In the model used for the exercises, the spatial grid dimensions are the x, y, and z directions. Predictions are defined for 10, 50, and 100 years of advective transport. Thus, the ADV Package calculates nine advective-transport predictions: the transport distances in the x, y, and z directions at 10, 50, and 100 years. For this run an observation also is defined for 200 years so that the full path is simulated. Two effective porosity parameters are defined. POR_1&2 is the porosity of the aquifers (layers 1 and 2 of the model) and POR_CB is the porosity of the confining bed. The values of POR_1&2 and POR_CB are set to 0.33 and 0.10, respectively.

Output from the MODFLOW-2000 simulation describing the movement of a particle originating at the proposed landfill location is shown in Figure 8.7.

Problem: Use the information about the particle path to answer all parts of Question 1. Plot the particle path on the model grid shown in Figure 8.7b. The X-position is measured along model grid rows: $X = 0$ at the left and increases to the right. The Y-position is measured along model grid columns: $Y = 0$ at the northern boundary and increases toward the south. The Z-position is measured vertically. In this application, $Z = -10$ at the bottom of layer 2 and increases in the upward direction. (MODFLOW users will note that in the ADV output the Z-axis definition is opposite to that used in MODFLOW, in which the model layer numbers increase

(a) ADVECTIVE-TRANSPORT OBSERVATION NUMBER 1
    PARTICLE TRACKING LOCATIONS AND TIMES:

LAYER	ROW	COL	X-POSITION	Y-POSITION	Z-POSITION	TIME
1	2	16	15500.	1500.0	100.00	0.0000
OBS #	12-	14	OBS NAME:	AD10x		
1	2	16	15156.	1609.3	89.366	0.31500E+09
1	2	15	15000.	1657.2	85.481	0.44658E+09
1	3	15	14085.	2000.0	69.953	0.11164E+10
1	3	14	14000.	2028.4	69.024	0.11668E+10
OBS#	15-	17	OBS NAME:	AD50x		
1	3	14	13269.	2341.2	62.686	0.15700E+10
1	3	13	13000.	2457.4	60.867	0.17072E+10
1	4	13	12076.	3000.0	56.119	0.21508E+10
1	4	12	12000.	3041.5	55.844	0.21813E+10
1	4	11	11000.	3817.0	52.850	0.25811E+10
1	5	11	10834.	4000.0	52.431	0.26481E+10
1	6	11	10022.	5000.0	50.679	0.29476E+10
1	6	10	10000.	5028.3	50.627	0.29548E+10
2	6	10	9804.5	5363.8	50.000	0.30232E+10
PARTICLE ENTERING CONFINING UNIT						
OBS #	18-	20	OBS NAME:	A100x		
2	6	10	9804.5	5363.8	46.239	0.31500E+10
2	6	10	9804.5	5363.8	40.000	0.33604E+10
PARTICLE EXITING CONFINING UNIT						
2	7	10	9552.7	6000.0	35.216	0.39200E+10
2	8	10	9375.4	7000.0	22.891	0.44052E+10
2	9	10	9379.0	8000.0	8.4270	0.45677E+10

FIGURE 8.7 (a) Part of MODFLOW-2000 List output file describing the movement of a particle that originates at the top of the cell (row 2, column 16, layer 1) containing the proposed landfill. Predictions AD10, AD50, and A100 are defined for times of 10, 50, and 100 years, respectively. (b) Diagram of the model grid for plotting the particle path (see Figure 2.1 for explanation of symbols).

with depth. The Z-axis is defined in this way so that the particle elevations are more intuitive and consistent with plotting routines.)

(b) *Determine the parameters that are important to the predictions using prediction scaled sensitivities and parameter correlation coefficients.*

This exercise addresses Question 2 by making the two comparisons that were presented in Figure 8.2 and discussed in Section 8.2.4. First, prediction and

8.8 EXERCISES

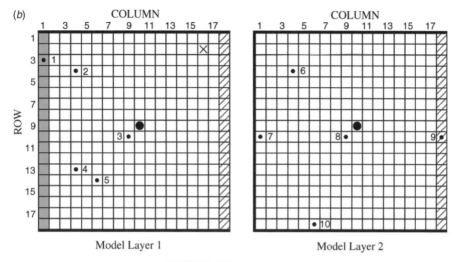

FIGURE 8.7 *Continued.*

composite scaled sensitivities are compared. Second, parameter correlation coefficients calculated using only the calibration observations are compared with those calculated with the addition of the predictions. These *pcc* are both calculated with prior information omitted.

For this problem, calculate *pss* that equal the percent change in the predicted quantity produced by a one-percent change in the parameter value (Eq. (8.2c)). Prediction scaled sensitivities calculated by UCODE_2005 are listed in data-exchange files. For MODFLOW-2000, the *pss* calculated by Eq. (8.2c) are in tables of dimensionless scaled sensitivities if the statistic for calculating prediction weights is set to the predicted value (the transport distance at a given time in a given direction) and STAT-FLAG is specified as 1. Composite scaled sensitivities calculated without the advective-transport prediction are listed in the _sc output file produced by MODFLOW-2000 and UCODE_2005 and are shown in Figure 7.5b of Exercise 7.1a for the optimal parameter estimates.

To obtain *pcc* that include only the calibration observations, the prior information needs to be omitted, which addresses one of the developer's concerns. These correlation coefficients can be calculated by using the optimized parameter values and completing a model run that produces the *pcc* without changing the parameter values.

To calculate *pcc* for the calibration observations plus the predictions, note first that the hydrologic conditions for the calibration are different from those for the predictions. In this model, the difference is the addition of pumpage. Therefore, correctly producing the *pcc* requires simulation of two conditions—one without pumpage and one with pumpage. The hydraulic-head and flow observations used for calibration occur during conditions without pumpage; advective-transport predictions occur during conditions with pumpage. Sensitivities related to

TABLE 8.3 Standard Deviations (in meters) Used to Calculate Weights[a] for the Advective-Transport Predictions

Direction	Time of Advective Travel		
	10 years	50 years	100 years
X	200	600	1000
Y	200	600	1000
Z	10	15	25

[a] The weights are needed to calculate parameter correlation coefficients and are determined using criterion 1 of Section 8.2.3.

calibration observations are calculated without pumpage, and those related to the predictions are calculated with pumpage. Both sets of sensitivities are used to compute the *pcc*.

As discussed in Section 8.2.3, the weighting specified for a prediction affects the calculated *pcc* (Eq. (8.3)). The standard deviations used to calculate the weights for the advective travel at specified times in the three coordinate directions are listed in Table 8.3.

The *pss* are plotted together with the *css* in Figure 8.8. In this figure, the scales are different for the *css* and *pss*, but this is not problematic because the analysis involves evaluating the relative values of each measure. The *pss* for effective porosity also are included in Figure 8.8. These parameters are not relevant to the model calibration with head and flow observations, but could be important in the calculation of advective transport times. The *pss* for effective porosity are computed with UCODE_2005 using central-difference perturbation. The *pcc* without and with predictions are shown in Tables 8.4 and 8.5, respectively.

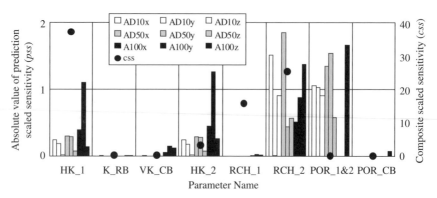

FIGURE 8.8 Composite and prediction scaled sensitivities for the calibrated parameters and for effective porosity. The *pss* for the *x*, *y*, and *z* grid directions are shown as the left, middle, and right columns, respectively, for each advective-travel time. The *pss* are defined as the percent change in advective travel caused by a one-percent change in the parameter value (Eq. (8.2c)).

8.8 EXERCISES

TABLE 8.4 Parameter Correlation Matrix Using Only the Hydraulic-Head and Flow Observations, with Prior Information Omitted, Using Final Parameter Values, and Calculated by MODFLOW-2000[a]

	HK_1	K_RB	VK_CB	HK_2	RCH_1	RCH_2
HK_1	1.00	−0.40	−0.90	−0.93	**0.96**	−0.90
K_RB		1.00	0.20	0.34	−0.32	0.32
VK_CB			1.00	**0.97**	**−0.97**	**0.97**
HK_2		Symmetric		1.00	**−0.99**	**0.996**
RCH_1					1.00	**−0.98**
RCH_2						1.00

[a] Correlation coefficients greater than 0.95 are in bold type.

TABLE 8.5 Parameter Correlation Matrix Using the Hydraulic-Head and Flow Observations and the Advective-Transport Predictions, with Prior Information Omitted, Using Final Parameter Values, and Calculated by MODFLOW-2000[a]

	HK_1	K_RB	VK_CB	HK_2	RCH_1	RCH_2
HK_1	1.00	−0.15	0.097	−0.15	0.70	0.36
K_RB		1.00	−0.62	−0.009	0.27	−0.090
VK_CB			1.00	0.32	−0.17	0.26
HK_2		Symmetric		1.00	−0.46	0.81
RCH_1					1.00	−0.23
RCH_2						1.00

[a] No correlation coefficients are greater than 0.95.

Problem

- Answer Question 2 above using the *css* and *pss* in Figure 8.8 and the *pcc* without and with predictions in Tables 8.4 and 8.5.
- Why are the *pss* in Figure 8.8 equal to or very close to zero for parameters K_RB, VK_CB, and RCH_1?
- Why is the *pss* for POR_CB equal to zero for all predictions except A100z? Consider the location of the particle at 100 years, shown in Figure 8.7a.
- Which of the effective porosity parameters should be included in analyses of prediction uncertainty conducted to further answer Question 2, and to answer Questions 3–5? Why?

(c) *Determine the parameters that are important to the predictions using the parameter–prediction statistic.*

This exercise addresses Question 2 using the parameter–prediction (*ppr*) statistic, calculated with the omission of existing prior information on parameters K_RB and VK_CB. As discussed in Section 8.2.5, the *ppr* statistic calculates parameter importance to predictions in a manner that includes the effects of parameter uncertainty and correlation as well as prediction sensitivities. The *pss* shown in Figure 8.8 include only the effects of prediction sensitivities.

The program OPR-PPR (Tonkin et al., in press) is used to calculate the *ppr* statistic for individual parameters. For these calculations, a 10-percent reduction in parameter standard deviation is specified. Thus, the *ppr* statistic represents the percent decrease in the standard deviation of a prediction that is produced by a 10-percent decrease in the standard deviation of one parameter. The effective porosity of the aquifer is also included in the *ppr* calculation, and prior information and weighting are used to realistically represent its uncertainty. The sensitivities of the heads and flows to POR_1&2 are zero; thus, its uncertainty would be infinite if prior information were not imposed. The weighting used for the prior information is calculated by forming a 95-percent confidence interval of 0.27 to 0.39 for the true effective porosity value (see Guideline 6 in Chapter 11).

Average *ppr* statistics for all advective-transport predictions are shown in Figure 8.9*a*. Figure 8.9*b* presents *ppr* statistics for predictions at 100 years, and Figure 8.9*c* shows the corresponding decreases in the prediction standard deviation. This intermediate result of the *ppr* statistic calculation is useful because it is important to evaluate whether a large percent reduction in a prediction standard deviation is associated with a very small change in the standard deviation. If so, then it might not be beneficial to use that particular *ppr* result for guiding field data collection, despite the large value of the statistic.

Problem
- Compare the results of Figure 8.9*a,b* with those in Figure 8.8. What are the differences in terms of which parameters rank as most important to predicted advective travel? For future data collection, which parameters would be most beneficial to further investigate according to the *ppr* results?
- Explain the different rankings of parameter importance by the *pss* and *ppr* results. Consider the parameter correlations shown in Table 8.4 in answering this question. Recall that because POR_1&2 is not applicable to the model calibration, its correlation with all other parameters is zero.
- Figure 8.9*c* shows the standard deviation decreases associated with the *ppr* values in Figure 8.9*b*. Are the standard deviation decreases for any of the predictions small enough to suggest that it might not be beneficial to collect additional data aimed at improving that prediction? Use distances traveled derived from Figure 8.7 and system dimensions of Figure 2.1.

OPR-PPR also is used to calculate the *ppr* statistic for all possible groups of two parameters. This analysis is applicable if field data collection will involve simultaneously obtaining information about two parameters. A 10-percent reduction in the standard deviation of each parameter is specified, and so the *ppr* statistic for a parameter pair represents the percent decrease in the prediction standard deviation that is produced by a 10-percent decrease in the standard deviation of each parameter in a group. The results for the advective-transport predictions at 100 years are shown in Figure 8.9*d*. The *ppr* statistics are similar for all pairs that include K_RB or POR_1&2 and are similar for all other pairs, so average values are shown for each of these groups of parameter pairs.

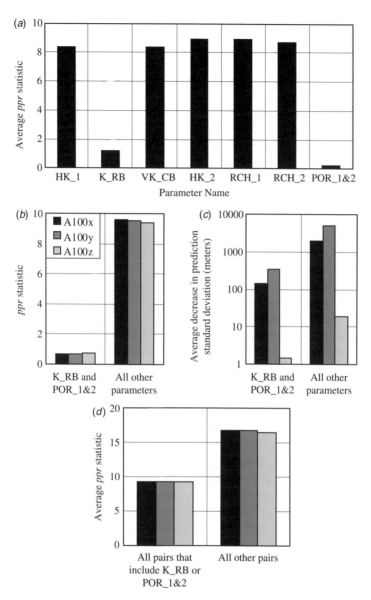

FIGURE 8.9 Parameter–prediction (*ppr*) statistic and intermediate calculations for evaluating the importance of model parameters to predicted advective transport. The statistic for each prediction is computed as the percent decrease in prediction standard deviation produced by a 10-percent reduction in the standard deviation of a parameter. (*a*) For each parameter, average *ppr* statistic for all predictions. (*b*) The *ppr* statistics for predicted transport in the *x*, *y*, and *z* directions at 100 years. The statistics for K_RB and POR_1&2 are similar, as are the statistics for all other parameters. Thus, the average value for each of these groups is shown. (*c*) Decreases in prediction standard deviation ($s'_{z_\ell} - s_{z_\ell(j)}$ in Eq. (8.8)) corresponding to the *ppr* results in (*b*). (*d*) Average *ppr* statistics for evaluating the importance of pairs of parameters to predicted advective transport at 100 years.

Problem: Which parameter pairs would be most beneficial to simultaneously investigate, according to the *ppr* results shown in Figure 8.9*d*?

(*d*) *Assess the importance of existing observations to the predictions using the observation–prediction (opr) statistic.*

This exercise addresses Question 3, by evaluating the relative contribution that the existing head and flow observations make toward reliably simulating the advective-transport predictions. This analysis is useful after initial model calibration to guide further field investigation of existing observations that rank as most important to the predictions with the goal of ensuring that their representation in the model is as accurate as possible. For example, for the most important head observations, field work might involve more accurately measuring the screened interval depth and areal location of the corresponding monitoring wells. This information then could be used to update the observation location in the model.

The program OPR-PPR (Tonkin et al., in press) is used to calculate the *opr* statistic for omitting individual observations. The *opr* statistic values equal the percent increase in the standard deviation of a prediction that is produced by omitting one observation. This analysis includes uncertainty in parameter POR_1&2, as described in Exercise 8.1c.

The results for the advective-transport predictions at 100 years are displayed in Figure 8.10*a*, which shows the *opr* statistics, and Figure 8.10*b*, which shows the corresponding increases in the prediction standard deviation. As discussed for the *ppr* statistic in Exercise 8.1c, this intermediate result of the *opr* statistic calculation is useful because it is important to evaluate whether a large percent increase in a prediction standard deviation is associated with a small increase in the standard deviation.

Problem
- Using the *opr* results presented in Figure 8.10*a*, identify the observations that rank as most important to the predictions.
- Observations can rank as important by the *opr* statistic if they are sensitive to parameters to which the predictions are sensitive, or if they are sensitive to parameters that are correlated with parameters to which the predictions are sensitive. Examine the dimensionless scaled sensitivities for the observations shown in Table 7.5 and the prediction scaled sensitivities shown in Figure 8.8. Do these sensitivities help explain the importance of observations head01.ss and flow01.ss?
- Observations also can rank as important if their removal substantially increases parameter correlations, because prediction uncertainty tends to increase as these correlations increase. In the base case for the *opr* statistic calculations, prior information on K_RB and VK_CB is omitted, and there are several parameter correlations that are very large in absolute value, as shown in Table 8.4. Table 8.6 summarizes the increases in these parameter correlations that occur when individual head observations are omitted in the *opr* calculations. Use the information in this table, along with knowledge of how omission of the

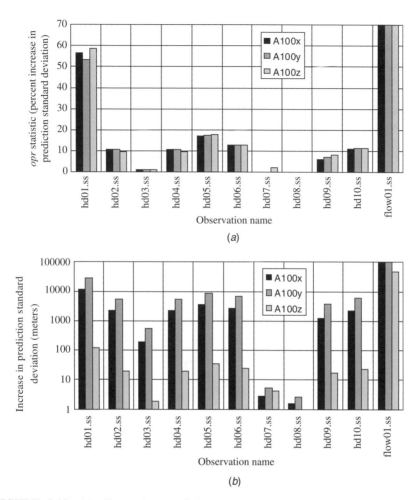

FIGURE 8.10 (*a*) Observation–prediction (*opr*) statistic calculated to evaluate the importance of existing head and flow observations to predicted advective transport in the *x*, *y*, and *z* directions at 100 years. The statistic is computed as the percent increase in prediction standard deviation produced by omitting an observation individually. The *opr* statistics for observation flow01.ss and predictions A100x and A100z equal 280,000 and 5404, respectively. (*b*) Increase in prediction standard deviation ($s_{z'_\ell} - s_{z'_\ell(\pm i)}$ in Eq. (8.11)) produced by omitting an observation. Increase in standard deviation for observation flow01.ss and prediction A100x equals 6×10^6 m. Note that this figure uses a logarithmic scale on the vertical axis.

flow observation affects parameter correlations (discussed in Exercises 4.1c, 5.1a, and 7.1f), to help explain the importance of observations head01.ss and flow01.ss by the *opr* analysis.

- Figure 8.10*b* shows the standard deviation increase associated with each *opr* statistic value in Figure 8.10*a*. Are the standard deviation increases for any

TABLE 8.6 Summary of Increases in Parameter Correlation Coefficients (*pcc*) Caused by Omitting Individual Observations[a]

Observation Name	Maximum Percent Increase in Any *pcc*	Number of *pcc* that Increase by More Than 1 percent
hd01.ss	5.8	3
hd02.ss	2.1	3
hd03.ss	0.0	0
hd04.ss	2.1	3
hd05.ss	2.6	3
hd06.ss	2.3	3
hd07.ss	0.0	0
hd08.ss	0.0	0
hd09.ss	0.5	0
hd10.ss	1.7	3
flow01.ss	7.5	6

[a] The percent increases and number of *pcc* shown in columns 1 and 2 of the table all occur for *pcc* that are greater than 0.90 in the base case calculation, in which no observations are omitted.

of the predictions small enough to suggest that further investigating observations for purposes of improving a particular prediction might not be warranted, despite relatively large *opr* statistics for the prediction?

(e) *Assess the likely importance of potential new observations to the predictions using dimensionless and composite scaled sensitivities and parameter correlation coefficients.*

This exercise addresses Question 4, by evaluating the likely importance to the predictions of potential new observations that would be collected under pumping conditions. The potential new observations include a hydraulic head in layer 1, row 9, column 18, and a streamflow gain or loss over all river cells in column 1 of the flow model. The potential new hydraulic-head observation location was chosen because the county has access to this property and county modelers believe that a location far from the river would provide substantial information about the model parameters.

The two potential observations have different units of measurement. Their relative importance can be evaluated using dimensionless scaled sensitivities, which are scaled using weights. As discussed in Chapter 3, Section 3.3.3, the statistics commonly used to determine the weights are variances, standard deviations, and coefficients of variation that reflect observation error. For potential measurements, it is preferable to use variances or standard deviations. Coefficients of variation can only be used to calculate the weights if a reasonable guess is specified for the anticipated observed value, because when coefficients of variation are used the weight is calculated as $\omega_{ii} = 1/[cv_i \times y]^2$, where y is the specified observed value. For this problem the variance of the potential head observation error is specified as 1.0025,

8.8 EXERCISES

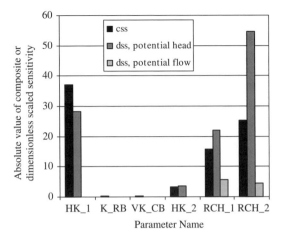

FIGURE 8.11 Composite scaled sensitivities (*css*) for the observations used in model calibration and dimensionless scaled sensitivities (*dss*) for two potential new observations. All sensitivities are calculated using the parameter values estimated by regression.

to be consistent with the variances of head observation errors used in the calibration. The standard deviation of measurement error for the flow is set to 0.44 m^3/s.

Parameter correlation coefficients with and without the potential new observations also need to be included in the analysis. They are used to evaluate if adding potential observations reduces the values of any problematic correlations, as discussed in Section 8.3.1.

The *dss* associated with the potential head and flow observations are plotted in Figure 8.11. Parameter correlation coefficient matrices with the potential observations are shown in Table 8.7.

Problem: Answer Question 4 by comparing the *dss* to the *css* for the existing observations and by comparing the *pcc* for this exercise with those evaluated in Exercise 8.1b.

(*f*) *Assess the likely importance of potential new observations to the predictions using the observation–prediction (opr) statistic.*

This exercise addresses Question 4, by using the *opr* statistic to calculate the decrease in prediction uncertainty caused by adding potential new observations collected under pumping conditions. First, *opr* statistics are calculated for the potential head and flow data described in Exercise 8.1e. Second, the *opr* statistic is calculated for the case of individually adding a new head observation in each cell of the model domain. This latter analysis identifies all areas of the domain that would be good candidates for new head observation locations, in terms of improving the advective-transport predictions. Both of these analyses can be completed using the program OPR-PPR (Tonkin et al., in press). Both analyses include uncertainty in parameter POR_1&2, as described in Exercise 8.1c.

TABLE 8.7 Parameter Correlation Matrices for Final Parameter Values Calculated by MODFLOW-2000 Using the Existing Hydraulic-Head and Flow Observations Under Conditions of No Pumping Together with, Under Pumping Conditions, (a) Only the Potential Flow Observation, (b) Only the Potential Hydraulic-Head Observation, and (c) Both Potential Observations[a]

	HK_1	K_RB	VK_CB	HK_2	RCH_1	RCH_2
	(a) Only the Potential Flow Observation					
HK_1	1.00	−0.42	−0.93	**−0.96**	**0.97**	−0.94
K_RB		1.00	0.20	0.34	−0.32	0.32
VK_CB			1.00	**0.97**	**−0.97**	**0.97**
HK_2		Symmetric		1.00	**−0.996**	**0.998**
RCH_1					1.00	**−0.99**
RCH_2						1.00
	(b) Only the Potential Hydraulic-Head Observation					
HK_1	1.00	−0.27	−0.56	**−0.98**	**0.96**	−0.93
K_RB		1.00	−0.38	0.18	−0.062	0.093
VK_CB			1.00	0.64	−0.61	0.63
HK_2		Symmetric		1.00	**−0.95**	**0.97**
RCH_1					1.00	**−0.97**
RCH_2						1.00
	(c) Both Potential Observations					
HK_1	1.00	−0.33	−0.43	**−0.96**	0.93	−0.89
K_RB		1.00	−0.42	0.21	−0.060	0.10
VK_CB			1.00	0.54	−0.50	0.53
HK_2		Symmetric		1.00	−0.91	**0.95**
RCH_1					1.00	**−0.95**
RCH_2						1.00

[a] Values greater than 0.95 are in bold type.

Results of the first analysis are presented in Figure 8.12, which shows the *opr* statistic calculated for each potential observation and each of the nine advective-transport predictions.

Problem
- Answer Question 3 by evaluating the *opr* statistics shown in Figure 8.12.
- Why does the potential head observation have larger values of the *opr* statistic for all predictions than does the potential flow observation? Consider the parameter correlation coefficients shown in Table 8.7.
- Why is the potential flow observation relatively unimportant to the predictions, in contrast to the very large importance of the existing flow observation as shown in Figure 8.10a.

8.8 EXERCISES

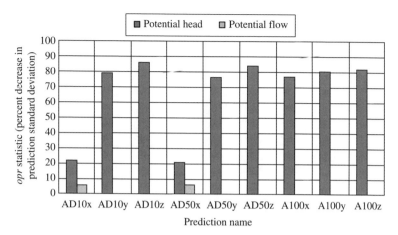

FIGURE 8.12 Observation–prediction (*opr*) statistic calculated to evaluate the importance of the potential head and flow observations to predicted advective transport in the *x*, *y*, and *z* directions at 10, 50, and 100 years. The statistic is computed as the percent decrease in prediction standard deviation produced by adding either the head or the flow observation.

Results of the second analysis are presented in Figure 8.13. To understand these results, examine the reduction in parameter correlation coefficients (*pcc*) caused by adding a potential hydraulic-head observation at any cell throughout the model domain. The reductions in *pcc* are summarized in Figure 8.14, which shows the maximum percent reduction in any *pcc* when a hydraulic-head observation is added at a particular cell center.

Problem
- Examine Figure 8.13 and identify the best locations for collecting additional hydraulic-head data, from the perspective of reducing the uncertainty in predicted advective transport at 100 years.
- How do the percent reductions in parameter correlations shown in Figure 8.14 help explain the results in Figure 8.13? In Figure 8.14, the maximum percent reduction in correlation is almost always associated with parameter VK_CB. Use your knowledge about the flow system to help explain why adding a hydraulic head most reduces the correlations for this parameter.

Exercise 8.2: Prediction Uncertainty Measured Using Inferential Statistics This exercise involves computing both linear and nonlinear confidence intervals on the predicted advective transport at 10, 50, and 100 years. Uncertainty in parameter POR_1&2 is included in these analyses, through prior information and weighting, as explained in Exercise 8.1c.

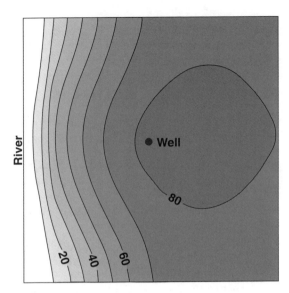

FIGURE 8.13 Observation-prediction (*opr*) statistics (Eq. (8.9)), showing the percent decrease in prediction standard deviation likely to be produced by collecting one new hydraulic-head observation under pumping conditions. New observations are located at each node in model layer 1. For each node, the new head is added to the existing 11 observations and the *opr* statistic is calculated. This procedure is repeated for all nodes and the resulting values are contoured. The contours range from 10 to 80 percent. The *opr* values plotted are averaged over the three advective transport directions for which the ADV Package produces results.

(*a*) *Calculate linear confidence intervals on the components of advective transport.*

This exercise addresses Question 5, using linear confidence intervals on the simulated advective-transport predictions to quantify their uncertainty. For students performing the simulations, instructions for calculating the intervals are available from the web site for this book listed in Chapter 1, Section 1.1.

Linear, individual 95-percent confidence intervals and linear simultaneous (Bonferroni) 95-percent confidence intervals in the x and y directions are shown on a map of model layer 1 in Figure 8.15*a,b*. Bonferroni simultaneous intervals are used instead of Scheffé simultaneous intervals because, for the finite number of intervals of interest here, the Bonferroni intervals are smaller (see discussion in Section 8.4.1). The intervals in the z direction are shown in Figure 8.16.

Problem
- Explain conceptually why the linear simultaneous intervals are larger than the linear individual intervals. Of these two linear confidence intervals, which might be the preferred representation of uncertainty? Why?
- Answer Question 4 above using the linear confidence intervals on the advective-transport predictions.

8.8 EXERCISES

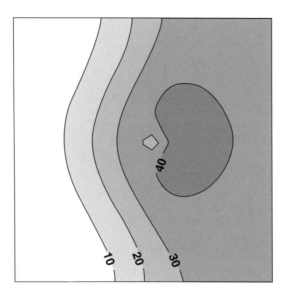

FIGURE 8.14 Maximum percent reduction in any parameter correlation coefficient (*pcc*), produced by adding one new hydraulic-head observation under pumping conditions. New observations are located at each node in model layer 1. For each node, the new head is added to the existing 11 observations and the *pcc* is calculated, and the percent decrease from the base case *pcc* (with only the existing 11 observations) is calculated. This procedure is repeated for all nodes and the resulting values are contoured. The contours range from 10 to 40 percent. The maximum percent reduction for each node is determined using only base case *pcc* that are greater than 0.90.

- Do the confidence intervals make sense? Is there anything surprising about them? Note that because they are linear, these intervals sometimes do not account for the physics of the problem. They can include values that are physically implausible, such as predicted advective-transport values that lie outside the model domain. This is typical of linear intervals.

(b) *Calculate nonlinear confidence intervals on the components of advective transport.*

This exercise revisits Question 5, using nonlinear individual and simultaneous (Scheffé $d = NP$) confidence intervals on the components of simulated advective transport. The nonlinear 95-percent confidence intervals in the x and y directions are shown on a map of model layer 1 in Figure 8.15c,d. The intervals in the z direction are shown in Figure 8.16. The nonlinear intervals were calculated using UCODE_2005.

In Figure 8.15c and 8.15d, there are dashed lines on the confidence intervals calculated for the particle position at 100 years. The interval limits involved are simulated using parameter values that cause the advective transport path to reach the well

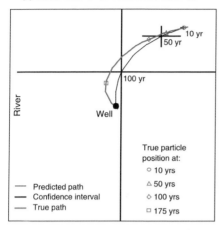
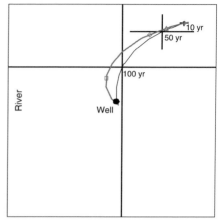
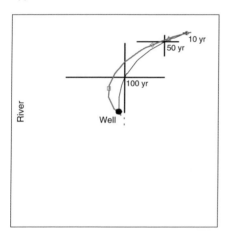
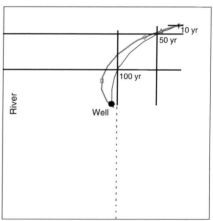

FIGURE 8.15 Plan view of the model grid showing the predicted advective-transport path from the proposed landfill and 95-percent confidence intervals in the x and y directions at simulated travel times of 10, 50, and 100 years (locations labeled on each map). The true path also is shown. (*a*) Linear individual confidence intervals. (*b*) Linear simultaneous (Bonferroni) confidence intervals. (*c*) Nonlinear individual confidence intervals. (*d*) Nonlinear simultaneous (Scheffé $d = NP$) intervals. (*a*) and (*b*) At 100 years, the linear intervals in the x and y directions extend outside the model domain. In (*a*), the upper limit in the x direction is 18,930 m and the limits in the y direction are $-17,680$ m and 28,390 m. In (*b*), the limits in the x direction are $-5,430$ m and 25,040 m and those in the y direction are $-33,110$ m and 43,820 m. (*c*) and (*d*) The dashed lines at 100 years reflect projections simulated by the ADV Package when a particle exits the model. See discussion in text about this and about the intervals in (*d*) at 50 and 100 years that extend to the river.

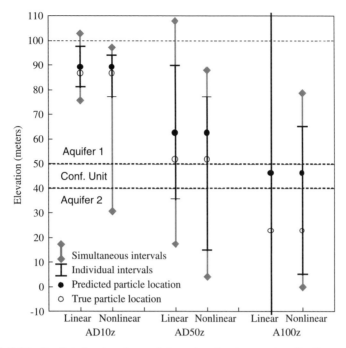

FIGURE 8.16 Predicted z locations of the advective-transport path, linear individual and simultaneous (Bonferroni) 95-percent confidence intervals, and nonlinear individual and simultaneous 95-percent confidence intervals at simulated travel times of 10, 50, and 100 years. True z locations also are shown. The bottom of the model domain lies at an elevation of -10 m. For prediction A100z, the limits of the linear individual intervals are -66 m and 152 m. The limits of the linear simultaneous intervals are -141 m and 234 m.

prior to 100 years. When a particle reaches a flow model boundary prior to the prediction time, the ADV Package projects the particle until the prediction time is reached, using the particle velocity when the particle exits the model (Anderman and Hill, 2001, p. 12). As they explain, this procedure is very useful when estimating parameter values because it makes sensitivities remain informative. In the model considered here, the large velocity at the pumping well causes the particle to be projected a considerable distance in the y direction after entering the well, and the resulting interval limit is not meaningful. A better approximation of the confidence interval limit in that direction is the location where the particle left the system. Here, that location is the well, as indicated by the solid lines in Figure 8.15c,d.

Other examples of projected particles are not apparent in Figure 8.15d, but affect the output files used to construct the figure. In Figure 8.15d, the confidence intervals in the x direction at 50 and 100 years extend to the river. These limits are simulated using parameter values that cause the advective transport path to reach the river prior to the prediction time of 50 or 100 years. The ADV Package projects the particle until the prediction time is reached. The interval limit printed in the output file falls outside the model domain, which is not meaningful. In this case, the appropriate limit at both 50 and 100 years is the location of the river, as shown in Figure 8.15d. Dashed lines are not needed because interval limits outside the model domain are not plotted.

Problem: Reevaluate the answer to Question 4 considering the nonlinear confidence intervals on the components of advective transport. Is the answer based on the nonlinear intervals different from that based on linear intervals?

9

CALIBRATING TRANSIENT AND TRANSPORT MODELS AND RECALIBRATING EXISTING MODELS

The methods presented in Chapters 3 to 8 are applicable to models of any system. However, there are special considerations when applying the methods to certain types of models. This chapter discusses three types of models that are of special interest to many scientific and engineering fields: transient models, transport models, and existing models that may need to be recalibrated.

9.1 STRATEGIES FOR CALIBRATING TRANSIENT MODELS

In many natural and engineered systems, conditions change with time. When simulating these transient systems, it is important to carefully consider (1) initial conditions, (2) representation and weighting of transient observations used for model calibration, and (3) definitions of model inputs and parameters added for the transient simulation.

9.1.1 Initial Conditions

For transient models, initial conditions define the system state at the beginning of the simulation. In some situations, such as for atmospheric systems, conditions change very quickly and there is no choice but to begin the simulation with initial conditions derived directly from measurements. In this case, the specified initial state of the system generally is inconsistent to some degree with the conservation equations

Effective Groundwater Model Calibration: With Analysis of Data, Sensitivities, Predictions, and Uncertainty. By Mary C. Hill and Claire R. Tiedeman
Published 2007 by John Wiley & Sons, Inc.

and properties of the model. That is, the initial simulated processes and properties result in a simulated system state that differs from the initially specified state. When the simulation is started, the simulated state changes from the initially specified state to become consistent with the simulated initial processes and properties and ongoing stresses. For a system in which conditions change rapidly, the inconsistencies between the initially specified state and the simulated initial processes and properties generally are not problematic because the ongoing stresses soon dominate the solution. As a result, comparing observed and simulated values becomes meaningful after a relatively short simulated time.

Other systems, like most groundwater flow systems, change slowly with time. For these systems it is preferable, and often possible, to begin the transient simulation with initial conditions produced by a steady-state simulation. This ensures initial conditions that are consistent with all simulated processes and properties except for the imposed transient stresses (Franke et al., 1987). This consistency is important, because inconsistencies in initial conditions can endure for long periods of simulated time in slowly changing systems. From the perspective of model calibration, enduring inconsistencies can affect model fit and, therefore, estimated parameter values and model design.

If initial conditions for slowly changing systems are derived directly from measurements, it is critical to evaluate how any enduring inconsistencies affect the model, the estimated parameters, and predictions of interest. Often, the effect of imposed initial conditions can be evaluated using a simulation in which transient stresses and processes are omitted from the simulation. For example, in a groundwater flow simulation, if the transient stress is pumpage and the predictions of interest are changes in head (drawdowns), it is important to simulate the system without any transient changes in pumpage to determine the head changes simulated as the model adapts to the imposed initial conditions. If the head changes are large relative to the head changes simulated with pumpage, comparisons of observed and simulated drawdowns can be meaningless, and attempts to calibrate the model using such comparisons can lead to a badly flawed model. Finally, for slowly changing systems it is important to reevaluate the effects of imposed initial conditions as regression proceeds and model construction and parameter values change.

9.1.2 Transient Observations

Four issues important to using transient observations are discussed in this book: the effects of imposed initial conditions, inconsistencies in observed and simulated temporal effects, weighting of transient observations, and temporal differencing. The first issue was discussed in Section 9.1.1. The other three are introduced here and are also addressed in Guidelines 4 and 6 of Chapter 11 and the field examples presented in Chapter 15.

Observed and Simulated Temporal Effects When temporally varying observations are used to calibrate models of transient systems, it is important to consider whether the transient effects contained in the observations are consistent with the

9.1 STRATEGIES FOR CALIBRATING TRANSIENT MODELS 215

simulated transient processes. For example, consider water-level data in wells collected on a daily or weekly basis by an automated recording device being used in a model for which simulated transient effects are limited to pumpage and recharge that vary seasonally. The data are likely to be affected by hydrologic processes that are not simulated in the model, such as water-table fluctuations due to short-term hydrologic events like storms or daily variations in pumping rates. Using the numerous high-frequency data as observations can require substantial data preparation effort and make it difficult to evaluate the aspects of model fit most important to the designed purpose of the model. It can also result in (a) optimized parameter estimates that are unreasonable because they are making up for unrepresented processes, (b) poor fit to observations and simulated values that are often consistently above or below the observations, and/or (c) problems with convergence of the parameter-estimation iterations.

Using observations that are consistent with the simulated processes eliminates the difficulties. We call these time-consistent observations. While it can be difficult to obtain completely time-consistent observations, substantial understanding of the data and the model can be derived from trying and generally sufficient consistency is achievable. In the example above, for which the data are far more frequent than the simulated processes, time-consistent observations can be obtained by temporally averaging observations or by selectively using observations that are thought to be representative for each season. Obviously, clearly describing the criteria used to obtain the observations from the data is critical to the integrity of any model.

If the data are affected by unrepresented, high-frequency transient processes but the data are infrequently collected, other possibilities need to be considered. In some cases, the time of year or time of day can be used to determine if the observation is likely to be different from the desired time-average quantity, and time-consistent observations can be obtained by making appropriate adjustments. More frequent data in nearby locations or similar circumstances sometimes can be used to support adjustments. If there is no way to know whether a value is too high or too low, this implies that no bias can readily be identified and the data can be used directly as an observation. In all cases, the weighting of the observations needs to reflect the analysis of errors conducted, as discussed in the next section.

A common situation is that data affected by transient processes are used to calibrate a steady-state model. This situation is discussed in Chapter 15, Section 15.2.1.

An example of using transient data to define the time discretization of simulated processes to obtain time-consistent observations is presented by Bravo et al. (2002). They used this approach to calibrate a model of steady-state groundwater flow and transient heat transport in a wetland system. Frequency domain analysis of water-level and temperature data revealed time periods when heat transport showed transient variation yet groundwater flow could be assumed to be at steady state. Calibration of the model over these time periods using the water-level and temperature observations allowed estimation of average rates of flow from the wetland to the groundwater system.

Weighting Transient Observations When determining weights for temporal observations, it is important to assess whether observation errors are correlated or are independent over time. Observation errors can be classified as presented in Appendix A, Section A.3 in the fifth assumption required for diagonal weights to be correct. Briefly, the classes are (1) constant over time (perfectly correlated), (2) correlated over time (autocorrelated), and (3) random over time (uncorrelated). For measurements of hydraulic head over time in a single monitoring well, examples of these types of error are as follows. Well elevation measurement error is constant in time, because the elevation is typically measured once and used to calculate hydraulic head whenever water levels in the well are measured. Temporally correlated errors include the transient drift of the reference point of a pressure transducer and the lag due to mechanical friction of float-type measuring device (Rosenberry, 1990). Random errors can be generated by procedures and devices used to measure the depth to water in a well.

Determining each of these types of error for a given observation requires careful consideration of measurements that contribute to each observation and the errors associated with each measurement. As discussed in Guideline 6 in Chapter 11, variances of these individual measurement errors can be added to obtain the total variance for the observation.

Temporal Differencing Differencing is a useful method for addressing errors that are constant over time. In groundwater modeling, differencing is commonly used because often the drawdown of hydraulic head caused by pumping is the observation of greatest interest. Drawdown equals the difference in hydraulic heads measured at two different times. In other scientific fields the system stress may be different, but often the change caused by some stress is of interest. In these circumstances it is desired not only to match the measured values, but also to match the changes in these values. Differences can be used to emphasize these changes and to achieve a simpler diagonal weight matrix. Differencing can produce a much more effective set of calibration observations in some circumstances. See Assumption 5 in Appendix A for more information.

9.1.3 Additional Model Inputs

A final consequence of a transient model is that, in most systems, the transition from steady state to transient simulation requires additional model inputs. In groundwater systems, the primary additional inputs include (1) storage coefficients; (2) stresses that vary over time, such as recharge and pumping rates; and (3) boundary conditions that vary over time, such as lake or river levels. These properties can be parameterized and estimated in the same manner as for system properties that are invariant in time, as suggested by the methods and guidelines presented in this book.

As for time-invariant model inputs, a common calibration issue for transient properties is that the potential number of parameters generally is large. For example, for a transient groundwater flow model with a total simulation time of 10 years and with recharge from precipitation that varies on a monthly basis, there are potentially

120 different recharge parameters even before spatial variability is considered. It is not likely that the observation data available for calibration would support independent estimation of all of these parameters. One strategy is to estimate an initial distribution of recharge over time, using site data on climate, geology, vegetation, and so on, and then define one or more multiplicative parameters that scale this initial distribution of recharge (e.g., Tiedeman and Gorelick, 1993). The parameters have initial values of 1.0, and regression is used to estimate their optimal values.

In some models of transient systems, only a subset of the simulated stresses vary in time. For example, in a transient model of a regional groundwater system in the arid environment of the Albuquerque Basin, Tiedeman et al. (1998b) simulated historical pumping rates that varied by season and used transient observations for calibration. Recharge and regional groundwater inflow were simulated as temporally constant, representing average annual values over the calibration period. This approach was reasonable because the typically large depth to the water table resulted in small seasonal recharge effects. A similar method was employed by Faunt et al. (2004).

9.2 STRATEGIES FOR CALIBRATING TRANSPORT MODELS

Predicting transport of introduced or naturally occurring constituents is of interest in many systems. Preferably transport observations, such as concentrations or concentrations summarized as advective-travel observations, moment observations, and so on, are available for model calibration. However, sometimes only observations related to the flow system are available for model calibration, and the model is used to predict some aspect of transport. This circumstance is difficult in that predictive capability can be compromised. However, models are still the only way to introduce conservation principles into the analysis of complicated natural environments, and, as discussed in Chapter 1, calibration is an important option in many circumstances. The difficulties involved in simulating concentration with models developed using only flow-system observations make the sensitivity analysis and uncertainty evaluation methods described in this book especially critical to pursue in these circumstances.

When calibrating transport models, it is important to carefully consider (1) what transport processes to include, (2) definition of the source, (3) scale issues, (4) numerical accuracy and execution time, (5) representation and weighting of transport observations used for model calibration, and (6) additional model inputs. These issues are addressed in the following sections. We also give some examples of how to obtain a tractable, useful model.

9.2.1 Selecting Processes to Include

The transport of constituents in natural and engineered systems is the result of many processes, generally including advection, transient changes in advection, dispersion, retardation, and chemical reaction with other dissolved or particulate transported constituents and the surrounding material (e.g., rocks and sediments in groundwater systems). In addition, there are processes such as density and temperature that can result in feedback effects on the flow field. Also, multiple phases such as nonaqueous phase liquids (NAPLs), alternate transport mechanisms such as colloids, and living particles such as viruses can be simulated. Execution times generally increase dramatically as more processes are included. The importance of thinking carefully about execution times is discussed in Chapter 15, Section 15.1.

The options for simulating transport are generally as follows:

Option 1. Use advection to approximate transport travel time and direction.

Option 2. Use advection and dispersion to simulate the arrival of low concentrations at the plume front, the concentration and time at the plume peak, and/or enduring low concentrations at the plume tail.

Option 3. Use advection, dispersion, reactions, and other mechanisms to account for additional processes that can affect arrival times, peak concentration times, and duration of low concentrations.

A model calibration effort need not use just one of the options mentioned. As suggested by Anderman et al. (1996), it can be advantageous to begin with just advective transport (option 1) to obtain the correct transport direction and timing. Progressively more processes can then be developed and tested against the available data and can be evaluated on the basis of the importance of the additional processes to predictions. This progression is consistent with Guideline 1 of Chapter 11.

Decisions about which transport processes to simulate depends on many things (e.g., see Zheng and Bennett, 2002, Chapter 7). Here we are concerned with how observations and parameters can be defined to achieve a tractable model calibration problem and a useful model. Relevant issues are discussed and examples are presented in Sections 9.2.2–9.2.7.

9.2.2 Defining Source Geometry and Concentrations

When simulating the transport of contaminants from a disposal source, the location at which the disposal occurred and the release history of contamination are important to simulated concentrations. To the extent that simulated source characteristics are incorrect, parameter estimates that produce a good match to measured concentrations might be unattainable or, even if a good match is achieved, may be unable to produce good predictions for other circumstances.

In some situations, the source characteristics are well known and can be used directly in the model. Even in these situations, however, difficulties can occur because local flow characteristics associated with source emplacement may not be simulated and/or the extreme concentration gradients at the source cause unrealistic numerical spreading of the plume at early time (e.g., LeBlanc and Celia, 1991; Zhang et al., 1998; Barlebo et al., 2004).

9.2 STRATEGIES FOR CALIBRATING TRANSPORT MODELS

In other situations, the source location and history are poorly known. An alternative to specifying the source characteristics is to simultaneously estimate its location and release history along with model parameter values. Wagner (1992) and Mahar and Datta (2001) present methods for this simultaneous estimation problem. Applications of their methods to synthetic examples show that the approach has promise, but that nonuniqueness can be problematic. Simultaneously estimated, unique values could only be achieved with constraints imposed on the parameter values and the source characteristics. When the source location is well known, but the source history is not, nonuniqueness can be less problematic: Sonnenborg et al. (1996) and Medina and Carrera (1996) successfully estimated source concentrations along with flow and transport parameters.

An advantage of defining parameters that represent the source characteristics is that sensitivity analysis methods can be used to investigate the importance of the parameters to model fit to observations and to model predictions.

9.2.3 Scale Issues

In addition to carefully considering which transport processes to simulate and which parameters to estimate, it also is important to recognize that features that are inconsequential to flow may be important to transport. This has been demonstrated for groundwater systems by a number of authors, including Poeter and Gaylord (1990), Zheng and Gorelick (2003), and De Marsily et al. (2005). Both unresolved smaller-scale features and misrepresented larger-scale features are of concern.

Calibration of transport models with unresolved and misrepresented features can cause problems with the regression and its results. For instance, the regression may estimate unreasonable parameter values to compensate for the inaccurate representation of system features, produce a poor fit to transport observations, and/or have difficulty converging to optimal parameter estimates.

Scale issues are problematic for groundwater models because (1) the variability consists of subsurface materials with properties that can vary by many orders of magnitude, (2) only a small portion of the subsurface material of any system is measured or measurable using current technology, and (3) the variability is important to the quantity and the quality of the groundwater needed by human and ecological communities.

This book does not address scale issues comprehensively, though the methods and guidelines provide important tools and ideas for addressing scale issues. Interesting methods that have been developed to deal with scale issues in transport problems include, for example, zonation (discussed further in Guideline 2, Chapter 11), the transitional probability method of Carle et al. (1998), the superparameter method of Tonkin and Doherty (2005), the constrained minimization method of Doherty (2003) and Moore and Doherty (2005, 2006), the representer method of Valstar et al. (2004), and the sequential indicator simulation method of Deutsch and Journel (1992, pp. 123–125, 148) and Gomez-Hernandez (2006), some of which were mentioned in Chapter 1 of this book. Scale issues are also addressed in many of the guidelines presented in Chapters 10 to 14. This is an area of active research and it has not yet matured to the point of achieving thorough comparison of methods for realistic problems. One can imagine that there might be a continuum between the solute spreading best represented by explicit representation of subsurface heterogeneity and solute spreading best represented by a simple representation of the heterogeneity combined with volume-averaged processes. An intermediate model would include some explicit representation of heterogeneity and some dispersion. It is not known where models along this continuum would be of most use in different practical problems. This seems to the authors of this book to be a rich area for future research.

9.2.4 Numerical Issues: Model Accuracy and Execution Time

Forward execution times of less than about 30 minutes or so are important to being able to explore the meaning of data and model processes regardless of the method used for model calibration (see Chapter 15, Section 15.1). Numerical accuracy is important to obtaining accurate sensitivities and parameter estimates. For transport problems, there is often a trade-off between accuracy and execution time—greater accuracy requires longer execution times. Also, as mentioned in Section 9.2.1, added processes can require much greater execution times.

The three options for simulating transport defined in Section 9.2.1 each have issues related to model execution time and accuracy of the numerical simulation.

Option 1 Simulate advective transport with particle tracking methods that require only slightly more execution time and computer storage than a flow model without any transport.

A problem with applying regression methods to models with advective-transport observations is that if a particle exits the grid when the model is solved for a particular set of parameters, its sensitivities cannot be calculated. Anderman and Hill (2001) address this problem by calculating a projected position for any particle that leaves the grid, using the particle velocity at the point where it exits the model. Sensitivities are then calculated using the projected position.

Advective-transport simulation accuracy can deteriorate as the model grid becomes coarse. This can be addressed by testing different grid refinements and possibly locally refining grids in selected parts of the model (e.g., Mehl and Hill, 2006). Zheng (1994) discusses two situations that can cause inaccurate particle tracking results when the model grid discretization becomes coarse: the presence of a weak sink or source, and a vertically distorted grid. He presents computational methods for minimizing the particle tracking errors that arise from these problems, which can be used when it is impractical to more finely discretize the grid. These methods are now used in many particle tracking codes.

Options 2 and 3 Simulating advection and dispersion requires substantially more execution time than does simulating advection only, and including reactions and other transport mechanisms generally requires even greater execution times. For

9.2 STRATEGIES FOR CALIBRATING TRANSPORT MODELS

these types of transport simulations, numerical accuracy issues are much more problematic than for simulation of advective transport. Concerns discussed here are time-step size coordination, problems with perturbation sensitivities when using Lagrangian methods, and the consequences of numerical dispersion.

Slight changes in time-step size can cause slight changes in the concentration simulated at a particular location and time. When calculating sensitivities by perturbation methods, these small changes in concentrations can cause large changes in sensitivities. This situation occurs commonly when the transport-step size is automatically calculated by the modeling software to satisfy, for example, the Courant number criterion (e.g., Zheng and Bennett, 2002, p. 187; Barth and Hill, 2005a). It is very likely that the transport-step size will be different in the two simulations needed to compute the perturbation sensitivities for each parameters, which adversely affects the sensitivities. This problem can be eliminated by executing the transport code for the starting parameter values, noting the model-calculated transport time step, and defining a somewhat smaller time step to calculate sensitivities and perform parameter estimation. In regression runs, the parameter values and simulated flow field changes from one regression iteration to the next. To ensure that the imposed transport-step size remains valid, occasionally use the updated parameter values in a forward run for which the model calculates the step size.

In Lagrangian methods of solute transport simulation, concentrations in a model cell or element are calculated from masses or concentrations associated with the particles present in the cell. Examples are the method of characteristics and the random-walk method. Simulated concentrations at one location tend to be accurate on average but are not smooth over time. Instead, they tend to oscillate about a smooth curve as particles leave and enter the cell. The oscillations are reduced as more particles are used, but enough particles to obtain a reasonably smooth curve often results in long execution times for practical problems. It is dangerous to use perturbation methods to calculate sensitivities based on oscillating concentrations. Depending on the oscillation captured in the two runs required to obtain perturbation sensitivities, the sensitivity can range from being much too small to much too large. Resulting sensitivities commonly do not vary smoothly from one time step to the next, or from one parameter estimation iteration to the next, which can cause substantial problems with convergence of the nonlinear regression procedure. Sonnenborg et al. (1996) minimized this problem by calculating the sensitivities using concentrations averaged over time periods that are longer than the model time steps. This approach can easily be implemented using universal inverse models such as UCODE_2005 and PEST.

Numerical dispersion is a common problem and its presence affects estimated parameters and model predictions. Mehl and Hill (2001) illustrated this by simulating a two-dimensional laboratory experiment constructed of discrete, randomly distributed, homogeneous blocks of five sands. They first demonstrated that when laboratory measurements of hydraulic conductivity and dispersivity values are used directly in the transport model, a poor fit to the measured breakthrough curve (BTC) is achieved (Figure 9.1a). Results of these simulations also show

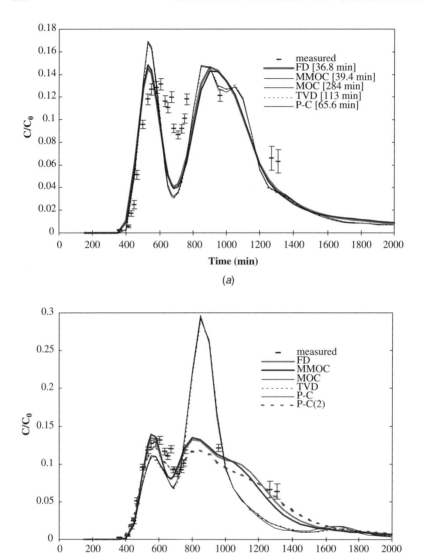

FIGURE 9.1 Measured and simulated breakthrough curves (BTCs) for a two-dimensional laboratory experiment of transport through saturated sands. For the measured concentration values, 95 percent confidence intervals are shown and reflect measurement error. (*a*) BTCs using measured hydraulic conductivities and dispersivities. Computation times are listed in brackets and are from a LINUX workstation, Pentium II-333, 64 Mb RAM. (*b*) BTCs using optimized hydraulic conductivities and measured dispersivities. The solution labeled P-C(2) uses dispersivity values increased to approximate the numerical dispersion common to the FD and MMOC methods of MT3DMS. (From Mehl and Hill, 2001.)

that, as coded in MT3DMS (Zheng and Wang, 1999; Zheng, 2005), the finite-difference (FD) method and the modified method of characteristics (MMOC) exhibit more numerical dispersion than the method of characteristics (MOC) and the total variation diminishing (TVD) method. A predictor-corrector (P-C) method added to MT3DMS for the study also had little numerical dispersion.

Sensitivities found using the different solution methods produced similar conclusions about what parameters were important, but regression estimates of dispersivity and hydraulic conductivity parameters were strongly affected by numerical dispersion. When dispersivity was set to laboratory-measured values, regression using FD and MMOC produced substantially different hydraulic-conductivity estimates than did MOC, TVD, and P-C. Better fits to measured BTCs were achieved for FD and MMOC (Figure 9.1b), which have more numerical dispersion. This suggests that the measured dispersivities were consistently too small and the estimated hydraulic conductivities were compensating for the bias in the measured dispersivities. When a single multiplicative dispersivity parameter and the five hydraulic-conductivity parameters were estimated, similar hydraulic-conductivity estimates and a similar fit were attained for all solution methods, and dispersivity estimates were larger for methods with little numerical dispersion.

9.2.5 Transport Observations

Three issues important to using transport observations are discussed: (1) simultaneous use of transport and flow-system observations, (2) weighting concentration observations, and (3) using point concentrations to determine other types of observations.

Simultaneous Use of Transport and Flow-System Observations Concentration observations are important to the estimation of both flow and transport parameters, typically providing substantial information about transmissive properties such as hydraulic conductivity and transmissivity, because (1) concentrations are sensitive to velocities and (2) in process-based models, velocity magnitude and direction depend on these properties.

Simultaneous use of concentrations and other types of data is likely to be more successful than a sequential estimation strategy, by which, for example, head and flow observations are used to estimate flow model parameter values, when these values are fixed and concentration data are used to estimate the transport parameters. Wagner and Gorelick (1987) were the first to develop a coupled estimation methodology and applied it to a synthetic example. Several later studies have shown that coupled estimation of flow and transport parameters produces parameter estimates that are more reasonable and have reduced uncertainty, compared to a sequential estimation strategy or a procedure whereby subsets of the observations (e.g., only heads or only concentrations) are used to estimate both flow and transport parameters (e.g., Gailey et al., 1991; Sonnenborg et al., 1996; Barlebo et al., 1998; Anderman and Hill, 1999). In some cases, a sequential estimation strategy might produce the same results as those from a coupled inverse procedure (e.g., Jacques et al., 2002). However, there is no guarantee of this, and thus the use of simultaneous estimation of flow and transport parameters is encouraged.

Weighting Concentration Observations As discussed in the context of Eq. (3.6) in Chapter 3, Section 3.3.3, the standard deviation of errors in concentrations often can be thought of as being proportional to the concentration. Under this circumstance, the standard deviation equals the product of a coefficient of variation and a concentration; both observed and simulated concentrations have been used in the literature, though there is some indication that simulated values are needed for unbiased parameter estimates (Anderman and Hill, 1999). Valstar et al. (2004) provide an example of using errors that are proportional to concentrations. Figure 9.2b shows how weighted residuals can vary depending on whether simulated or observed values are used to calculate the weights.

When applying parameter-estimation methods to transport models, a numerical problem can occur when the range of concentration observations spans more than about four orders of magnitude (Barth and Hill, 2005a,b). Such a large range occurs for many constituents, such as dissolved aqueous species or pathogens. The difficulty arises when observation uncertainty is represented as being proportional

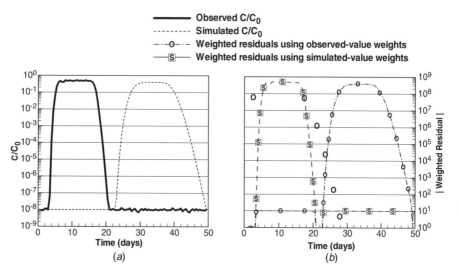

FIGURE 9.2 (*a*) Simulated breakthrough curves trailing and not overlapping the observed breakthrough curve, and (*b*) the resulting weighted residuals employing weights calculated using coefficients of variation and either observed or simulated concentrations (referred to as observed- and simulated-value weights in the legend). Observed- and simulated-value weights produce weighted residuals of similar magnitude. However, decreasing the transport rate so that the simulated breakthrough curve shifts to the right and beyond the period of observations decreases the sum of squared weighted residuals with observed-value weighting while having no significant effect on weighted residuals with simulated-value weighting. (From Barth and Hill, 2005a.)

9.2 STRATEGIES FOR CALIBRATING TRANSPORT MODELS

to the concentration. Applying a constant coefficient of variation for all observations can result in enormous weighted residuals for small concentrations. This can occur even if the concentrations are log-transformed. A solution is to place a lower bound on the statistic, and thus an upper bound on the weight, as suggested by, for example, Keidser and Rosbjerg (1991). Barth and Hill (2005a) show that it can be important to approach the upper bound gradually.

Alternatives to Using Point Concentration Measurements as Observations Often when calibrating transport models, use of point concentrations as measures of goodness of fit can be problematic, because (1) concentration measurements can vary over many orders of magnitude, (2) concentration measurements are often scarce spatially, and (3) simulated point concentrations depend on the particular representation of heterogeneity in the model, though the predictions of interest may be averaged quantities that do not have this dependence. Thus, alternative measures of goodness of fit might be preferable when calibrating transport models. Here, we cite four alternatives to using point concentration measurements.

1. To calibrate a model of natural-gradient tracer transport in an extremely heterogeneous aquifer, Feehley et al. (2000) divide the model domain into six zones along the flow direction and compare simulated and observed masses within each zone.
2. For calibrating a model of a different natural-gradient tracer test in the same aquifer studied by Feehley et al. (2000), Julian et al. (2001) compare the maximum simulated concentration from all model layers at a given areal location with the maximum observed value from all vertical sampling points at that location.
3. Barth and Hill (2005a,b) use moments of the concentration distribution.
4. Anderman et al. (1996) use concentration measurements to derive advective-transport observations.

9.2.6 Additional Model Inputs

Simulation of transport often brings additional observations to the model calibration effort, and also brings additional system characteristics that typically are determined at least in part using additional estimated parameters. If only advection is considered, effective porosity is the sole additional system characteristic required (as was the case for simulation of advective-transport predictions in Exercise 8.1). If dispersive processes are included, dispersivity in up to three spatial directions is needed. If multicomponent reactive transport is considered, there are potentially a large number of additional system characteristics (e.g., see Parkhurst, 1995; Prommer et al., 2003). All of the new system characteristics can vary spatially. As for transient models, the additional model inputs can be parameterized and parameters estimated as suggested by the methods and guidelines presented in this book.

Problems with insensitivity and correlation, as discussed in Chapter 4, can be troublesome in transport models. When only advection is simulated and the flow field is steady state, hydraulic conductivity, recharge, and effective porosity tend to be intercorrelated. Insensitivity and correlation can be severe for multicomponent reactive transport models, because each component is potentially characterized by several separate transport properties.

9.2.7 Examples of Obtaining a Tractable, Useful Model

First, we mention two examples presented in more detail in Chapter 15 with very different scales and modeling objectives. In a regional model of the Death Valley groundwater system with 1500-meter grid spacing, advective transport was chosen to simulate transport predictions. In a site-scale model of the Grindsted landfill in Denmark with grid spacing as small as 1 meter, advection and dispersion were included to calibrate with concentration observations. See Chapter 15 for additional information.

Three recent applications of nonlinear regression to multicomponent reactive transport models achieved tractable problems by carefully selecting which parameters to specify rather than estimate, or by simplifying the simulated processes without sacrificing the ability of the model to reasonably represent the simulated system.

1. To calibrate a model of vapor phase hydrocarbon transport, Gaganis et al. (2002) grouped sets of individual hydrocarbon constituents with similar thermodynamic properties into composite constituents. They then defined thermodynamic parameters associated with these composite constituents, and thus substantially reduced the total number of model parameters. These transport parameters were estimated using concentrations of the composite constituents as the calibration observations.

2. Ghandi et al. (2002b) calibrated a model of cometabolic trichloroethylene biodegradation by retaining the full model complexity and large number of parameters, and fixing several insensitive parameters at values obtained from laboratory experiments or previous modeling studies.

3. Essaid et al. (2003) made two simplifications when applying regression to a model of hydrocarbon dissolution and biodegradation. First, they represented biodegradation by using first-order reactions rather than Monod kinetics, because of high correlations between the Monod kinetics parameters. Second, they defined a single dissolution rate parameter for all hydrocarbon components, because of high correlation between the dissolution rate parameter and biodegradation rate parameter for each component. This approach was supported by independent experiments. Even with these simplifications, the model retained a considerable amount of complexity, and the observation data supported estimation of a large number of

parameters including individual first-order anaerobic biodegradation rates for all of the hydrocarbon components.

9.3 STRATEGIES FOR RECALIBRATING EXISTING MODELS

Models frequently are recalibrated as additional observations become available or as other new information is obtained. For groundwater systems, examples of other new information include new observations, possibly affected by different stresses such as pumpage or drought; new geologic interpretations; and new information on the distribution of areal recharge.

Recalibrated models can be developed and evaluated using the methods described in previous chapters of this book. While methods that allow building on previous regression results, such as the Kalman filter (Drécourt and Madsen, 2002), could be used, for nonlinear models using nonlinear regression with the new information is more straightforward. Negative consequences such as greater execution time often are not serious enough to warrant using the more complicated methods.

Model results can help determine when the possibly considerable additional investment in model recalibration is needed. While such decisions are often based largely on data and policy criteria, the model (or alternative models) generally provides the best available representation of system processes and can provide important insight. Table 9.1 presents the issues likely to be of concern and methods useful in addressing each issue. The guidelines presented in this book are also likely to be useful.

If a model is recalibrated, the following issues need to be considered.

- Does the recalibrated model produce predictions that differ significantly from those produced using previous models?

To address this issue, compare predictions simulated using the recalibrated model to those from previous models.

- Is the uncertainty of the predictions greater or smaller than previously calculated?

To address this issue, compare linear and possibly nonlinear confidence intervals for predictions produced by the recalibrated model to intervals produced by previous models. Generally, it is expected that model uncertainty will be reduced when data are added, given the same number of parameters, but this may not always be the case, because of model nonlinearity. Even if the model construction is unchanged, the new estimated parameter values will cause the sensitivities to be different, and the effect of this difference on the calculated uncertainty may be greater than the effect of the information provided by the new observations. Often new observations motivate modifications to the model construction, such as changes in where parameters apply, what processes are included, and perhaps how many parameters are defined and/or estimated.

TABLE 9.1 Issues to Consider When Deciding if Model Recalibration Is Needed[a]

Issue	Method	Section or Comment	Guideline
Do new observations suggest the model is incorrect or suggest an alternative model?	Observation residuals and weighted residuals.	6.2.1	4, 6, 9
	Use graphical analysis to compare model fit to new observations with the fit to observations used in model calibration.	6.4.1 to 6.4.4	
Do new system data suggest the model is incorrect or suggest an alternative model?	Compare new system data with model input and parameter values.	Usually requires GIS or 3D visualization;	2, 6, 10
	Framing new data as prior information and calculating prior residuals and weighted residuals is sometimes useful.	6.2.1, 6.4.1 to 6.4.4	
Is the information provided by new observations likely to affect estimated parameter values or parameter uncertainty?	dss, css, pcc	4.3	3, 11
	Leverage statistics	4.3.6, 7.5.2	
	Influence statistics:		
	Cook's D	7.5.2	
	$DFBETAS$	7.5.2	
Is the information provided by new observations likely to affect predictions of interest?	opr	8.2.2	12
Is new information about parameter values or model construction likely to affect predictions of interest?	ppr	8.2.1	12

[a] Validity of results depends on the accuracy of the model; consider analyzing results from the perspective of simplifications and approximations made in model construction.

9.4 EXERCISES (OPTIONAL)

In Exercises 9.1–9.8 the model that was developed, calibrated, and evaluated in Exercises 2.1 through 8.2 is recalibrated using new hydraulic-head and streamflow

9.4 EXERCISES

gain observations obtained during a long-term transient aquifer test. Water-supply wells have been completed in both aquifers at the areal location shown in Figure 2.1a and the aquifer test was conducted using these wells. During the test, groundwater was withdrawn for 283 days at a rate of about 1.0 m^3/s from each of the two aquifers (total pumping rate of about 2.0 m^3/s). Because of fluctuations in the pumping rate during the aquifer test, the true average pumping rate is uncertain.

The model used in the recalibration simulates steady-state groundwater flow without pumping (the conditions used in the previous exercises), and then uses that solution as initial conditions for a transient simulation of the aquifer test. Observations used for the recalibration include the observations used in the previous exercises and the observations of hydraulic head and river discharge collected during the aquifer test. In these exercises, the term "transient model" refers to the combined simulation that includes steady-state flow without pumping and transient flow during the aquifer test. The term "steady-state model" refers to the model without pumping that was used in previous exercises.

Figure 9.3 shows the volumetric budget for the transient flow system with pumping, calculated using the true parameter values. Simulated heads at selected times are shown in Figure 9.4a–d.

Exercises 9.1 and 9.2: Simulate Transient Hydraulic Heads and Perform Preparatory Steps Exercises 9.1 and 9.2 involve initial MODFLOW-2000 simulations of the transient model. Instructions for performing these simulations are available from the web site for this book listed in Chapter 1, Section 1.1. Students who are not performing the simulations may skip these exercises.

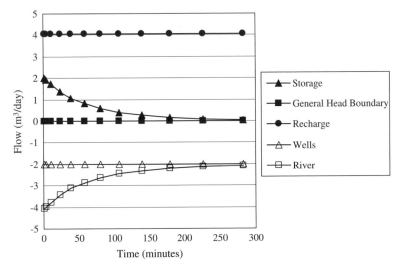

FIGURE 9.3 Transient budget showing simulated flows in the true system with pumping. Inflows are positive in sign; outflows are negative.

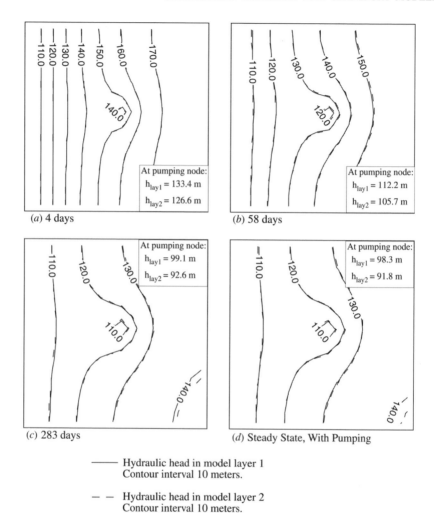

FIGURE 9.4 Simulated hydraulic heads for the true system with pumping, after (*a*) 4 days, (*b*) 58 days, (*c*) 283 days, and (*d*) at steady state.

Exercise 9.3: Transient Parameter Definition Parameters needed for the transient model that were not applicable to the steady-state model are the pumping rate and the specific storage for each model layer. The pumping rate is treated as a potentially estimated parameter because of fluctuations in the pumping rate during the aquifer test. The regression is used here to estimate the constant rate that is most consistent with the observed drawdown.

The names and starting values of storage and pumping parameters are given in Table 9.2. In addition, parameters HK_1, VK_CB, HK_2, K_RB, RCH_1, and RCH_2 are defined as for the steady-state model (see Exercise 3.1).

9.4 EXERCISES

TABLE 9.2 Parameter Names and Starting Values for Properties that Are Only Applicable in the Transient System

Flow-System Property	Parameter Name	Starting Value
Specific storage in layer 1	SS_1	2.6×10^{-5}
Specific storage in layer 2	SS_2	4.0×10^{-6}
Pumping rate in each of model layers 1 and 2, in m^3/s	Q_1&2	-1.1

All work for this exercise involves modifying computer files, as described in the instructions on the web site for this book listed in Chapter 1, Section 1.1. Students who are not performing the simulations may skip this exercise.

Exercise 9.4: Observations for the Transient Problem This exercise involves defining observations and their weights for the transient model. The observations are listed in Tables 9.3 and 9.4 and include steady-state hydraulic heads, drawdowns during pumping, and steady-state and transient discharge to the river.

The head and drawdown observations are in the same locations used for the steady-state model (Figure 2.1*b*). As discussed for the steady-state model, the hydraulic-head observations were generated by including random error. The error in the elevation of each observation well has a mean of 0.0 and a variance of 1.0, and each water-level measurement error has a mean of zero and a variance of 0.0025. The total variance of the error in each hydraulic-head observation is 1.0025.

There are three observations of groundwater discharge to the river, which include one observation for the steady-state conditions without pumping, and two observations during the aquifer test (Table 9.4). As in the steady-state model, the reach over which flow is measured extends the entire length of the river.

(a) *Define observations of hydraulic head, drawdown, and flow.*

Use the information in Tables 9.3 and 9.4 to define the observations of hydraulic head, drawdown, and flow in the appropriate input files, and simulate a forward model run. Instructions for performing this simulation are available from the web site for this book listed in Chapter 1, Section 1.1. Students who are not performing the simulations may skip this exercise.

(b) *Calculate weights on observations for the transient model.*

Problem
- Use the information provided on head observation error, the discussion of transient observations in Section 9.1.2, and the discussion of weighting in Guideline 6 and Appendix A to explain the variance on drawdowns listed in Table 9.3.

TABLE 9.3 Transient Hydraulic-Head Observations[a]

Observation Name	Lay	Row	Column	Time of Observation, After Beginning of Simulation (days)	Reference Stress Period in MODFLOW	Time of Observation, After Beginning of Reference Stress Period (days)	Observed Head (m)	Observed Drawdown (m) (Calculated by MODFLOW)	Variance of Head Observation Error (m^2)	Variance of Drawdown Observation Error (m^2)
hd01.ss	1	3	1	0.0	1	0.0	101.804	0.000	1.0025	—
hd01.1	1	3	1	1.0088	1	1.0088	101.775	0.029	1.0025	0.005
hd01.283	1	3	1	282.8595	1	282.8595	101.675	0.129	1.0025	0.005
hd02.ss	1	4	4	0.0	1	0.0	128.117	0.000	1.0025	—
hd02.1	1	4	4	1.0088	1	1.0088	128.076	0.041	1.0025	0.005
hd02.4	1	4	4	4.0353	1	4.0353	127.560	0.557	1.0025	0.005
hd02.108	1	4	4	107.6825	1	107.6825	116.586	11.531	1.0025	0.005
hd02.283	1	4	4	282.8595	1	282.8595	113.933	14.184	1.0025	0.005
hd03.ss	1	10	9	0.0	1	0.0	156.678	0.000	1.0025	—
hd03.1	1	10	9	1.0088	3	0.0	152.297	4.381	1.0025	0.005
hd03.283	1	10	9	282.8595	5	272.7713	114.138	42.540	1.0025	0.005
hd04.ss	1	13	4	0.0	1	0.0	124.893	0.000	1.0025	—
hd04.1	1	13	4	1.0088	3	0.0	124.826	0.067	1.0025	0.005
hd04.283	1	13	4	282.8595	5	272.7713	110.589	14.304	1.0025	0.005

hd05.ss	1	14	6	0.0	1	0.0	140.961	0.000	1.0025	—
hd05.1	1	14	6	1.0088	3	0.0	140.901	0.060	1.0025	0.005
hd05.283	1	14	6	282.8595	5	272.7713	119.285	21.676	1.0025	0.005
hd06.ss	2	4	4	0.0	1	0.0	126.537	0.000	1.0025	—
hd06.1	2	4	4	1.0088	3	0.0	126.542	−0.005	1.0025	0.005
hd06.283	2	4	4	282.8595	5	272.7713	112.172	14.365	1.0025	0.005
hd07.ss	2	10	1	0.0	1	0.0	101.112	0.000	1.0025	—
hd07.1	2	10	1	1.0088	3	0.0	101.160	−0.048	1.0025	0.005
hd07.283	2	10	1	282.8595	5	272.7713	100.544	0.568	1.0025	0.005
hd08.ss	2	10	9	0.0	1	0.0	158.135	0.000	1.0025	—
hd08.1	2	10	9	1.0088	3	0.0	152.602	5.533	1.0025	0.005
hd08.283	2	10	9	282.8595	5	272.7713	114.918	43.217	1.0025	0.005
hd09.ss	2	10	18	0.0	1	0.0	176.374	0.000	1.0025	—
hd09.1	2	10	18	1.0088	3	0.0	176.373	0.001	1.0025	0.005
hd09.283	2	10	18	282.8595	5	272.7713	138.132	38.242	1.0025	0.005
hd10.ss	2	18	6	0.0	1	0.0	142.020	0.000	1.0025	—
hd10.1	2	18	6	1.0088	3	0.0	142.007	0.013	1.0025	0.005
hd10.283	2	18	6	282.8595	5	272.7713	122.099	19.921	1.0025	0.005

[a]All observations are located at cell centers.

TABLE 9.4 Data for Transient Flow Observations

Observation Name	Time of Observation, After Beginning of Simulation (days)	Reference Stress Period	Time of Observation, After Beginning of Reference Stress Period (days)	Observed Gain in River Flow (m^3/s)	Coefficient of Variation of River Gain	Standard Deviation of River Gain (m^3/s)
flow01.ss	0.0	1	0.0	4.4	0.10	—
flow01.10	10.0882	5	0.0	4.1	—	0.38
flow01.283	282.8595	5	272.7713	2.2	—	0.21

9.4 EXERCISES   235

- Calculate the weights on the hydraulic-head, drawdown, and flow observations for the transient model using information in Tables 9.3 and 9.4. Compare your results with the square roots of the weights shown in Figure 9.5.

Exercise 9.5: Evaluate Transient Model Fit Using Starting Parameter Values   In this exercise, the initial fit is evaluated for the forward transient model run. Tables of observed and simulated hydraulic heads and flows are shown in Figure 9.5.

Problem: Comment on the model fit achieved with the starting parameter values. Are there any residuals that are clearly outliers? How do the residuals compare to the weighted residuals?

Exercise 9.6: Sensitivity Analysis for the Initial Model   This exercise involves evaluating sensitivities for the transient model. For students performing the simulations, instructions for calculating the sensitivities are available from the web site for this book listed in Chapter 1, Section 1.1.

(a) *Evaluate contour maps of one-percent scaled sensitivities for the transient flow system.*

Contour maps of one-percent scaled sensitivities after 4, 58, and 283 days of pumpage, for parameters HK_1, HK_2, VK_CB, K_RB, SS_1, and SS_2 are shown in Figures 9.6–9.8. One-percent scaled sensitivities for parameters RCH_1 and RCH_2 at all times are the same as those for the steady-state flow system without pumpage (shown in Figure 4.4).

These maps reflect the flow system dynamics and are useful for understanding the effect of each parameter on the simulated hydraulic heads. Thus, they could be used to help guide collection of additional head observations that would provide information about individual parameters. However, note that for the transient model, there are a large number of maps, and that a location or time important to one parameter might not be important to another. It is difficult to use these maps to clearly identify locations that would be most beneficial, for example, to improving a set of parameter estimates, and they do not address the issue of improving predictions. Limitations on the use of the one-percent scaled sensitivity maps are discussed further in Chapter 4, Section 4.3.7.

This exercise focuses on using the physics of the groundwater flow system to understand the one-percent scaled sensitivities. In a confined flow system like the one considered here, the principal of superposition applies. This occurs because hydraulic head is a linear function of the applied fluxes, spatial dimension and time, as discussed in Chapter 1, Section 1.4.1. As a result of these linear relationships, hydraulic head calculated for the transient flow system, with applied fluxes of areal recharge and pumpage, is equal to the sum of the hydraulic head calculated for the flow system with areal recharge only and the drawdown calcu-

DATA AT THE HEAD LOCATIONS

OBS#	OBSERVATION NAME	OBSER-VATION *	SIMUL. EQUIV. *	RESIDUAL	WEIGHT**.5	WIEGHTED RESIDUAL
1	hd01.ss	102.	100.	1.58	0.999	1.58
2	dd01.1	-0.290E-01	-0.153E-04	-0.290E-01	14.1	-0.410
3	dd01.tr2	-0.129	-0.906E-01	-0.384E-01	14.1	-0.543
4	hd02.ss	128.	139.	-11.2	0.999	-11.2
5	dd02.tr1	-0.410E-01	-0.949E-02	-0.315E-01	14.1	-0.446
6	dd02.tr2	-0.557	-0.276	-0.281	14.1	-3.97
7	dd03.tr3	-11.5	-13.0	1.43	14.1	20.3
8	dd02.tr4	-14.2	-18.8	4.59	14.1	64.9
9	hd03.ss	157.	174.	-17.7	0.999	17.7
10	dd03.tr1	-4.38	-3.67	-0.715	14.1	-10.1
11	dd03.tr2	-42.5	-56.2	13.7	14.1	194.
12	hd04.ss	125.	139.	-14.4	0.999	-14.4
13	dd04.tr1	-0.670E-01	-0.163E-01	-0.507E-01	14.1	-0.718
14	dd04.tr2	-14.3	-18.8	4.54	14.1	64.3
15	hd05.ss	141.	157.	-16.2	0.999	-16.2
16	dd05.tr1	-0.600E-01	-0.368E-01	-0.232E-01	14.1	-0.328
17	dd05.tr2	-21.7	-28.5	6.79	14.1	96.0
18	hd06.ss	127.	140.	-13.1	0.999	-13.1
19	dd06.tr1	0.500E-02	-0.125E-01	0.175E-01	14.1	0.247
20	dd06.tr2	-14.4	-19.2	4.82	14.1	68.2
21	hd07.ss	101.	103.	-1.76	0.999	-1.75
22	dd07.tr1	0.480E-01	-0.109E-02	0.491E-01	14.1	0.694
23	dd07.tr2	-0.568	-1.38	0.813	14.1	11.5
24	hd08.ss	158.	174.	-15.8	0.999	-15.8
25	dd08.tr1	-5.53	-5.81	0.277	14.1	3.91
26	dd08.tr2	-43.2	-57.3	14.0	14.1	199.
27	hd09.ss	176.	190.	-13.9	0.999	-13.9
28	dd09.tr1	-0.992E-03	-0.506E-01	0.496E-01	14.1	0.702
29	dd09.tr2	-38.2	-49.5	11.3	14.1	159.
30	hd10.ss	142.	157.	-15.0	0.999	-15.0
31	dd10.tr1	-0.130E-01	-0.436E-02	-0.864E-02	14.1	-0.122
32	dd10.tr2	-19.9	-26.1	6.21	14.1	87.8

DATA FOR FLOWS REPRESENTED USING THE RIVER PACKAGE

OBS#	OBSERVATION NAME	MEAS. FLOW	CALC. FLOW	RESIDUAL	WEIGHT**.5	WEIGHTED RESIDUAL
33	flow01.ss	-4.40	-4.86	0.461	2.27	1.05
34	flow01.10	-4.10	-4.72	0.618	2.63	1.63
35	flow01.283	-2.20	-2.86	0.663	4.76	3.16

FIGURE 9.5 Part of MODFLOW-2000 LIST output file showing initial model fit and weights for the head, drawdown, and flow observations.

9.4 EXERCISES

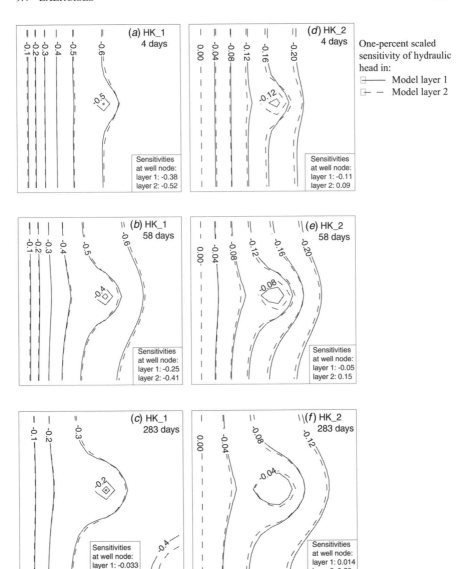

FIGURE 9.6 Contour maps of one-percent scaled sensitivity of hydraulic head to (*a*)–(*c*) parameter HK_1 [($\partial h/\partial$HK_1) × (HK_1/100)] and (*d*)–(*f*) parameter HK_2 [($\partial h/\partial$HK_2) × (HK_2/100)] after 4, 58, and 283 days of pumping in the transient flow model, calculated using the starting parameter values. Contour labels apply to contours for both model layers.

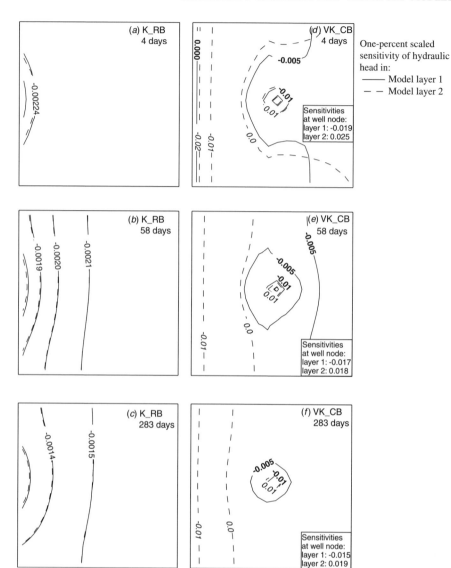

FIGURE 9.7 Contour maps of one-percent scaled sensitivity of hydraulic head to (a)–(c) parameter K_RB $[(\partial h/\partial \text{K\_RB}) \times (\text{K\_RB}/100)]$ and (d)–(f) parameter VK_CB $[(\partial h/\partial \text{VK\_CB}) \times (\text{VK\_CB}/100)]$ after 4, 58, and 283 days of pumping in the transient flow model, calculated using the starting parameter values. In (a)–(c), contour labels apply to contours for both model layers; in (d)–(f), bold contour labels apply to model layer 1 and italic contour labels apply to model layer 2.

9.4 EXERCISES

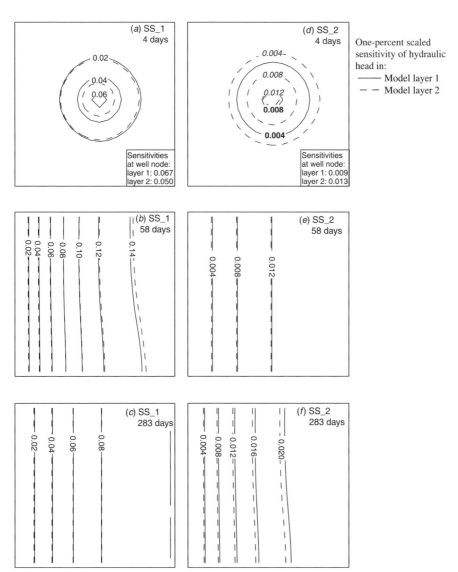

FIGURE 9.8 Contour maps of one-percent scaled sensitivity of hydraulic head to (a)–(c) parameter SS_1 [$(\partial h/\partial \text{SS\_1}) \times (\text{SS\_1}/100)$] and (d)–(f) parameter SS_2 [$(\partial h/\partial \text{SS\_2}) \times (\text{SS\_2}/100)$] after 4, 58, and 283 days of pumping in the transient flow model, calculated using the starting parameter values. In (a)–(c) and (e) and (f), contour labels apply to contours for both model layers; in (d), bold contour labels apply to model layer 1 and italic contour labels apply to model layer 2.

lated for the flow system with pumpage only. Because taking the derivative is a linear process, the principle of superposition also applies to sensitivities. The one-percent scaled sensitivities of hydraulic head for the flow system with areal recharge only are those calculated for the steady-state model, and are shown in Figure 4.4. The one-percent sensitivities of hydraulic head to the hydraulic-conductivity parameters HK_1, HK_2, K_RB, and VK_CB for the transient flow system without areal recharge (with pumpage only) are shown in Figures 9.9 and 9.10.

Problem: Explain the one-percent scaled sensitivity maps for the transient system by answering the following questions using your knowledge of the flow system and the principle of superposition:

- Why are the one-percent scaled sensitivities of hydraulic head to HK_1 and HK_2 positive for the flow system with pumpage only?
- Explain the distribution of the one-percent scaled sensitivities for HK_1 at 4 days and at 283 days.
- Use the sensitivities for HK_1 in Figure 4.4a, Figure 9.6a–c, and Figure 9.9a–c to convince yourself that the principle of superposition can be used to calculate sensitivities for this model.
- In the steady-state model, the one-percent scaled sensitivities for K_RB are the same throughout the model domain (Figure 4.4b). Why do the sensitivities for K_RB in the transient system vary over the model domain (Figure 9.7a–c and Figure 9.10a–c)?
- Why are the one-percent scaled sensitivities for SS_1 and SS_2 (Figure 9.8) concentric around the pumping wells at early time and nearly parallel to the river at late time?

(b) *Use composite scaled sensitivities to evaluate the information observations provide about the defined parameters.*

In preparation for performing nonlinear regression, examine the composite scaled sensitivities for the parameters of the transient model, shown in Figure 9.11. Use these *css* to help decide which parameters to estimate by regression.

Problem
- Which parameters have the smallest and largest composite scaled sensitivities?
- Using suggestions from Chapter 4, Section 4.3.4 about evaluating relative and individual *css* values, determine which parameters are likely to be estimated by the regression, given the information provided by the observations.

9.4 EXERCISES

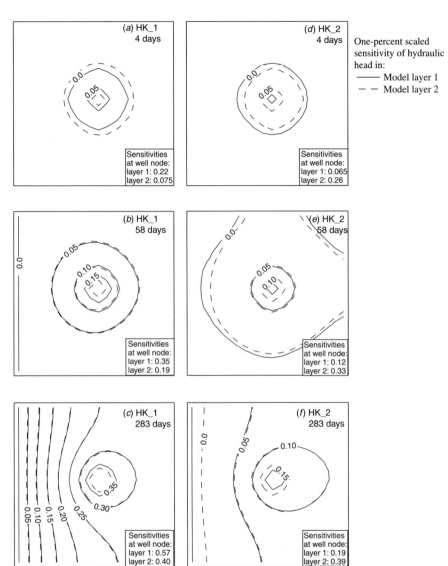

FIGURE 9.9 Contour maps of one-percent scaled sensitivity of hydraulic head to (*a*)–(*c*) parameter HK_1 [$(\partial h/\partial \text{HK\_1}) \times (\text{HK\_1}/100)$] and (*d*)–(*f*) parameter HK_2 [$(\partial h/\partial \text{HK\_2}) \times (\text{HK\_2}/100)$] after 4, 58, and 283 days of pumping in the transient model without areal recharge (with pumpage only), calculated using the starting parameter values. Contour labels apply to contours for both model layers.

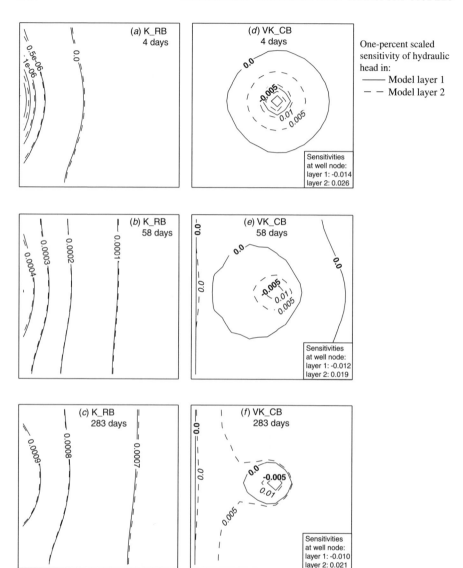

FIGURE 9.10 Contour maps of one-percent scaled sensitivity of hydraulic head to (a)–(c) parameter K_RB [($\partial h/\partial$K_RB) × (K_RB/100)] and (d)–(f) parameter VK_CB [($\partial h/\partial$VK_CB) × (VK_CB/100)] after 4, 58, and 283 days of pumping in the transient model without areal recharge (with pumpage only), calculated using the starting parameter values. In (a)–(c), contour labels apply to contours for both model layers; in (d)–(f), bold contour labels apply to model layer 1 and italic contour labels apply to model layer 2.

9.4 EXERCISES

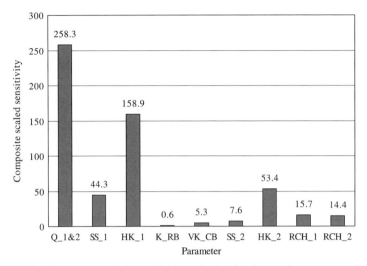

FIGURE 9.11 Composite scaled sensitivities calculated at the starting parameter values for the transient model.

(c) *Evaluate parameter correlation coefficients.*

As discussed for the steady-state regression, it is important to use parameter correlation coefficients for the initial model to assess the likelihood of uniquely estimating all flow system parameters given the available observation data. The correlation coefficients calculated by MODFLOW-2000 are shown in Table 9.5.

Problem
- Are any of the correlation coefficients greater than 0.95 in absolute value? Are any greater than 0.90 in absolute value?
- What do the correlation coefficients indicate about the likelihood of estimating all of the parameters independently using the head, drawdown, and flow data?

Exercise 9.7: Estimate Parameters for the Transient System by Nonlinear Regression In most applications, the problems of sensitivity and uniqueness identified by the analysis above would lead to first trying to estimate the more sensitive parameters and then, using the updated values, attempt the regression with the less sensitive parameters as well. Here, however, model execution times are relatively short, so it is feasible to try estimating all the parameters in the first regression run. Generally, execution times for parameter estimation can be long and tend to be longer when using UCODE_2005 than when using MODFLOW-2000 because of the perturbation sensitivity calculations performed by UCODE_2005. Approximate execution times can be calculated as described in Chapter 15, Section 15.1.

TABLE 9.5 Correlation Coefficient Matrix for Starting Parameter Values Using the Hydraulic-Head, Drawdown, and Flow Observations, Calculated for the Transient Problem by MODFLOW-2000[a,b]

	Q_1&2	SS_1	HK_1	K_RB	VK_CB	SS_2	HK_2	RCH_1	RCH_2
Q_1&2	1.00	−0.91	**−0.99**	−0.057	−0.67	−0.41	**−0.96**	−0.66	−0.83
SS_1		1.00	0.88	−0.078	0.80	0.043	0.89	0.58	0.75
HK_1			1.00	−0.029	0.68	0.41	0.92	0.67	0.82
K_RB				1.00	−0.36	0.38	0.22	0.051	0.055
VK_CB					1.00	−0.23	0.61	0.43	0.55
SS_2		Symmetric				1.00	0.41	0.30	0.35
HK_2							1.00	0.62	0.81
RCH_1								1.00	0.16
RCH_2									1.00

[a]The matrix produced using UCODE_2005 is nearly identical.
[b]Correlation coefficients greater than 0.95 in absolute value are in bold type.

Performing nonlinear regression for the transient model involves modifying computer files and simulating either MODFLOW-2000 or UCODE_2005. For students performing the computer simulations, instructions are available from the web site for this book listed in Chapter 1, Section 1.1. For students not performing the simulations, Figure 9.12 summarizes the results of the regression run.

Problem
- Examine the results shown in Figure 9.12a. What happened during this regression run?
- The parameter values calculated for all regression iterations are shown in Figure 9.12b. On the basis of this figure, which parameter does the regression have the most difficulty estimating? Is the answer consistent with the information about the regression behavior shown in Figure 9.12a?
- The starting, estimated, and true parameter values are shown in Table 9.6. Why do the estimated values differ from the true values?
- In this problem the starting parameter values are close to the final parameter values. Given the information provided about objective functions in Chapters 4 and 5, what problems might be expected given starting parameter values that are progressively further from the optimal values?

Exercise 9.8: Evaluate Measures of Model Fit This exercise evaluates model fit for the transient model regression performed in Exercise 9.7, using statistics shown in Figure 9.13.

Problem
- What conclusion about model fit might be drawn from the result that s^2 (the calculated error variance of Figure 9.13) is less than 1.0?

9.4 EXERCISES

(a)

SELECTED STATISTICS FROM MODIFIED GAUSS-NEWTON ITERATIONS

MAX. ITER.	PARAMETER CALC. PARNAM	CHANGE MAX. MAX. CHANGE	CHANGE ALLOWED	DAMPING PARAMETER
1	VK_CB	0.868519	2.00000	1.0000
2	K_RB	1.43887	2.00000	1.0000
3	K_RB	0.894564	2.00000	1.0000
4	K_RB	0.346358	2.00000	1.0000
5	K_RB	0.559047E-01	2.00000	1.0000
6	K_RB	0.182992E-02	2.00000	1.0000

SUMS OF SQUARED WEIGHTED RESIDUALS FOR EACH ITERATION

SUMS OF SQUARED WEIGHTED RESIDUALS

ITER.	OBSERVATIONS	PRIOR INFO.	TOTAL
1	0.13469E+06	0.0000	0.13469E+06
2	625.19	0.0000	625.19
3	38.638	0.0000	38.638
4	26.242	0.0000	26.242
5	23.892	0.0000	23.892
6	23.846	0.0000	23.846
FINAL	23.841	0.0000	23.841

*** PARAMETER ESTIMATION CONVERGED BY SATISFYING THE TOL CRITERION ***

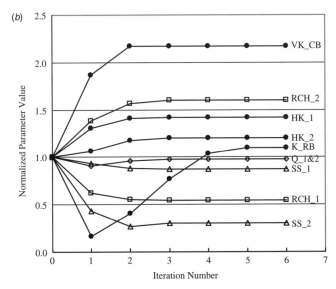

FIGURE 9.12 (a) Selected statistics from the modified Gauss–Newton iterations from Exercise 9.7. This is a fragment from the global output file of MODFLOW-2000. (b) Normalized parameter values at the end of each iteration of the transient regression. For each parameter, the graphed values are normalized by the starting value of the parameter (see Table 9.6).

TABLE 9.6 Starting, Estimated, and True Parameter Values for the Transient Model

Parameter Name	Starting Value	Estimated Value in Steady-State Regression	Estimated Value in Transient Regression	True Value
Q_1&2	−1.10	—	−1.07	−1.00
SS_1	2.6×10^{-5}	—	2.3×10^{-5}	2.0×10^{-5}
HK_1	3.0×10^{-4}	4.6×10^{-4}	4.3×10^{-4}	4.0×10^{-4}
K_RB	1.2×10^{-3}	1.2×10^{-3}	1.3×10^{-3}	1.0×10^{-3}
VK_CB	1.0×10^{-7}	9.9×10^{-8}	2.2×10^{-7}	2.0×10^{-7}
SS_2	4.0×10^{-6}	—	1.2×10^{-6}	2.0×10^{-6}
HK_2	4.0×10^{-5}	1.5×10^{-5}	4.8×10^{-5}	4.4×10^{-5}
RCH_1	63.072	47.45	34.10	31.536
RCH_2	31.536	38.53	50.44	47.304

```
LEAST-SQUARES OBJ FUNC (DEP.VAR. ONLY)----- =  23.841
LEAST-SQUARES OBJ FUNC (W/PARAMETERS)------ =  23.841
CALCULATED ERROR VARIANCE--------------- =   0.91697
STANDARD ERROR OF THE REGRESSION---------- =   0.95758
CORRELATION COEFFICIENT----------------- =   0.99999
        W/PARAMETERS--------------------- =   0.99999
ITERATIONS------------------------------ =        6

MAX LIKE OBJ FUNC   =  -35.070
AIC STATISTIC----   =  -17.070
BIC STATISTIC----   =   -3.0715
```

FIGURE 9.13 Selected statistics related to overall model fit, from the modified Gauss–Newton iterations of the regression run in Exercise 9.7. This is a fragment from the global output file of MODFLOW-2000

- Construct a confidence interval for the true error variance (Eq. (6.2)) to determine if the deviation of s^2 from a value of 1.0 is significant. How does this result affect the answer to the question in the previous bullet?
- Calculate the fitted standard deviation for heads and for drawdowns. Do the fitted standard deviations suggest that the model provides a good fit to these data?

Exercise 9.9: Perform Graphical Analyses of Model Fit and Evaluate Related Statistics In this exercise, the fit of the transient model to the head, drawdown, and flow data is evaluated using graphical methods and associated statistics.

9.4 EXERCISES

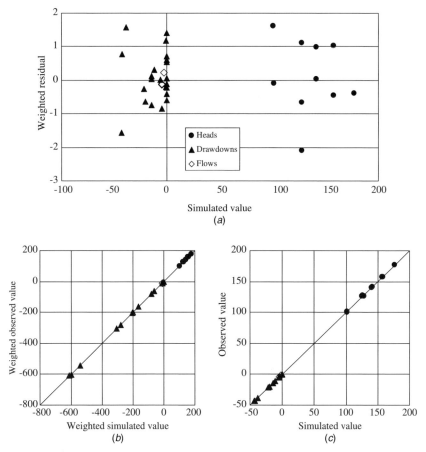

FIGURE 9.14 (a) Weighted residuals versus simulated values, (b) weighted observed values versus weighted simulated values, and (c) observed versus simulated values for the transient regression.

(a) *Evaluate graphs of weighted residuals and weighted and unweighted simulated and observed values.*

Graphs for analyzing model fit are shown in Figure 9.14.

Problem
- Do the weighted residuals appear to be randomly distributed with respect to the simulated values?
- Comment on the utility of the graphs in Figure 9.14b,c for analyzing the model fit to the data.
- Does the correlation R between the weighted simulated values and weighted observed values, shown in Figure 9.13, provide evidence that there is a good fit of the model to the data?

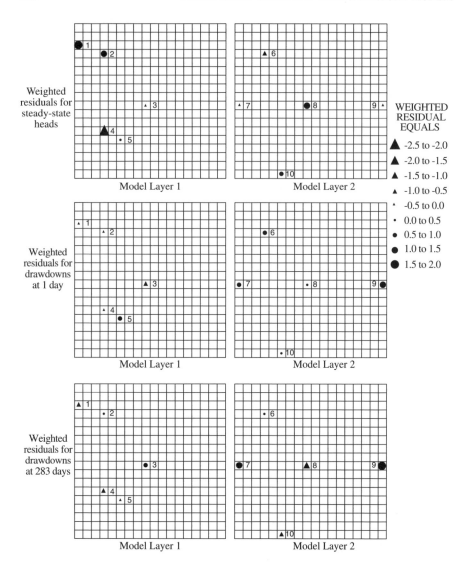

FIGURE 9.15 Weighted residuals for the transient regression plotted on maps of the two model layers.

(b) *Evaluate graphs of weighted residuals against independent variables and the runs statistic.*

In this exercise, the randomness of the hydraulic head and drawdown weighted residuals is evaluated (a) graphically with respect to their spatial location in the model, and (b) by applying the runs statistic to residuals as ordered in the input files containing the observations. The graphical analysis is conducted using

```
# RESIDUALS >=0.: 18
# RESIDUALS <0. : 17
NUMBER OF RUNS  : 17 IN 35 OBSERVATIONS

INTERPRETING THE CALCULATED RUNS STATISTIC VALUE OF -0.339
NOTE: THE FOLLOWING APPLIES ONLY IF
       #RESIDUALS >=0. IS GREATER THAN 10 AND
       #RESIDUALS <0. IS GREATER THAN 10
THE NEGATIVE VALUE MAY INDICATE TOO FEW RUNS:
  IF THE VALUE IS LESS THAN -1.28, THERE IS LESS THAN A 10 PERCENT
     CHANCE THE VALUES ARE RANDOM,
  IF THE VALUE IS LESS THAN -1.645, THERE IS LESS THAN A 5 PERCENT
     CHANCE THE VALUES ARE RANDOM,
  IF THE VALUE IS LESS THAN -1.96, THERE IS LESS THAN A 2.5 PERCENT
     CHANCE THE VALUES ARE RANDOM.
```

FIGURE 9.16 Runs statistic and critical values from the regression run in Exercise 9.7. This is a fragment from the global output file of MODFLOW-2000.

Figure 9.15, which shows the weighted residuals plotted on maps of the model domain. For this model, there are only three times (steady state, 1 day, and 283 days) at which observations are available at all wells, and plotting weighted residuals at these times is feasible. For transient models with observations at many times, such maps can only be evaluated for a subset of the times. The evaluation using the runs statistic is shown in Figure 9.16.

Problem
- For each of the observation times, does the spatial distribution of the weighted residuals in Figure 9.15 appear to be random? Why is the distribution of the weighted residuals for the steady-state head observations very similar to that in the steady-state regression (Figure 6.9)?
- For the transient regression, are there enough positive and negative weighted residuals so that the runs statistic can be evaluated using the critical values printed in the output file? Use Figure 9.16.
- Does the runs statistic indicate that the residuals are random with respect to the order in which the calibration observations are listed?

(c) *Assess independence and normality of weighted residuals.*

A normal probability graph of the weighted residuals from the transient problem and the associated statistic R_N^2, the correlation coefficient between the ordered weighted residuals and the normal order statistic, are shown in Figure 9.17.

Problem
- Does the normal probability graph indicate that the weighted residuals are independent and normally distributed?
- Are the results of this analysis consistent with the evaluation of R_N^2?

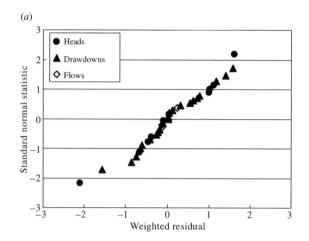

(b)

```
CORRELATION BETWEEN ORDERED WEIGHTED RESIDUALS AND NORMAL ORDER
   STATISTICS FOR OBSERVATIONS = 0.969
-------------------------------------------------------------------
COMMENTS ON THE INTERPRETATION OF THE CORRELATION BETWEEN WEIGHTED
RESIDUALS AND NORMAL ORDER STATISTICS:

The critical value for correlation at the 5% significance level is 0.943

IF the reported CORRELATION is GREATER than the 5% critical value,
ACCEPT the hypothesis that the weighted residuals are INDEPENDENT AND
NORMALLY DISTRIBUTED at the 5% significance level. The probability that
this conclusion is wrong is less than 5%.

IF the reported correlation IS LESS THAN the 5% critical value REJECT
the hypothesis that the weighted residuals are INDEPENDENT AND NORMALLY
DISTRIBUTED at the 5% significance level.

The analysis can also be done using the 10% significance level.
The associated critical value is 0.952
-------------------------------------------------------------------
```

FIGURE 9.17 (a) Normal probability graph of weighted residuals from the transient regression. (b) R_N^2 statistic and critical values from the regression run in Exercise 9.7. This is a fragment from the global output file of MODFLOW-2000.

- Should normal probability graphs of independent and correlated random numbers be prepared to determine acceptable deviations of the weighted residuals from independence and normality?

Exercise 9.10: Evaluate Estimated Parameters This exercise evaluates the sensitivity, correlation, uncertainty, and reasonableness of the estimated parameter values for the transient model.

9.4 EXERCISES

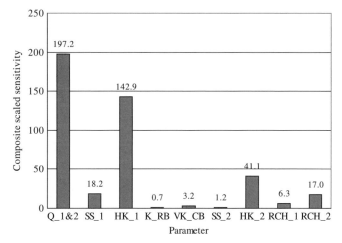

FIGURE 9.18 Composite scaled sensitivities calculated at the final parameter values for the transient regression.

(a) *Composite scaled sensitivities.*

Composite scaled sensitivities for the final parameter estimates are shown in Figure 9.18.

Problem: Compare the composite scaled sensitivities for the final parameter estimates with those for the initial parameter values (Figure 9.11). Why are the two sets of values different? Are they so different that conclusions drawn from the composite scaled sensitivities for the initial parameter values (Exercise 9.6b) have changed?

(b) *Parameter estimates and confidence intervals.*

The starting parameter values, true parameter values, optimal parameter estimates, and approximate, linear, individual, 95-percent confidence intervals are shown in Figure 9.19. Each of these quantities is plotted as a percentage of the estimated parameter value, by dividing each value or interval limit by the corresponding parameter estimate and then multiplying the result by 100.

Problem
- Compare the parameter estimates for the transient regression with those for the steady-state regression (Figure 7.7). Which parameter estimates are closer to their true values in the transient regression than in the steady-state regression? Why? Which parameter estimates are closer to their true values in the steady-state regression than in the transient regression? Why?
- Why are the confidence intervals on most hydraulic-conductivity and recharge parameters smaller for the transient regression than for the steady-state regression? Why is the confidence interval on K_RB not significantly smaller?

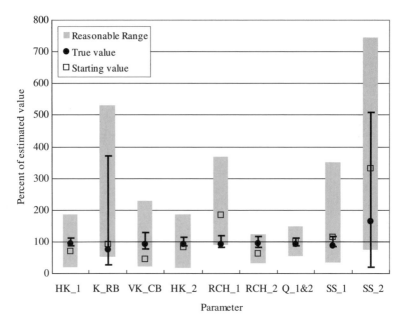

FIGURE 9.19 Starting and true parameter values, limits of approximate, linear, individual 95-percent confidence intervals (black bars), and limits of reasonable ranges of parameter values (grey bars), expressed as percentage of the estimated values, for the transient regression run. The estimated value for each parameter is represented by the thick line at a value of 100.

(c) *Reasonable parameter ranges.*

The optimal parameter estimates and linear confidence intervals are compared to reasonable parameter ranges in Figure 9.19.

Problem
- Are all the estimated parameter values within their respective reasonable ranges?
- Some of the linear confidence intervals extend outside the reasonable ranges. Using the analysis suggested in Chapter 7, Section 7.6, does this indicate any significant problems with the model construction or with the calibration observations?

(d) *Parameter correlation coefficients.*

Recall that some of the initial parameter correlation coefficients for the transient model (Table 9.5) were very close to 1.0. The parameter correlations calculated by MODFLOW-2000 at the optimal estimates are shown in Table 9.7.

Problem: Are any correlations larger than 0.95? How do the correlations compare to those for the initial parameter values?

9.4 EXERCISES

TABLE 9.7 Correlation Coefficient Matrix for Final Parameter Values Using the Hydraulic-Head, Drawdown, and Flow Observations, Calculated for the Transient Problem by MODFLOW-2000[a,b]

	Q_1&2	SS_1	HK_1	K_RB	VK_CB	SS_2	HK_2	RCH_1	RCH_2
Q_1&2	1.00	−0.75	**−0.99**	−0.089	−0.50	−0.056	0.95	−0.17	−0.91
SS_1		1.00	0.74	−0.19	0.82	−0.60	0.70	0.12	0.68
HK_1			1.00	0.0003	0.51	0.057	0.91	0.18	0.90
K_RB				1.00	−0.38	0.42	0.28	0.005	0.095
VK_CB					1.00	−0.70	0.43	0.090	0.44
SS_2		Symmetric				1.00	0.078	0.021	0.065
HK_2							1.00	0.14	0.88
RCH_1								1.00	−0.23
RCH_2									1.00

[a]The matrix produced by UCODE_2005 is nearly identical.
[b]Correlation coefficients greater than 0.95 in absolute value are in bold type.

Given the very large correlation between parameter pair HK_1 and Q_1&2, it is important to test whether the parameter estimates are unique. Accomplish this by running the regression with a few sets of different parameter starting values. Try values of the hydraulic-conductivity and storage parameters that are 10 times larger and smaller than the starting parameter values, and pumping and recharge parameters that are two times larger and smaller than the starting values. If any set of different starting parameters causes the regression to have difficulty converging, try less extreme starting values.

Problem: Does the regression converge to the same estimates as in Exercise 9.7? What does this mean in terms of uniqueness and optimality of the estimated parameter values?

Exercise 9.11: Test for Linearity This exercise uses the modified Beale's measure to assess the linearity of the transient model. For students performing the simulations, instructions for calculating this measure are provided on the web site for this book listed in Chapter 1, Section 1.1. The primary output from this computation is shown in Figure 9.20.

```
USING FSTAT = 2.2700, BEALES MEASURE = 84.181
IF BEALES MEASURE IS GREATER THAN 0.44, THE MODEL IS NONLINEAR.
IF BEALES MEASURE IS LESS THAN 0.40E-01, THE MODEL IS
EFFECTIVELY LINEAR, AND LINEAR CONFIDENCE INTERVALS ARE FAIRLY
ACCURATE IF THE RESIDUALS ARE NORMALLY DISTRIBUTED.
```

FIGURE 9.20 Part of the BEALE-2000 output file for the transient regression, showing the value of the modified Beale's measure.

Problem: Does the modified Beale's measure indicate that the model is effectively linear? What are the implications of this result with regard to the parameter confidence intervals shown in Figure 9.19? What options are there for obtaining more accurate measures of parameter uncertainty?

Exercise 9.12: Predictions The recalibrated model can now be used to update the predictions of advective transport, as requested by the landfill developer who had argued that it was inappropriate to use the calibrated steady-state model for evaluating the predictions. The landfill developer agrees that the transient model addresses many of his concerns with the steady-state model. Although direct observations of transport are still not available, the model has been calibrated with transient observations of head and flow that were collected under the same stress conditions that will be in place when the landfill is operating and that will be used for predicting advective transport. In addition, these data were sufficient to calibrate the model without including prior information. The landfill developer also is pleased that the uncertainty of most flow system parameters has been reduced, compared to the parameter uncertainty calculated using the steady-state model.

A thorough analysis of the potential transport from the landfill still requires an advective-dispersion transport model, but, as for the steady-state model, advective transport alone will be considered first. Including dispersion is not needed for this problem if the advected particle goes to the well and the time of arrival is not a concern.

The advective travel paths (Exercise 9.12a) and their uncertainty (Exercise 9.12c) can be analyzed under steady-state conditions with pumping, because the landfill will be operating under the condition of approximately steady pumping from the proposed supply wells near the center of the aquifer. The evaluation of the model's ability to simulate the predictions (Exercise 9.12b) also uses a transient simulation that includes the calibration observations. All analyses of the predictions use models with parameter values from the calibrated transient model.

The results of the simulations are used to address questions similar to those posed in Exercises 8.1 and 8.2 for the original steady-state model and predictions. Instructions for these simulations are available from the website for this book listed in Chapter 1, Section 1.1.

(a) *Predict advective transport with the updated steady-state model.*

First, predict the advective-transport path from the landfill location using the ADV Package in a forward MODFLOW-2000 run of a steady-state model. Model output that describes the movement of the advective path is shown in Figure 9.21.

Problem: Plot the particle path on a copy of the model grid of Figure 2.1*b*, and answer the following questions.

- Does an advective path from the landfill go to the well or the river? How long does it take to get there?

9.4 EXERCISES

```
ADVECTIVE-TRANSPORT OBSERVATION NUMBER  1
          PARTICLE TRACKING LOCATIONS AND TIMES:
 LAYER    ROW    COL   X-POSITION   Y-POSITION   Z-POSITION      TIME
-----------------------------------------------------------------------------
    1      2     16      15500.       1500.0       100.00       0.0000
 OBS #     1-     3         OBS NAME: AD10
    1      2     16      15178.       1575.8        85.940      0.31500E+09
.............................................................................
    1      2     15      15000.       1615.4        79.690      0.47394E+09
    1      2     14      14000.       1875.5        56.849      0.12269E+10
    1      3     14      13600.       2000.0        51.405      0.14794E+10
    2      3     14      13469.       2037.2        50.000      0.15518E+10
 PARTICLE ENTERING CONFINING UNIT
.............................................................................
 OBS #     4-     6         OBS NAME: AD50
    2      3     14      13469.       2037.2        48.862      0.15700E+10
.............................................................................
    2      3     14      13469.       2037.2        40.000      0.17114E+10
 PARTICLE EXITING CONFINING UNIT
    2      3     13      13000.       2167.8        34.419      0.20230E+10
    2      3     12      12000.       2539.7        25.685      0.26478E+10
.............................................................................
 OBS #     7-     9         OBS NAME: A100
    2      3     12      11165.       2909.6        20.380      0.31500E+10
.............................................................................
    2      3     11      11000.       2988.7        19.436      0.32485E+10
    2      4     11      10980.       3000.0        19.336      0.32603E+10
    2      4     10      10000.       3609.3        14.987      0.38208E+10
    2      5     10       9464.0      4000.0        13.057      0.41490E+10
    2      5      9       9000.0      4426.0        11.385      0.44536E+10
    2      6      9       8497.7      5000.0        10.083      0.48233E+10
    2      7      9       8046.1      6000.0         8.1157     0.53184E+10
.............................................................................
 OBS #    10-    12         OBS NAME: A175
    2      7      9       8018.8      6524.4         6.9647     0.55200E+10
.............................................................................
    2      7      8       8000.0      6988.7         6.1411     0.56728E+10
    2      8      8       7999.0      7000.0         6.1113     0.56810E+10
    2      8      9       8000.0      7001.1         6.1068     0.56817E+10
    2      9      9       8384.8      8000.0         3.0823     0.59752E+10
    2      9     10       9000.0      8186.7         1.6827     0.60413E+10
```

FIGURE 9.21 Part of the MODFLOW-2000 List output file (from the steady-state model with pumping) describing the path of a particle that originates at the top of the cell containing the proposed landfill.

- Does the path differ from that predicted using the model calibrated with only steady-state observation data?

(b) *Evaluate the model's ability to simulate predictions using composite and prediction scaled sensitivities, and parameter correlation coefficients.*

To assess the model's ability to simulate the predictions, evaluate whether the observation data support parameters that are important to the predictions. This evaluation uses the classification in Figure 8.2 and involves (1) comparing prediction and composite scaled sensitivities (*pss* and *css*), and (2) comparing parameter correlation coefficients (*pcc*) calculated using only the calibration observations with *pcc* calculated with the addition of the predictions.

The *pss* are calculated as the percent change in simulated value caused by a one-percent change in the parameter value (Eq. (8.2c)). The *css* are calculated using the final parameter values for the transient calibration (Figure 9.18). The *pcc* using only the calibration observations are shown in Table 9.7. The *pcc* with the addition of the predictions are calculated using a simulation with (1) steady-state conditions with pumpage, during which advective transport is predicted, (2) steady-state conditions without pumpage, during which the steady-state head and flow observations are simulated, and (3) transient conditions with pumpage, during which the transient head and flow observations are simulated.

Weights for the advective-travel predictions are needed to calculate the *pcc* with predictions. Weights at 10, 50, and 100 years are the same as those for Exercise 8.1b, shown in Table 8.3. Weights for the prediction at 175 years are calculated using a standard deviation of 1300 m in the *x* and *y* directions, and a standard deviation of 50 m in the *z* direction.

The *css* are plotted together with the *pss* in Figure 9.22. The *pcc* with the predictions are shown in Table 9.8.

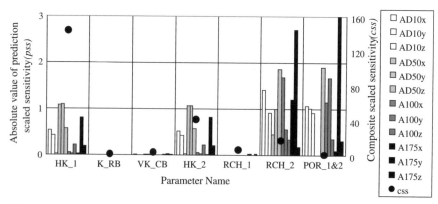

FIGURE 9.22 Final composite scaled sensitivities (*css*) from the calibrated transient model, and advective-transport prediction scaled sensitivities (*pss*) calculated using a model with parameters from the transient calibration. The *pss* represent the percent change in advective travel that would be caused by a one-percent change in the parameter value.

9.4 EXERCISES

TABLE 9.8 Parameter Correlation Coefficient Matrix Calculated by MODFLOW-2000 Using Parameter Estimates and Observations from the Transient Calibration, and Using the Advective-Transport Predictions[a]

	Q_1&2	SS_1	HK_1	K_RB	VK_CB	SS_2	HK_2	RCH_1	RCH_2
Q_1&2	1.00	−0.65	**−0.99**	−0.066	−0.40	−0.035	−0.92	−0.37	−0.84
SS_1		1.00	0.63	−0.26	0.80	−0.71	0.58	0.22	0.53
HK_1			1.00	−0.050	0.42	0.036	0.84	0.38	0.82
K_RB				1.00	−0.43	0.42	0.32	0.016	0.076
VK_CB		Symmetric			1.00	−0.75	0.30	0.15	0.32
SS_2						1.00	0.063	0.028	0.047
HK_2							1.00	0.31	0.79
RCH_1								1.00	−0.17
RCH_2									1.00

[a] Correlation coefficients greater than 0.95 in absolute value are in bold type.

Problem: Using Figure 9.22 and Tables 9.7 and 9.8, determine what parameter values are most important to predicted advective transport, and assess how well these parameter values are supported by the observation data used in the transient calibration. Does this analysis suggest that the model does a good job of simulating the predictions?

(c) *Evaluate prediction uncertainty using inferential statistics.*

In this exercise, linear and nonlinear, simultaneous, 95-percent confidence intervals on the advective-transport predictions are computed. Equation (8.12) is used for the linear intervals and Eq. (8.14) is used for the nonlinear intervals. The results of Exercise 9.12b showed that most of the advective transport predictions have relatively large sensitivities to the aquifer effective porosity, parameter POR_1&2. Because of this result, uncertainty in effective porosity is considered when calculating the confidence intervals on advective transport, using prior information and weighting, as was done for the calibrated steady-state model in the exercises of Chapter 8. The weighting used for the prior information is calculated by forming a 95-percent confidence interval of 0.27 to 0.39 for the true effective porosity value.

The linear and nonlinear, simultaneous confidence intervals in the x and y directions are shown on an aerial map of the model domain in Figure 9.23a,b. Intervals in the z direction are shown in Figure 9.23c.

In Figure 9.23b, dashed lines form the upper limits of the nonlinear confidence intervals for the predictions in the x and y directions at 175 years and in Figure 9.23c, the lower limit of the interval in the z direction at 175 years extends below the bottom of the system. These limits are simulated using parameter values that cause the advective-transport path to reach the well in less than 175 years. As explained in Exercise 8.2b, when a particle reaches a flow model boundary such as a well prior to the prediction time, the ADV Package of MODFLOW-2000 projects the particle until the prediction time is reached (Anderman and Hill, 2001, p. 12). The dashed line in Figure 9.23b,c

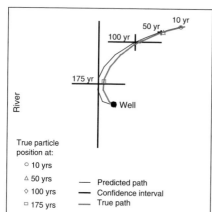
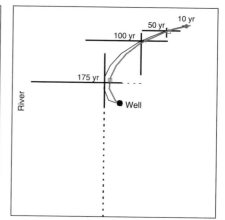
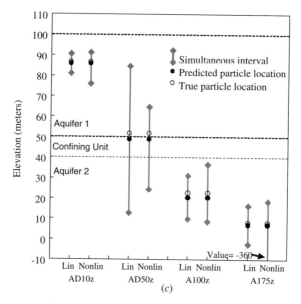

FIGURE 9.23 (*a*) and (*b*) Plan view of model grid showing predicted advective-transport path from the proposed landfill and 95-percent confidence intervals in the *x* and *y* directions at simulated travel times of 10, 50, 100, and 175 years. The true path is also shown. These results are produced using the parameter estimates from the transient calibration. (*a*) Linear, simultaneous confidence intervals. (*b*) Nonlinear, simultaneous confidence intervals. (*c*) Cross-section showing predicted vertical particle positions and linear and nonlinear confidence intervals. True *z* locations are shown. At 175 years, the interval limits with dashed line in (*b*) and the nonlinear interval in (*c*) are affected by projections simulated by the ADV Package when a particle exits in the model, as discussed in the text. Smaller intervals at late time reflect that the pumpage limits the potential location at late time.

9.4 EXERCISES

shows the limits formed by the projected particles. A more realistic approximation of the limits in the x and y directions is the location of the well, and in the z direction it is the aquifer depth at which the particle enters the well. These limits are represented by the solid lines on Figure 9.23b,c for the nonlinear intervals in the x, y, and z directions at 175 years.

Problem
- Are the predictions calculated in this exercise more or less certain than those calculated using the original steady-state model (Figure 8.6)? Why? What are the potential effects of model nonlinearity?
- What are the differences in size and symmetry between the linear intervals (Figure 9.23a) and nonlinear intervals (Figure 9.23b)?
- How do the predicted path and particle locations at 10, 50, 100, and 175 years compare to the true advective-transport path? Why is the prediction closer to the true path and particle locations than for the calibrated steady-state model (compare Figure 9.23 with Figures 8.15 and 8.16)?

10

GUIDELINES FOR EFFECTIVE MODELING

This chapter introduces and summarizes a set of guidelines for effective modeling of natural and engineered systems. These guidelines show how data, models, and the methods presented in Chapters 3 through 9 can be used together to gain insight into the simulated system, and to successfully attain goals related to calibrating and evaluating the simulated system. The guidelines are summarized in Table 10.1 and are explained in Chapters 11 through 14. The guidelines are organized into four topics: (a) Guidelines 1 through 8 for model development (presented in Chapter 11), (b) Guidelines 9 and 10 for model testing (Chapter 12), (c) Guidelines 11 and 12 for evaluating potential new data (Chapter 13), and (d) Guidelines 13 and 14 for evaluating prediction uncertainty (Chapter 14).

In Figure 1.1, the terms "system information" and "observations" are used for what are sometimes called "soft" and "hard" data, respectively. For a groundwater system, the system information includes hydrologic and hydrogeologic data; observations include hydraulic heads, streamflow gains and losses, and concentrations used directly or used interpretively to define advective-travel observations. In the guidelines, the terms "system information" and "observations" are used instead of "hard" and "soft" data because we believe they describe the data more clearly. For example, prior information generally is derived from system information, but because it appears in the regression objective function in the same manner as observations, it is sometimes classified as hard data. Using the terms system information and observations reduces the confusion.

Effective Groundwater Model Calibration: With Analysis of Data, Sensitivities, Predictions, and Uncertainty. By Mary C. Hill and Claire R. Tiedeman
Published 2007 by John Wiley & Sons, Inc.

GUIDELINES FOR EFFECTIVE MODELING

TABLE 10.1 Guidelines for Effective Development and Use of Models

Guideline[a]	Description and Suggested Actions[b]
Preliminary Steps (Not covered by the guidelines. See, for example, Anderson and Woessner, 1992)	
Define purpose	Design the model to meet the modeling objectives.
Develop conceptual models	Select processes and system characteristics. Identify ways to attain a tractable model and aspects that are uncertain.
Choose code	Use modular codes that allow easy inclusion and exclusion of processes.
Model Development Guidelines (Chapter 11)	
1. Apply the principle of parsimony	• Start simple and add complexity as warranted by the hydrology and hydrogeology, the inability of the model to reproduce observations, and the complexity that can be supported by the available observations.
2. Use a broad range of system information (soft data) to constrain the problem	• Identify spatial and temporal structure. Use it to represent the system well using few parameters. • Do not add features or parameters to improve model fit if they contradict system information. • Possibly use geographic information systems (GIS) and 3D database and visualization methods to organize, analyze, interpret, and present data.
3. Maintain a well-posed, comprehensive regression problem	• Maintain a well-posed regression: define few parameters. • Maintain a comprehensive model: represent many aspects with parameters. • To be both well-posed and comprehensive, seek simple models that represent important system dynamics. • Detect ill-posed regressions with *css* and *pcc*.
4. Include many kinds of data as observations (hard data) in the regression	• Add different kinds of observations; this can be critical to obtaining a reasonably accurate model. In groundwater flow model calibration, it is very important to include information about flows. • Use *opr* to evaluate which observations dominate the predictions.
5. Use prior information carefully	• Begin with no prior information to investigate the observations. • Insensitive parameters (e.g., small *css*): include with prior information or exclude to reduce run time. Include for Guidelines 11–14. • Sensitive parameters: do not use prior information to make unrealistic optimized parameter values realistic. See Guideline 10.
6. Assign weights that reflect errors	• Assign weights that equal $1/\sigma_i^2$.
7. Encourage convergence by making the model more accurate and evaluating the observations	• If nonlinear regression does not converge (can occur even when *css*, *pss*, and so on indicate observations are sufficient to estimate the parameters), work to make the model represent the system more accurately and make sure observations are interpreted correctly.

(Continued)

TABLE 10.1 *Continued*

Guideline[a]	Description and Suggested Actions[b]
8. Consider alternative models	• Use model fit, *dss*, *css*, *pss*, and system information to determine what to change. • Develop alternative models using deterministic or stochastic methods. • Judge models based on better fit and more realistic parameter estimates.
Model Testing Guidelines (Chapter 12)	
9. Evaluate model fit	• Use standard error, AIC_c, and other statistics from Chapter 6 to assess overall model fit. • Use weighted and unweighted residuals to assess details of model fit.
10. Evaluate optimized parameter values	• Unreasonable estimated parameter values can indicate model error. • Perhaps combine parameters with overlapping confidence intervals, divide parameters with large *css*.
Potential New Data Guidelines (Chapter 13)	
11. Identify new data to improve simulated processes, features, and properties	• Use fit-independent statistics *dss*, *css*, *pcc*, *leverage* to identify potential important new observations. • Use *css* and *pcc* to identify parameters for which existing and potential observations contain substantial information. Consider representing the associated system characteristics using additional estimated parameters.
12. Identify new data to improve predictions	• Identify observations and parameters important to predictions using fit-independent statistics *dss*, *css*, *pss*, *pcc*, *ppr*, *opr*.
Prediction Uncertainty Guidelines (Chapter 14)	
13. Evaluate prediction uncertainty and accuracy using deterministic methods	• Use regression to determine whether predicted values of interest (such as regulatory guidelines) contradict the observations. • Use postaudits to test prediction accuracy.
14. Quantify prediction uncertainty using statistical methods	• Use statistical inference—linear and nonlinear. Includes uncertainty intervals. • Use designed and random sampling—omit poor-fit realizations. • Include parameters not estimated by regression, perhaps with prior information. • Consider alternative models by including the probability of each.

[a] The guidelines generally are used iteratively, not just once in sequence.
[b] *dss*, *css*, *pss*, dimensionless, composite, and prediction scaled sensitivities, respectively; *pcc*, parameter correlation coefficients; *ppr*, parameter–prediction statistic; *opr*, observation–prediction statistic; 3D, three-dimensional; σ_i^2 is the best approximation of the observation error variance. See text for discussion of weight matrices. Fit-independent statistics are italicized.

This chapter explains the purpose of the guidelines, discusses them in the context of previous work and other modeling approaches, and provides suggestions for effectively implementing them during modeling.

10.1 PURPOSE OF THE GUIDELINES

An entire modeling protocol is presented by Anderson and Woessner (1992, pp. 4–9), which spans the modeling procedure from defining the purpose of the model and selecting a code, through predictive analysis and postaudits. The guidelines presented in Chapters 10 to 14 fit into that protocol, enhancing the sensitivity analysis, calibration, prediction, and uncertainty evaluation phases. The guidelines also emphasize investigation of different conceptual models. The guidelines do not address the preliminary steps of the protocol. For example, there are no guidelines for the important steps of defining the modeling objectives and selecting or programming a code with the appropriate capabilities.

The guidelines are closely tied to the modeling process represented in Figure 1.1. As discussed in Chapter 1, Section 1.1, Figure 1.1 shows how the model, with its defined parameters, quantitatively links the system information and the observations to the predictions of interest and measures of prediction uncertainty. Figure 1.1 emphasizes the direct links the model provides between the triad composed of *observations*, *parameters*, and *predictions*. The methods and statistics presented in Chapters 3 through 8 take advantage of these links. Selected statistics that connect each element of the triad are listed in Table 10.2. The guidelines show how modelers can use these links and associated methods and statistics advantageously during model development, testing, and evaluation of predictions and their uncertainty.

TABLE 10.2 Statistics[a] from Chapters 4, 7, and 8 that Indicate the Importance of Observations to Parameters, Parameters to Predictions, or Observations (Through the Parameters) to Predictions

Observations–Parameters	Parameters–Predictions (Chapter 8)
dss, css, pcc, leverage (Chapter 4) Parameter standard deviations, coefficients of variation, confidence intervals, DFBETAS, Cook's *D* (Chapter 7)	*pss, ppr*
Observations–Parameters–Predictions (Chapter 8)	
opr Prediction standard deviations, coefficients of variation, confidence intervals. Cross-validation, jackknifing, bootstrapping (only mentioned briefly in the text).	

[a]*dss, css, pss*, dimensionless, composite, and prediction scaled sensitivity, respectively; *pcc*, parameter correlation coefficient; *ppr*, parameter–prediction statistic; *opr*, observation–prediction statistic. Fit-independent statistics are italicized.

In the context of the entire modeling process shown in Figure 1.1, the ideas suggested in the guidelines are aimed at facilitating effective use of the system information and the observations to constrain the model and at making the model and model development more transparent. The goal is to produce a model that represents the simulated system more accurately, compared to modeling procedures that use these data less effectively, and to encourage clear testing.

10.2 RELATION TO PREVIOUS WORK

The guidelines are presented in the context of groundwater modeling problems but are applicable to other fields. Many aspects of the approach have had a long history in a variety of fields. The idea of parsimony—starting simple and building complexity slowly—is emphasized in Guideline 1 and has been discussed by Popper (1982), Cooley et al. (1986), Constable et al. (1987), Backus (1988), Cooley and Naff (1990), and Parker (1994). The importance of conceptual models is discussed by many authors, including Bredehoeft (2003, 2005). Most of the graphical analyses of Guideline 8 were suggested for application to groundwater problems by Cooley and Naff (1990) as derived from Draper and Smith (1981). Very similar approaches were tested using simple and complex synthetic test cases in Poeter and Hill (1996, 1997) and in Hill et al. (1998). Alternative guidelines have been presented by Refsgaard and Henrikson (2004). Hill et al. (2004) provide a review.

From the perspective of stochastic inverse methods (e.g., Kitanidis, 1997), many aspects of the approach presented here can be applied directly. This is accomplished by considering the parameters of the stochastic model to be analogous to the parameters discussed in this work, and calculating sensitivities appropriately.

Alternatively, the approach presented here can be thought of as a strategy to approximate mean, or effective, values. Stochastic methods generally require that the mean of any spatially distributed quantity, such as hydraulic conductivity, be constant, a simple function, or known. Unfortunately, geologic media often defy these limitations. A model developed using the guidelines presented here can be used to evaluate whether the mean is constant, and, if not, to provide an estimate of what could be a very complex spatial distribution, often with sharp contrasts. Once large-scale variations are established, stochastic methods can be used to assess the influence of small-scale variations. To date, methods to characterize large- and small-scale variations mostly have been considered separately, and integration is sorely needed. One goal of such work can be thought of as identifying the aspects of a given problem that can most profitably be regarded as deterministic, and the aspects that can be most profitably be regarded as stochastic, given the information available (perhaps using the ideas in Guidelines 2 and 4) and the objectives of the work (such as the predictions considered in Guideline 12).

10.3 SUGGESTIONS FOR EFFECTIVE IMPLEMENTATION

Although the guidelines are presented roughly in the order along which most studies proceed, flexible application is important to their success. We encourage modelers to

TABLE 10.3 Common Questions, Useful Statistics and Graphs, and Related Figures, Tables, and Guidelines[a]

Questions	Statistic or Graph	Figures (**F**) or Tables (**T**)	Guideline
Evaluate Information that Observations Provide on Parameters			
Which of the defined parameters can be estimated? What types of additional parameters can be estimated	*css*	**F**: 4.3, 7.5, 9.11, 9.18, 11.3 **T**: 5.5	1, 2, 3, 5, 11
	leverage	**F**: 7.2; **T**: 13.1	
	DFBETAS	**T**: 7.3	
Can the defined parameters be estimated uniquely?	*pcc*	**F**: 4.2, 5.4, 9.5, 11.4 **T**: 4.2, 4.3	3
Which observations contain the most information?	*dss*	**F**: 8.1, 13.1; **T**: 4.1, 7.5, 13.1	3, 4, 11
	Cook's *D*	**F**: 15.6; **T**: 7.2	
	one-percent-scaled sensitivities	**F**: 4.4, 9.6–9.10	
Evaluate Model Fit			
Are the observations correctly interpreted, simulated, and weighted?	Extreme values of residuals and weighted residuals	**F**: 6.8	9
What is the overall fit for all observations and for each type of observation?	Standard error, fitted error statistics, AIC_c, BIC	**F**: 6.6, 9.13, 11.6	9
How is the fit to individual observations? Are the weighted residuals random?	Weighted residuals vs. [weighted] simulated values	**F**: 6.1, 6.2, 6.7, 9.14, 11.7, 15.3	9
	Graphs using independent variables	**F**: 6.9, 9.15, 12.2, 12.3, 15.4, 15.5	
	Runs test	**F**: 6.4, 6.10, 9.16, 12.4	
Are the weighted residuals normally distributed?	Normal probability graphs	**F**: 6.5, 6.11, 6.14, 9.17	—
	R_N^2	**F**: 6.12, 9.17	

(*Continued*)

TABLE 10.3 Continued

Questions	Statistic or Graph	Figures (**F**) or Tables (**T**)	Guideline
How can misfit be obscured? (Question is included to discourage use of this graph.)	[Weighted] observed vs. [weighted] simulated values	**F**: 6.3, 6.7, 9.14, 15.3	9
Evaluate Estimated Parameter Values			
Are the parameter estimates reasonable?	Estimated parameter values, confidence intervals, reasonable ranges	**F**: 7.4, 7.7, 7.8, 7.19, 12.5, 15.15 **T**: 7.9	
Are the parameter estimates unique?	*pcc*	**T**: 7.4, 7.6, 7.8, 9.7	10
Evaluate Predictions			
How accurate are the predictions?	Check for bias using model fit and estimated parameter values	As for model fit and estimated parameters	13
How precise are the predictions?	Confidence intervals, Monte Carlo	**F**: 8.15, 8.16, 9.23, 14.1	14
How reliable are the predictions?	Accuracy and precision → reliability	As for previous two questions	
What parameters are most important to the predictions?	*css*, *pss*, *pcc* *ppr*	**F**: 8.1, 8.8, 9.22, 15.8 **T**: 8.4, 8.5, 9.8 **F**: 8.9, 15.8	12
What observations are most important to the predictions?	*dss*, *css*, *pss*, *pcc* *opr*	**F**: 8.1, 8.11; **T**: 8.7 **F**: 8.10, 8.12, 8.13, 15.9, 15.10	12

[a] *dss*, *css*, *pss*, dimensionless, composite, and prediction scaled sensitivities, respectively; *pcc*, parameter correlation coefficients; *ppr*, parameter–prediction statistic; *opr*, observation–prediction statistic; [weighted], could be unweighted instead. Fit-independent statistics are italicized.

10.3 SUGGESTIONS FOR EFFECTIVE IMPLEMENTATION

follow the guidelines out of order if warranted by the individual modeling situation, or to revisit some of the guidelines during the course of model development, calibration, and evaluation. For example, analyses of prediction uncertainty discussed in Guideline 14 often are useful in guiding data collection, which is the topic of Guidelines 11 and 12.

Sun (1994, p. 210) recognized the need for flexible application of modeling steps. He noted that there is an inherent difficulty associated with the optimal design of data collection for nonlinear problems: the solution for the optimal design depends on the values of the unknown parameters, which in turn depend on the data. In addition, new data may cause the conceptual model to evolve and may challenge previous conceptual models and result in changes to many aspects of the model, including the optimized parameter values. Sun (1994) presents some elegant methods of addressing this problem that are generally very computationally intensive. The methods presented in this book tend to be simpler and less computationally intensive, while still being useful in many situations. The methods presented here may be used alone or may serve as preliminary steps to a more computationally intensive evaluation.

The guidelines do not suggest formally considering the predictions or using the model to evaluate potential new data until Guidelines 11 and 12. This is because it is expected that a reasonably accurate model is needed for a quantitative evaluation of predictions. The placement of predictions in the guidelines is not intended to diminish the importance of considering prediction issues throughout data collection and model development. Indeed, as predictions differ from observations significantly in terms of location, depth, time, type, or system stresses, it becomes increasingly important to simulate the predictions as calibration proceeds, as emphasized in Figure 1.1. This allows the modeler to understand how the assumptions and simplifications being made during model calibration affect the predictions. However, the results need to be considered cautiously until the model is reasonably accurate, which is the reason for the order of the guidelines. In addition, ethics require that the model not be designed to obtain desired predictions. If assessing predictions during calibration in any way endangers the integrity of the model, delay the simulation of predictions until the end of model development.

Many statistical and graphical analyses related to inverse modeling methods were presented in Chapters 3 to 8. In the guidelines, additional examples of using these statistics and graphs are presented. To aid cross-referencing, Table 10.3 lists most of the statistics and graphs discussed in Chapters 3 to 8 and shows the figures and guidelines in which they are presented and discussed in Chapters 11 to 14. These are presented in the context of typical questions that arise during model sensitivity analysis, calibration, and evaluation.

As the methods described in this work are used in the context of the guidelines, modelers may devise new methods or apply these to new situations. Thoughtful innovation is welcome and essential in this immature field.

11

GUIDELINES 1 THROUGH 8— MODEL DEVELOPMENT

Eight guidelines focus on model development: (1) follow the principle of parsimony in all model development endeavors; (2) use system information effectively; (3) use as few parameters as possible to represent as many important aspects of the system as possible; (4) include observations that cover a broad range of system dynamics; (5) use prior information when appropriate; (6) specify weighting that represents errors; (7) encourage convergence of the regression by using results from failed regressions to guide model improvements; and (8) consider alternative conceptual models.

GUIDELINE 1: APPLY THE PRINCIPLE OF PARSIMONY

> The methods of science depend on our attempts to describe the world with simple theories. Theories that are complex become unstable, even if they happen to be true. Science may be described as the art of oversimplification: the art of discerning what we may with advantage omit.
>
> —Popper (1982)

The principle of parsimony calls for keeping the model as simple as possible while accounting for the system processes and characteristics that are evident in the observations and are important to the predictions, and while respecting all system information. In many fields, including groundwater hydrology, the known

---

Effective Groundwater Model Calibration: *With Analysis of Data, Sensitivities, Predictions, and Uncertainty.* By Mary C. Hill and Claire R. Tiedeman
Published 2007 by John Wiley & Sons, Inc.

complexities of the simulated systems often seem overwhelming, and applying parsimony in model development can require substantial restraint.

Of greatest concern are two- or three-dimensional spatial fields that also may vary in time. Literally an infinite number of parameters could be defined. In numerical models, the possible number of parameters is finite because of the discretization of the numerical grid or mesh, but it is still far more than can be supported using observations of the simulated system, and probably more than is useful.

G1.1 Problem

Keeping a model simple is important because though more complex models generally fit the observations more closely compared to simpler models, they can have greater prediction error. For example, consider the situation shown in Figure 11.1a,b, where the true model is linear. A more complicated model (Figure 11.1b) clearly produces a better fit to the observations, but much of the improved fit is achieved by matching the observation error rather than the system processes. In this example the predictions are less accurate in the more complicated model than in the simpler model. Figure 11.1c displays the general situation, in which there is a trade-off between model fit and prediction accuracy with respect to the number of parameters. All model-fit statistics used for model discrimination include a penalty as the number of parameters increases to account for this effect; see Chapter 6, Section 6.3.2.

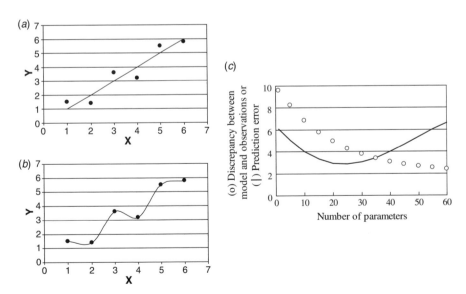

FIGURE 11.1 (a) Data with a true linear model. (b) The same data with an overly complex model with little predictive capability. (c) Schematic graph showing conceptually the trade-off between model fit to observations and prediction accuracy with an increasing number of parameters.

In practice, we do not know many of the characteristics of the underlying model. We do not know, therefore, when we are using observations advantageously to characterize the processes of concern, and when we are fitting errors and are probably degrading the predictive capabilities of the model. Guideline 1 suggests approaching this problem from the left in Figure 11.1c—that is, by starting with simple models and building complexity carefully. By starting with a simple model, the modeler can more easily understand the effect that added complexity, such as additional simulated processes or using more parameters to represent system features, has on model fit, optimal parameter estimates, predicted values, and prediction uncertainty. This helps keep the behavior of the model as a whole in perspective, compared to narrowly focusing on a small portion of the spatial or temporal domain of the model. This "big picture" view is consistent with the sparse data available for characterizing many systems. In many cases, it also is consistent with the detail needed to obtain useful predictions.

There has been an active discussion in the Earth science literature about the advantages and disadvantages of using models with different levels of complexity. For example, Parker (1994) and Smith et al. (1999) address these issues with regard to a geophysical investigation and suggest the utility of simple models; Murray (2002, 2003), Bras et al. (2003), and Harry (2003) discuss general numerical modeling issues, the first two with an emphasis on geomorphic modeling; de Marsily et al. (2005) stress the importance of detailed hydraulic-conductivity structure in simulations of groundwater transport (an issue also mentioned in Chapter 9, Section 9.2.3); Hill (2006) and Gomez-Hernandez (2006) debate simplicity and complexity of groundwater models. Yeh and Sun (1990) suggest a stepwise approach. Oreskes (2000) discusses the paradox of complex models and the importance of refutability and transparency. Refutability means the model is constructed such that different assumptions can be tested; transparency means the model dynamics are understandable. Both refutability and transparency suffer as a model becomes more complicated, and this loss needs to be weighed against perceived advantages gained. One goal of the guidelines is to increase refutability and transparency.

G1.2 Constructive Approaches

Applying the principle of parsimony to all aspects of model development is important. For example, only include the processes needed for the system being simulated. The most useful models are designed modularly so that different combinations of processes can easily be included to test their relevance and unused capabilities do not interfere by increasing execution time or computer storage, or affecting simulated values. This approach also allows execution time to be managed efficiently, as noted in Chapter 15, Section 15.1.

To represent a system adequately with relatively few parameters, as suggested in Guideline 1, the model and parameters need to be defined carefully to capture the important processes and system features. When considering changes that decrease the complexity of a model, it is important to test whether system information and/or observations contradict the changes and the importance of the changes to

GUIDELINE 1: APPLY THE PRINCIPLE OF PARSIMONY

the predictions. The remaining model development guidelines suggest how to obtain a parsimonious, useful model.

The remainder of this section investigates two difficulties commonly encountered that need to be managed well to obtain a useful model. The first is nonlinearities of the forward model and the second is variability of system properties. Both are presented in the context of groundwater models, but the basic ideas are directly applicable to other fields.

Managing Forward Model Nonlinearities Part of managing execution time efficiently is managing forward model nonlinearities efficiently. Replacing nonlinear forward problems with linear approximations as much as possible can dramatically reduce execution time. If designed wisely, this can be achieved without substantially diminishing model accuracy. Basically this comes down to managing the nonlinearity to best serve the purpose of the model.

In groundwater flow simulations, for example, unconfined and convertible layers (as they are called in MODFLOW96 and MODFLOW-2000, respectively) can be replaced by confined layers with approximate defined thicknesses during model calibration. Using confined layers is always good practice for steady-state simulations because in the final calibrated model the saturated thicknesses are expected to conform to observed hydraulic heads. Allowing saturated thicknesses to vary as the parameters vary during calibration can be a numerical nightmare and produces no advantage. The result will be about the same regardless of whether confined layers are used during calibration; we suggest choosing the easier option and spending the effort on more worthwhile endeavors. A sensible approach is to maintain saturated thicknesses that are consistent with the observed heads and, therefore, the expected simulated heads in the final calibrated model. This can save a tremendous amount of time and aggravation.

The approach also can be useful in transient simulations depending on how much de-watering or saturation occur over time. It is important to consider the proportion of the pumped strata that becomes dewatered, which may not be the same as the proportion of a pumped model layer. For example, if a well intersects permeable material that is 50 m thick and is represented by ten 5-m thick model layers, it is the proportion of the 50-m thickness that needs to be considered.

Once a model is close to being calibrated, two options can be pursued.

First, the importance of the linear approximation can be evaluated. It is easy to evaluate the model inaccuracy that results from defining layers as confined with approximate specified thickness instead of as unconfined or convertible. This involves simply comparing a forward simulation that includes the water table and the convertible layers to a forward simulation with approximate layer thicknesses and confined layers.

Second, if needed, there are several options for integrating the nonlinearity into the simulation. For example, the top of the system can be updated using the topmost simulated hydraulic heads iteratively until little change occurs between iterations. Also, the water table can be explicitly simulated using, for example, the wet/dry capability of MODFLOW.

The likely consequence of using confined layers can be evaluated by considering the effect of dewatering on the transmissivity (hydraulic conductivity times saturated thickness) versus the effect of the possible range of hydraulic-conductivity values. For example, dewatering half the saturated thickness of a model layer or hydrogeologic unit reduces transmissivity by a factor of two, whereas possible variation in hydraulic conductivity for the layer easily can be an order of magnitude or more. In this situation, it would be reasonable to set the thickness and define model layers as confined, at least for preliminary regression runs. As noted, the model layers can then be represented as unconfined or convertible layers for final regression runs.

Managing Variability Often the variability of data is the driving force behind increasing model complexity. Scheibe and Chien (2003), however, present a study that suggests that increased model complexity is not always advantageous. They investigated an extensive groundwater data set using numerical simulations with transport predictions and different levels of detail used in representing the hydraulic-conductivity (K) field. The data are used to construct predictive models that were not calibrated. Scheibe and Chien (2003) draw the following conclusion important to Guideline 1: "model conditioning to local (effectively point support) data, even hundreds of such data, provides little benefit for prediction and may even provide misleading results. One would expect that conditioning data would improve predictions overall and decrease model uncertainty (narrow the range of variations in predicted behavior). However, the average summary performance metric for the simulations conditioned to borehole flowmeter measurements of K... was not significantly improved over the homogeneous base case."

However, Scheibe and Chien (2003) also found that conditioning on larger-scale hydraulic-conductivity data, consisting of estimates based on geophysical tomography, did significantly improve the predictions. Thus, this study found that adding very detailed complexity on the basis of local-scale measurements was not beneficial to predictive analyses, but that adding less detailed complexity was beneficial. These conclusions were possible because the investigators started with a very simple model, which served as a base case for objectively assessing whether or not the additional complexities were advantageous.

GUIDELINE 2: USE A BROAD RANGE OF SYSTEM INFORMATION TO CONSTRAIN THE PROBLEM

In most scientific and engineering modeling studies, there is system information that is related to model inputs. Effective use of this information in building conceptual models of the system can mean the difference between a model that represents the system well and one that does not. This applies whether or not the model is constructed in a parsimonious manner. In developing the parsimonious models encouraged by these guidelines, we try to use system information to define simplifications and approximations that produce a model with just enough of the right detail, and no more (paraphrasing Albert Einstein). The goal, of course, is for the resulting model to be as useful as possible, which requires as much transparency and refutability as possible (see Guideline 1).

G2.1 Data Assimilation

Guidelines 2 and 4, when taken together, emphasize what is sometimes referred to as data assimilation or data fusion. Examples of approaches for incorporating different types of system information in groundwater flow and transport model development are reported by Rubin et al. (1992), McKenna and Poeter (1995), Poeter and McKenna (1995), Eppstein and Dougherty (1996), Woodbury and Ulrych (2000), Barrash and Clemo (2002), and Chen and Rubin (2003). Many of these studies provide site-specific applications of the methods. A general framework for hydrologic data assimilation that is not limited to groundwater systems is described by McLaughlin (2002). Koltermann and Gorelick (1996) divide approaches of using field data to construct groundwater models into three categories: structure imitating, process imitating, and descriptive. It is becoming common to use diverse types of data in model construction, and methods for integrating these data have much potential for further development.

G2.2 Using System Information

System information can be used to define model structure, including the choice of processes to simulate, or to directly provide information on parameter values through determination of reasonable ranges (Chapter 7, Section 7.6), parameter limits (Chapter 5, Section 5.5), or prior information (Chapter 3, Sections 3.1 and 3.4.1). Here we focus on using system information to constrain the structure of groundwater models. This reflects the problem-specific nature of using system information to constrain model structure. We expect that providing concrete approaches for a specific field will be more useful than a general presentation.

If a groundwater model is to have any credibility, the simulated hydraulic-conductivity distribution needs to be consistent with the known hydrology and hydrogeology. Most groundwater investigations consider relatively shallow systems for which substantial surficial and subsurface information can be determined. This is in contrast to many fields of geophysics and other Earth sciences in which the great depths of interest preclude substantially constraining the calibration with known geology. Indeed, Carerra et al. (2005) state the following in relation to groundwater models: "when available, geologic information about parameter variability is so compelling (in the sense that it can be included deterministically) that it overcomes the advantages of conventional geostatistics."

For groundwater systems, hydrogeologic data often indicate that faults, fractures, and/or depositional processes have produced sharp contrasts in the hydraulic-conductivity distribution. These contrasts sometimes need to be explicitly represented in the model to simulate the system accurately. While zones of constant value provide an unrealistically uniform distribution of hydraulic conductivity within the zone, they are very useful when large-scale contrasts overwhelm the importance of smaller scale features. Even when smaller scale features are important, they often need to be characterized within the structure provided by such zones. At large or small scales, depositional conditions may suggest a gradual refining or coarsening in horizontal or vertical directions such that interpolation methods instead of zonation are most useful. When many interpolation points are used, as in pilot point methods, advantages of zonation can sometimes be captured. Representing hydraulic-conductivity variations is also discussed in Chapter 9, Section 9.2.3. D'Agnese et al. (1997, 1999) provide a good example of analyzing three-dimensional hydrologic and hydrogeologic data to construct a groundwater flow model of a complex system. This system is discussed in Chapter 15.

Commonly, the information used to constrain a problem as described in this guideline also is used to support the prior information on parameters discussed in Guideline 5. For example, the results of aquifer tests may be used to determine that two hydrogeologic units have similar hydraulic-conductivity values and probably can be combined to form one parameter in the regression. This information may be an important constraint on the problem. Later, the same results might be used to determine a prior information value for the combined or individual hydrogeologic units.

G2.3 Data Management

Evaluating, integrating, and using different types of system information for model development can require sophisticated data management capabilities. The level of sophistication required depends on the complexity of the system investigated. For example, in groundwater models, the system hydrogeology might be represented as homogeneous, as layered, or as a complex heterogeneous distribution. Homogeneous models require a simple data management structure, layered models generally require standard GIS (geographic information systems) capabilities, and more complex representations could require sophisticated methods found mostly in either fairly expensive software packages such as GMS (Environmental Modeling Systems, 2006) or more expensive software packages such as StratWorks 3D (Landmark, 2006), Earthvision (Dynamic Graphics, 2006), and GOCAD (Earth Decision, 2006). D'Agnese et al. (1999) and associated publications describe the software used to develop a complex groundwater flow model of the regional Death Valley groundwater system, and part of their discussion is presented in Chapter 15. Examples applied to glacial sediments are presented by Frind et al. (2002) and Ross et al. (2005). Many aspects of their approaches are directly applicable to studies of other types of systems.

For fully three-dimensional systems, the methods available for data organization and analysis are not very mature and can be very expensive. Recent and continuing advances in computer capabilities and standardization of technology related to sophisticated visualization and databases are likely to result in greatly improved methods in the coming years.

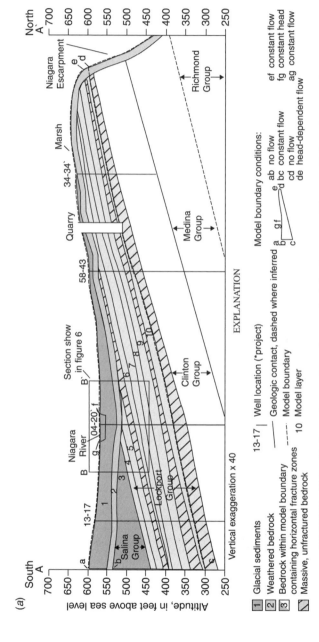

FIGURE 11.2 Hydrogeology of a fractured dolomite near Niagara Falls, New York. (*a*) Generalized section showing extent of model layers and model boundary conditions. (*b*) Relationships among stratigraphy, regional water-bearing zones, and model layers. Use of proposed nomenclature for the Lockport Group does not constitute formal acceptance by the U.S. Geological Survey. (From Yager, 1996, Figures 2 and 5.)

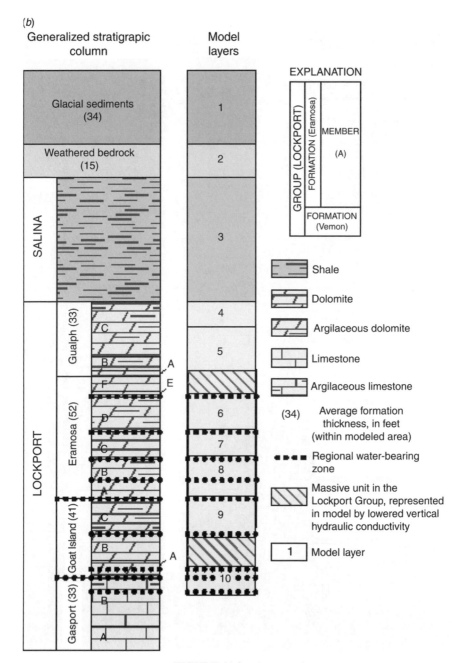

FIGURE 11.2 *Continued.*

GUIDELINE 3: WELL-POSED, COMPREHENSIVE REGRESSION   277

Generally, the constraints imposed by the system information are never enough to fully characterize the system. Computational methods have been proposed for identifying the structure of a model and can be useful as long as the system data are respected, such as geologic and hydrogeologic information for groundwater systems. Parameter structure identification methods are presented by Tsai et al. (2003a,b), who also provide a comprehensive review of past work on this topic applied to groundwater models. A continuing challenge is the integration of these methods with the constraints imposed by the system information.

G2.4 Application: Characterizing a Fractured Dolomite Aquifer

Yager (1996) simulated a groundwater flow system near Niagara Falls, New York, that is dominated by fractured dolomite (Figure 11.2). Definition of parameters for this system appeared problematic until a simple and powerful relation was derived from aquifer-test results available as part of the system information. Two factors contributed to the simplification.

1. The dominant regional fractures in the dolomite are roughly horizontal and along bedding planes. Fractures between these planes are dominated by vertical flow.
2. Transmissivities calculated for different aquifer tests were approximately proportional to the number of bedding-plane fractures intersected by the pumped well.

These two factors led to the assumption that each fracture has equal transmissivity, so that model-layer transmissivity is proportional to the known number of fractures in each layer. This relationship allows the entire heterogeneous horizontal hydraulic-conductivity distribution to be realistically represented using a single hydraulic-conductivity parameter and multiplication arrays that indicate the number of fractures in each model layer. Multiplication arrays are available in MODFLOW-2000 and possibly other models.

GUIDELINE 3: MAINTAIN A WELL-POSED, COMPREHENSIVE REGRESSION PROBLEM

The first part of this guideline suggests that the regression problem be well posed. For the purposes of these guidelines, a regression is well posed if it converges to an optimal set of reasonable parameter values given reasonable starting parameter values. In Earth systems, available observations are commonly sparse, so the requirement of maintaining a well-posed regression usually produces rather simple models with relatively few estimated parameters. Thought of in another way, only this simple level of model complexity can be supported by the observations, and the regression is providing an assessment of the information contained in the data. Thus, determining the greatest possible level of model complexity while maintaining a well-posed regression can be considered an objective analysis of the information provided by the observations. Prior information and regularization can be used to support additional complexity (see Guideline 5). However, it is important to model transparency for the modeler to know and to communicate to others what complexity is supported by the observations, what is supported by other types of information, and what is pure speculation.

The second part of this guideline suggests that the regression problem be comprehensive. This means characterizing as many aspects of a given system as possible using defined parameters. Being comprehensive is important for two reasons.

1. Preconceived notions about various aspects of the system only can be quantitatively tested against the observations by defining parameters, calculating sensitivities, and attempting estimation by regression. Such testing and resultant reevaluation of system characteristics is a key advantage of using regression methods, as discussed, for example, by Poeter and Hill (1997).
2. Many methods evaluate model uncertainty using the parameter variance–covariance matrix. As more aspects of the system are represented by defined parameters, more aspects are represented in the uncertainty evaluation using these methods.

The two parts of this guideline represent a fundamental tension faced by modelers of most natural systems. For most modelers, a comprehensive regression problem is easier to achieve than a well-posed regression problem. That is, it is easier to add complexity than to be simple, even if a simple design could be found that represents the system well.

A number of the statistics discussed in this book and listed in Table 10.2 can be used to encourage a well-posed problem. In the initial stages of model development, it can be advantageous to use fit-independent statistics. Sections G3.1 and G3.2 discuss the utility of two of the most useful fit-independent statistics: composite scaled sensitivities (css), presented in Chapter 4, and parameter correlation coefficients (pcc), presented in Chapters 4 and 7. Dimensionless scaled sensitivities (dss) also are useful, particularly for better understanding css values.

G3.1 Examples

The css and pcc, along with the system information discussed in Guideline 2, can be used to define parameters and to decide which parameters to estimate using regression. The css and pcc are well suited for this purpose because they are fit-independent, as discussed in Chapter 4. This means that they depend only on the observation sensitivities and weights; they are independent of the model fit to observed values. When evaluated at the starting parameter values, these fit-independent statistics can be used to determine what sets of parameters are likely to be estimated successfully given a model and a set of observations.

Composite Scaled Sensitivities (css) Composite scaled sensitivities were used to help achieve a well-posed regression for the three-layer model of the Death

GUIDELINE 3: WELL-POSED, COMPREHENSIVE REGRESSION

Valley regional groundwater flow system (DVRFS) (see Chapter 15). The bar chart of *css* for the initial, uncalibrated model used by D'Agnese et al. (1997, 1999) (Figure 11.3a) indicates that the K4 and RCH parameters are likely to be easily estimated by regression, whereas the ANIV1 and ETM parameters are not likely to be

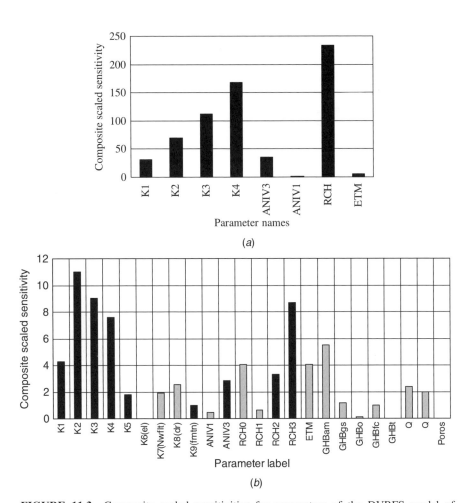

FIGURE 11.3 Composite scaled sensitivities for parameters of the DVRFS model of D'Agnese et al. (1997, 1999) for (*a*) the initial model and (*b*) the final model. In (*b*), parameters estimated by regression have black bars; parameters defined but not estimated by regression have gray bars. Parameters represent the following: K*, hydraulic conductivity; ANIV*, vertical anisotropy; RCH*, areal recharge; ETM, maximum evapotranspiration parameter; GHB*, conductance of head-dependent boundaries used to represent springs; Q, pumping (the two Q parameters apply to different areas); POROS, effective porosity. The observations provide no information for POROS, but this parameter is important to the transport predictions of interest. Together these parameters define all aspects of the system except the lateral and bottom boundary conditions.

easily estimated. In general, the available observations appear to contain substantial information about the K (hydraulic conductivity) and RCH (areal recharge) parameters, and less information about the ANIV (vertical anisotropy) and ETM (maximum evapotranspiration) parameters. Composite scaled sensitivities were calculated often during the calibration of this model and were used to determine what new parameters to introduce and whether previously excluded parameters should be included.

The *css* for the final model are shown in Figure 11.3*b*. Note that there are additional K and RCH parameters, and that most of these were estimated by regression. This is consistent with the initial evaluation showing that the data contained substantial information for these types of parameters. An important aspect of this analysis is that the basic conclusions from the initial and final evaluations are the same, despite the model nonlinearity and the substantial model and parameter-value changes made during calibration. This stability is typical and makes this method useful. If problems are too nonlinear to be stable, the utility of the composite scaled sensitivity method is diminished and possibly absent altogether.

Chapter 4, Section 4.3.4 states that if any parameters have *css* values that are less than one-percent of the largest *css* values, problems with convergence can be expected in regression. In Figure 11.3*b*, there are a few parameters with very small values, and these were not estimated. There are others with larger values that were not estimated for other reasons. For example, consider the one new type of parameter, GHB*, which are the hydraulic conductivity of the head-dependent boundary conditions used to represent groundwater supported springs. None of the GHB parameters were estimated in the regression for the final model because they tended to produce a good match solely to the flow of the spring or set of springs at which they were applied. This was evident because the dimensionless scaled sensitivities for each of these parameters generally were large for only one observation. Any error in the spring-flow measurements would have been fit by the model through adjustment of the GHB parameters. The values of the GHB parameters were determined primarily on the basis of hydrogeologic arguments and a few preliminary regression results.

Recall that the *pcc* indicate whether the estimated parameter values are likely to be unique. For the parameters of Figure 11.3*b*, all *pcc* were less than 0.95, suggesting that the estimates are unique. These simulations used a parsimonious model and the sensitivity-equation sensitivities of MODFLOWP (Hill, 1992) and MODFLOW-2000 (Hill et al., 2000), so potential problems with *pcc* accuracy, discussed in Chapter 4, Section 4.4.2, were not expected.

Parameter Correlation Coefficients (*pcc*) Anderman et al. (1996) and Anderman and Hill (1998) used *pcc* to investigate what types of observations were needed to achieve a well-posed regression for a model of groundwater flow in a shallow aquifer on Cape Cod, Massachusetts. The system had a lake and a well-monitored sewage plume. Three different sets of observation data were considered: (1) hydraulic heads

GUIDELINE 3: WELL-POSED, COMPREHENSIVE REGRESSION

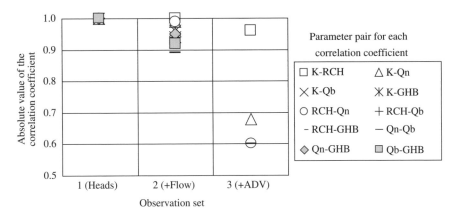

FIGURE 11.4 Parameter correlation coefficients for five parameters for three data sets from the Cape Cod sewage plume model as reported by Anderman et al. (1996), evaluated for the final parameter values. For values close to 1.00, parameter estimates are likely to be nonunique. For values less than 0.95, unique values are expected. The five parameters are K, hydraulic conductivity; RCH, areal recharge; Qb, sewage discharge sand-bed flux; Qn, northern boundary flux; GHB, conductance of the lake bottom. For data set 3, correlations not shown are $<|0.5|$. (From Anderman and Hill, 1998.)

only, (2) hydraulic heads and a lake seepage value, and (3) hydraulic heads, lake seepage, and an advective-transport observation derived from the monitored concentrations of the sewage plume. Figure 11.4 shows the *pcc* calculated at final parameter values for five model parameters, for each of the three observation data sets. This figure clearly shows that with only hydraulic heads (data set 1), all parameters are completely correlated (the absolute values of all correlation coefficients equal 1.00), so that any parameter estimates found by the regression are not unique. Adding one lake seepage measurement (data set 2) reduced correlations somewhat, but correlations remain very large. Only the observation set with the advective-travel observation (data set 3) could uniquely estimate all of the parameters.

G3.2 Effects of Nonlinearity on the *css* and *pcc*

As discussed above and in Chapter 4, Section 4.4, the utility of many of the statistics presented in this book can be affected by model nonlinearity. Some (like *pcc*) also are affected by even slightly inaccurate sensitivities. Here, we consider *css* and *pcc* because the authors have the most experience with these statistics; the discussion can largely be extended to other statistics. In general, difficulties with nonlinearity and inaccurate sensitivities occur when the conclusions drawn from the statistics are in error. The most common problems, their consequences, and suggested resolutions are displayed in Table 11.1.

TABLE 11.1 Consequences and Suggested Actions When Nonlinearity and Inaccurate Sensitivities Plague Parameter Correlation Coefficients and Nonlinearity Plagues Composite Scaled Sensitivities

Problem	Effect	Consequences	Actions
		Parameter Correlation Coefficients	
Nonlinearity	Correlation coefficients calculated using different sets of parameter values result in very different conclusions about whether the observations contain enough information to estimate parameters uniquely. (See Figure 4.2.)	Incorrect conclusions might be drawn about whether the observations are adequate to estimate the defined parameters uniquely.	In nonlinear systems, correlation coefficients can vary substantially for different sets of parameter values. The primary concern is whether the optimized values are unique.[a] That is, are any parameters extremely correlated at the optimized parameter values? If the answer is inconclusive because of problems with nonlinearity or inaccurate sensitivities, consider further tests: (a) Check whether the estimates and objective-function values are similar[b] in regressions with different starting values. (b) Reduce problems related to small css by combining parameters.[c] (c) If correlation is suspected, change the parameter values substantially in a way that reflects the suspected correlation. If the resulting objective-function value is similar,[b] extreme correlation is confirmed. Note: Parameter correlation cannot be detected by the progress or convergence of a single regression. (See Exercise 5.1b.)
Inaccurate calculated sensitivities (most problematic if css is small)[c]	Extremely correlated parameters do not have correlation coefficients equal to 1.00 or −1.00, as needed to detect correlation.		

282

Composite Scaled Sensitivities (css)

Nonlinearity	css values calculated using different sets of reasonable parameter values result in very different conclusions about whether the observations contain enough information to estimate certain parameters.	(a) Judgments about the feasibility of estimating a parameter may be in error. (b) Parameterizations and other aspects of model structure designed using css values results in unexpected model performance. For example, a system characteristic changes from being important to unimportant when the parameterization is changed.	(a) Judgments about likely success or failure of estimation are preliminary. If results are not as expected (e.g., parameters with relatively large css cannot be estimated), recalculate css for updated sets of reasonable parameter values. (b) Keep evaluating css as the model evolves. If problems persist, try to figure out why. Appropriate strategies depend on the situation. In nonlinear systems, parameter changes can fundamentally modify system dynamics. Use model fit to observations, understanding of the mathematical model, and the system information to evaluate the system dynamics for the different parameter values and pameterizations. If the problems continue to occur for plausible parameter values and system dynamics, consider using a global search technique (Chapter 5, Section 5.2).

---

[a]The focus here is nonuniqueness caused by parameter correlation, but further test (a) checks for local minima.
[b]"Similar" for parameter values means differences are small relative to the calculated parameter standard deviations (see Exercise 7.1e). "Similar" for objective-function values means within a few percent.
[c]This is discussed in Chapter 4, Section 4.4.2.

GUIDELINE 4: INCLUDE MANY KINDS OF DATA AS OBSERVATIONS IN THE REGRESSION

Guideline 4 stresses the importance of using as many kinds of observations as possible. Guidelines 2 and 4, when taken together, emphasize data assimilation. References on data assimilation are listed in Section G2.1.

Different systems offer different observation possibilities. For example, in groundwater flow problems it is important to augment commonly available hydraulic-head observations with flow observations. Flows often constrain solutions much more than do hydraulic heads, which tend to be easier to match. Using observations that reflect the rate and/or direction of groundwater flow, therefore, tends to promote the development of more accurate models. In many settings, measurements of groundwater flow are difficult to obtain. Often concentrations of contaminants are used. Groundwater age dates and geochemical measurements are alternative types of data that also can provide valuable information on flow rates and directions (see Section G4.4). Many studies have shown that regression results improve when transport observations are included in regression models of flow and/or transport, compared to use of only hydraulic heads and flows. Studies illustrating this finding are cited in Chapter 9, Section 9.2.5.

The observations that can be used with MODFLOW-2000, UCODE_2005, and PEST are described in Chapter 2, Table 2.1. Detailed analyses of the importance of different observation types are presented by, for example, Anderman et al. (1996), Poeter and Hill (1997), and in this book.

Three issues about observations are discussed next: the use of interpolated observations, clustered observations, and observations that are inconsistent with model construction. Lastly, three applications are presented.

G4.1 Interpolated "Observations"

In some circumstances, it is appealing to use interpolated values to increase the number of "observations" available for the regression. Interpolated values are obtained by interpolating the actual observations. If interpolated "observations" are used, then the errors of the interpolated "observations" are correlated with each other and with the errors in the actual observations. Thus, the weight matrix needs to be full (see Chapter 3 and Guideline 6). In groundwater examples, Clifton and Neuman (1982), Neuman (1982), Neuman and Jacobson (1984), and Carrera and Neuman (1986) kriged hydraulic-head measurements to generate interpolated hydraulic heads and used them as observations in the regression. When kriging is used, the associated kriging variances and variogram can be used to calculate the variance–covariance matrix on hydraulic-head observation errors that is needed to calculate the full weight matrix for the observations.

The disadvantage of interpolation methods is that the interpolated "observations" generally are not consistent with the processes governing the simulated system. For example, Cooley and Sinclair (1976) show that for groundwater systems, interpolated hydraulic heads are not necessarily based on the physics of groundwater flow. Thus, the interpolated values are expected to respect the underlying processes

GUIDELINE 4: INCLUDE MANY KINDS OF DATA AS OBSERVATIONS   285

represented in the model only to the extent that (1) they are constrained by the actual observations and (2) the actual processes happen to be represented well by the chosen interpolation scheme. The problems associated with item (2) can be severe if there are abrupt spatial variations in aquifer properties. Abrupt spatial variations can result in regions of steep and flat gradients that are only partially captured by the actual observations and, therefore, by the interpolation. This could cause the "observed" hydraulic-head distribution to be unrealistically smooth. Alternatively, it could cause a steep gradient evident in some of the actual observations to be unrealistically extrapolated.

These problems can be avoided if the observations are used directly in the regression. Then, the processes built into the model do the "interpolation," and lack of fit focuses attention on the quality of the observations and the model, which is where the focus of model calibration needs to be.

G4.2 Clustered Observations

In many circumstances data are clustered in limited areas or within short time intervals. For example, in groundwater systems, wells are often clustered in areas of high hydraulic conductivity where yields are highest and/or near population centers served by groundwater. As a result of the clustering of wells, hydraulic-head observations tend to be clustered spatially. Also, hydraulic-head observations are often limited to recent times, long after the start of substantial pumping.

There are two problems in regression related to clustered observations.

First, the presence of a large number of observations may make it difficult to evaluate whether the model is reproducing basic system characteristics represented by the observations. If the system characteristics are well represented by a subset of the observations, it can be productive to use the subset in at least the initial stages of regression. Alternatively, the observations can be averaged over defined areas or times. The error used to weight an averaged observation would not necessarily be much smaller than the error of the individual observations, because many components of error are caused by effects that averaging does not eliminate.

Second, clustered data can be problematic if they dominate regression methods such that observations are ignored elsewhere in the system or at other times. This would produce poor fit to sparse observations. If poor model fit occurs where observations are sparse, the possibility that clustering is the problem can be tested by grossly increasing the weights on some of the sparsely distributed observations in some regression runs. If this does not significantly improve the fit to those observations, then clustered data probably are not the problem. Instead, problems such as conceptual model error should be considered—that is, some aspect of model construction may be preventing any set of parameter values from producing a match at the location(s) in question. This is discussed in Section G4.3.

Data clustering is not a major problem in many groundwater models because most of the data clusters are hydraulic heads in areas of high hydraulic conductivity where significant quantities of water are pumped. In these areas, sensitivities of hydraulic heads to most parameters tend to be relatively small and the clustered wells will not adversely affect the regression.

G4.3 Observations that Are Inconsistent with Model Construction

Model simplifications and assumptions can result in observation data that reflect processes that are impossible to simulate using the constructed model. This can occur by design or can be unintended.

Situations that occur by design include those discussed for transient groundwater models in Chapter 9, Section 9.1.2. Consequences of designed inconsistencies depend on how the inconsistency is expected to affect model fit. If the omitted processes are expected to create a bias, such that the observations are consistently higher or lower than values simulated by an accurate model, the observations need to be adjusted to eliminate the bias. If the omitted processes are not expected to create a bias, so that the observations may be higher or lower than values simulated by an accurate model, the observation weights need to be adjusted to reflect the expected error introduced by omitting the processes.

Unintended inconsistencies can be more difficult. An example is presented by Hvilshøj and Jensen (2000) in which an analytic solution for a dipole test in a well was inconsistent with measured heads. This was resolved by simulating the test with a heterogeneous numerical model. If matching observations is troublesome or estimated parameter values are unrealistic, try to identify contributing model inconsistencies.

First, inspect the model for characteristics that could prohibit fitting the observation(s) in question using reasonable parameter values. In groundwater models when the problem is with fitting hydraulic heads, check for nearby head-dependent boundaries or specified heads. When the problem is with fitting flows, think about how the model simulates flow approaching or leaving the location in question. Compare the simulated flow paths involved to the flow paths conceptualized based on field information to ensure there are no unintentional simulated barriers or sources of water. Also, consider if there are any processes that occur in the true system and affect the observed values but that the model does not simulate.

Second, the existence, and sometimes the cause, of model inconsistencies can be identified by assigning the observation(s) in question enormous weight(s) (equivalent to a tiny variance, standard deviation, or coefficient of variation) and proceeding with a regression run. There are two possibilities.

1. *The fit to observation(s) does not improve.* This is a clear indication of inconsistency between the model and the observation(s). The model is constructed in such a way that the observation cannot be matched. Consider the options presented in the preceding paragraph.
2. *The fit to observation(s) improves.* Two results need to be checked.
   (a) *Model fit to other observations.* If the fit to the observation(s) is achieved by making the fit to other observations much worse, note what other observations those are. Determine if the original difficulty in fitting the observation(s) was because of other inconsistent observations, and resolve any inconsistencies.

GUIDELINE 4: INCLUDE MANY KINDS OF DATA AS OBSERVATIONS   287

(b) *Optimized parameter values.* If the fit to the observation(s) is achieved with clearly unrealistic parameter values, evaluate the associated simulated features, such as the geometry of hydrogeologic units in groundwater models. Is the simulated geometry or some other aspect of the features possibly in error?

G4.4 Applications: Using Different Types of Observations to Calibrate Groundwater Flow and Transport Models

Imaginative modelers continually explore, including testing new types of observations in the development of environmental models. Here, we describe several examples from the groundwater literature in which innovative types of observations were included. Generally, the use of such observations requires the use of a universal inverse code such as UCODE_2005 or PEST.

Age and Geochemistry Observations It can be especially difficult to obtain measurements of flows to or from groundwater systems in arid environments, where the water-table elevation is often far below surface-water bodies. To overcome the lack of flow data when calibrating a regional flow model of the Middle Rio Grande Basin aquifer system in central New Mexico, Sanford et al. (2004a,b) used an extensive set of carbon-14 and other geochemical data to estimate hydraulic-conductivity and recharge parameters.

Carbon-14 measurements were used to infer age dates, which in the regression were compared to simulated ages calculated by backward particle tracking with MODPATH (Pollock, 1994).

Geochemistry data were used to infer whether or not groundwater originated from the Rio Grande River. These data were used in the regression by defining nine hydrochemical zones. In each zone, an observation was defined as the percentage of groundwater that originated in the river, which was determined using geochemistry samples collected from all wells in the zone. The corresponding simulated percentage of water in the zone that originated in the river was calculated by backtracking a large number of particles from model layer 2 to any recharge location, and calculating the percentage of these particles that were tracked back to the river. The advantage of defining the observations in this way is that they were then continuous instead of discrete numbers, so that sensitivities could be calculated and the observations fit into readily available regression methods.

Use of these two data sets in addition to hydraulic-head data allowed estimation of 59 model parameters with no large correlation coefficients for any parameter pairs. The regression also produced recharge estimates that were more consistent with independent data than were recharge values estimated by previous groundwater models of the basin that used very few flow data.

Temperature Observations If groundwater flow is sufficient for advective heat transport to occur, then groundwater temperature data can provide information about parameters that govern groundwater flow.

Manning and Ingebritsen (1999) describe several studies in which temperature data were used to develop models of heat transport in relatively deep groundwater flow systems through, for example, sedimentary basins and mountainous terrain. Although most of the studies cited did not use formal inverse modeling techniques, they illustrate the substantial benefit of temperature data for constraining crustal permeability estimates in numerical models.

Anderson (2005) recently reviewed the use of heat as a tracer in groundwater systems and cites numerous studies in which temperature data are used to constrain hydraulic-conductivity estimates. For example, temperature measurements have been widely used in recent years in unsaturated zone models to constrain estimates of the vertical hydraulic conductivity of streambeds (e.g., Burow et al., 2005; Niswonger et al., 2005). This constraint leads to more accurate estimates of the flux between streams and groundwater systems. Only a small number of studies have incorporated temperature observations in a formal inverse modeling context; two such studies are described next.

Woodbury and Smith (1988) demonstrate the advantages of temperature data to estimating parameters of a cross-sectional model of groundwater flow and heat transport through and beneath a large landslide. They showed that when only hydraulic-head data were used for calibration, the estimated recharge and hydraulic-conductivity parameters were perfectly correlated. With temperature observations, these parameters could be uniquely estimated, as could additional parameters such as thermal conductivity and basal heat flux. The latter parameters had a fairly high degree of uncertainty, but the estimates were consistent with results from other studies.

Bravo et al. (2002) simulated groundwater flow and heat transport through a wetland system characterized by a fluvial sedimentary layer underlain by sandstone. Subsurface temperatures varied on a daily and seasonal basis. Application of inverse modeling to a flow and heat transport model using both head and temperature observations yielded parameter estimates with much smaller uncertainty and produced fewer problems with convergence of the regression procedure, compared to calibration of a flow model using only hydraulic-head observations.

GUIDELINE 5: USE PRIOR INFORMATION CAREFULLY

Prior information allows measurements related to defined parameters to be included in the regression (see Eq. (3.1)). The measurements involved are a subset of the system information of Guideline 2 and are related to model input.

G5.1 Use of Prior Information Compared with Observations

It can be argued that prior information should be treated differently than observations for two reasons. First, experience has shown that in many systems observations often can be measured more accurately than prior information. Second, the relationship between observed and simulated values is usually more direct than is the relationship between prior information and model parameter values. Often both problems result from what could be called scale issues: local variability makes it difficult to measure many model input quantities at a scale that is consistent with the model input. Resulting measurements can be grossly in error relative to what is appropriate in the model.

To the extent that measurements of the model input values represented by parameters are accurate as well as applicable to the scale of the model, model calibration may become unnecessary or less important. This book addresses problems in which model calibration is important, which implies that the measurements related to model inputs are inadequate in some way.

We suggest that for problems in which observations are more accurate and well understood, they be emphasized more than prior information that is less accurate and poorly understood. For systems with accurate measurements that directly relate to some or all of the parameters, the prior information might be more strongly emphasized and perhaps used in a manner closer to that suggested here for the observation data. In the applications described below, geophysical data are used in that way.

To encourage understanding of the information that is directly available from the observations alone, Guideline 5 suggests initially omitting from the regression any prior information on parameters. Two reasons generally motivate the subsequent inclusion of prior information.

First, if the parameter is insensitive, as indicated by a small composite scaled sensitivity (css), regression that includes the parameter often will not converge. Problematic parameters can be identified as those with the largest fractional changes calculated by Eq. (5.7), which are printed by most nonlinear regression programs (e.g., see Exercise 5.2b). Two options generally exist for dealing with these problematic parameters: (1) specify prior information for the parameter or (2) set the parameter value so that it is not changed during the regression.

In the regression, specifying prior information on an insensitive parameter usually results in a parameter estimate that is close to the specified prior value. Thus, the estimated parameter value generally is equal to or close to the prior information value regardless of which option is chosen. Model execution time is less when the parameter value is set, because this eliminates the need to calculate sensitivities for the parameter. Thus, option (2) often works well for model calibration. This will continue to be a good option as long as the parameter remains insensitive. Sensitivities can be checked by occasionally calculating css for all defined parameters.

Other groundwater studies, in which prior information was used either because of insensitive parameters or to explore its effect on parameter or prediction uncertainty, include those by Parker and Islam (2000), Christensen and Cooley (1999), Heidari and Ranjithan (1998), Christensen (1997), Bentley (1997), and Cooley (1983a).

The second common reason for using prior information occurs when a parameter value estimated by the regression is unreasonable. This problem is discussed in

Chapter 5, Section 5.5. As noted there, the most productive response to this problem depends on the amount of information the observations provide on the parameter in question.

1. If little information is provided, the problem falls into the category of insensitive parameters. Detection and resolution of such problems are discussed above.
2. If substantial information is provided, the unrealistic estimated parameter value may indicate problems with the model or the data, as discussed in Chapter 7, Section 7.6 and Guideline 10 in Chapter 12. These problems need to be resolved to achieve an accurate model.

In both of these situations, imposition of prior information during model calibration is not the best way to proceed.

Weiss and Smith (1998) also suggest cautious use of prior information and present methods of identifying parameters for which specification of prior information would be most beneficial. Their methods are based on analyzing attributes of scaled objective-function surfaces and parameter confidence regions. One method identifies parameters for which imposition of prior information will most stabilize the regression in terms of making it better posed. This method is likely to produce similar results as would be obtained by analyzing composite scaled sensitivities to determine parameters about which the observations provide little information.

G5.2 Highly Parameterized Models

In some situations, a modeler purposely defines more parameters than can be directly supported by the data, to represent potential variability in system properties. When very large numbers of parameters are defined, the model is considered highly parameterized. To obtain a tractable regression problem, such models require the use of prior information; the associated weights generally result in it being classified as regularization (see Guideline 6). The regularization is used to penalize parameter distributions that violate certain requirements. Commonly, the requirement is simply that the parameter values be close to specified values. This approach was considered in a simple way by Hill et al. (1998), and in more sophisticated ways by Valstar et al. (2004), among others. Alternatively, neighboring estimated parameter values that differ from one another are penalized so that high frequency variations are discouraged. Resulting distributions tend to be smooth. This is one of the approaches presented by Tikhonov and Arsenin (1977) and has been used extensively (e.g., Eppstein and Dougherty, 1996; Moore and Doherty, 2005, 2006). This approach is available through the regularization capability of PEST (Doherty, 1994, 2005).

Thus, parsimonious and highly parameterized models are two end-member approaches to obtaining tractable regression problems for complex groundwater systems. Parsimonious models are the focus of much of this book, though the methods presented have potential utility for highly parameterized models as well.

For example, parameter correlations can be revealed using single parameters multiplying the highly parameterized fields. Extreme correlations would suggest that though the variability produced by the highly parameterized model may be important, the overall mean depends on the value of other parameters. Thus, despite a good fit to observations, the model may poorly simulate results sensitive to the values of the individual system properties. The issue of scale discussed in Chapter 9, Section 9.2.3 also is relevant. Parsimonious methods are not necessarily limited to characterizing large-scale variations (e.g., Carle et al., 1998). Highly parameterized models generally represent variability at the scale of the grid or, in the case, for example, of pilot points, at a larger scale. Methods for sub-grid scale effects exist (e.g. Anderman and Hill, 2001; Rubin et al., 2003), but, to the authors' knowledge, have not yet been used in highly parameterized models.

G5.3 Applications: Geophysical Data

For models of groundwater systems, geophysical data are commonly used to define model layer thickness and define parameterizations as discussed in Guideline 2. A few investigations have used geophysical data more directly; selected studies are listed in Table 11.2. Most commonly, the geophysical data are used to support prior information, which is why geophysical data are included here under Guideline 5. However, in some of these studies the geophysical data is classified as a type of observation because equations relating the geophysical data to hydraulic conductivity are included as a model equation.

GUIDELINE 6: ASSIGN WEIGHTS THAT REFLECT ERRORS

Chapter 3, Section 3.1 of this book shows how weights and weight matrices appear in the objective functions minimized in nonlinear regression; Section 3.3.3 presents the purpose and theoretical requirements of weighting; and Section 3.4.2 suggests that, except for limited testing, it is useful to define weights that equal one divided by the variance of the errors in observations and prior information, or a weight matrix that equals the inverse of the variance–covariance matrix of errors.

Under Guideline 6, we first show how weights can be determined using common field data and assumptions and we then discuss selected issues related to weighting. The discussion reveals the importance of assigning weighting in a way that respects its intended role in the regression. Seven points emphasized in the discussion are:

1. The strategy of defining weighting based on likely error is supported by theory, has a strong intuitive appeal, and provides practical advantages. A chief advantage is that the strategy provides a formal mechanism for including an analysis of errors in model development.

TABLE 11.2 Selected Investigations Showing the Use of Geophysical Data as Observations or to Support Prior Information in Groundwater Model Calibrations

Reference	System	Parameter Estimation Method[a]	Geophysical Data Type[b]	Method of Including in Model Calibration[b]	Observed and Prior Values[b]	Parameters Estimated[b]	Results
Rubin et al. (1992) Copty et al. (1993)	2D groundwater flow; synthetic	Geostatistical inversion ML	τ	Function relates SVL to K. Processing τ to obtain SVL is independent of groundwater model calibration.	Hydraulic head, K	Mean and covariance of log K probability distribution function	Including τ data produced a much more accurate K field.
Hyndman et al. (1994)	2D groundwater flow, solute transport; synthetic	NLR	τ	Coinversion of τ, solute arrival time, and concentrations.	Solute arrival time, concentrations, τ	K, dispersivity, SSL field used to delineate K zones	Combining seismic and tracer data can provide high-resolution estimates of aquifer zonation and properties.
Hyndman et al. (1996)	3D groundwater flow, solute	NLR	τ	Used two steps with feedback: delineate K zones, estimate parameters.	Solute arrival time, drawdown, τ		
Hyndman et al. (2000)	transport; Kesterson, CA	NLR	τ	Process τ to yield SSL. Generate SSL fields. Function relates SSL to K.	Concentrations, drawdown, τ	Dispersivity, SLK	

Reference	Model	Regression method	Geophysical data	Role of geophysical data	Observations	Estimated parameters	Comments
Dam and Christensen (2003)	2D groundwater flow; synthetic	NLR	ER	Function relates ER to K. Processing to obtain ER data is independent of groundwater model calibration.	Hydraulic head, prior K, prior ER	ER, K, ERK	With ER data, a greater number of more accurate K parameters were estimated, compared with no ER data.
Day-Lewis et al. (2006)	3D groundwater flow, solute transport; Mirror Lake, NH	NLR	Radar	Postcalibration evaluation of alternative models.	Solute arrival times, concentrations, drawdown	K, effective porosity, dispersivity	Used radar data to resolve alternative conceptual models that were nonunique based on groundwater model calibration.

[a] ML, maximum likelihood; NLR, weighted least-squares nonlinear regression (in most circumstances these are equivalent; see Appendix A).
[b] τ, seismic travel time; ER, electrical resistivity; K, hydraulic conductivity; SVL, seismic velocity; SSL, seismic slowness; ERK, parameters relating electrical resistivity to K; SLK, parameters relating seismic slowness to K.

2. Additional insight into model fit can be achieved if the weighting relates directly to observation error instead of relating to it proportionately, as required by theory.
3. Theoretical considerations and available data can be used to determine weighting that is adequate, partly because regression results are not very sensitive to modest variations in weighting. The requirement that the weights reasonably represent observation error provides a sufficiently restrictive framework for determining weights.
4. The terminology used by different parameter-estimation methods causes part of the confusion that plagues weighting.
5. Some kinds of model errors can be included in determining weighting as long as the expected value of each error represented by the weighting equals zero.
6. Large weights on selected observations or prior information can be useful to the following interrelated goals: (a) ensuring the data are not being ignored, (b) determining whether a plausible solution exists with a given model construction, and (c) identifying model construction errors. This is related to the analyses proposed in Sections G4.2 and G4.3.
7. For uncertainty analyses to be meaningful, all observations and prior information need to be weighted based on likely errors.

G6.1 Determine Weights

Substantial guidance for determining weights is provided by the idea that weights need to equal one divided by the variance of the observation error and that weight matrices need to equal the inverse of the error variance–covariance matrix. Even if alternate weighting is chosen, it is important to evaluate errors in observations and prior information as discussed here to ensure the data are used appropriately. Using this strategy to determine the weights provides a formal mechanism for including analysis of errors in model development.

For problems with one kind of observation (e.g., all hydraulic heads) measured and simulated with errors of apparently equal variance, it is common to set all weights equal to 1.0. For example, see the Theis problem of Figure 5.3. The calculated standard error of the regression (defined after Eq. (6.1)) can be compared to the expected standard deviation of the errors to evaluate the likely model error (see Chapter 6, Section 6.3.2).

For commonly used diagonal weight matrices, the weight is defined to be equal to one divided by the variance of the errors, σ_i^2, as discussed in Chapter 3, Section 3.4.2. More readily understood quantities are σ_i, the standard deviation, and σ_i/y_i or σ_i/y_i', the coefficient of variation, where y_i is an observed value or prior estimate and y_i' is an equivalent simulated value. Variances are readily calculated from these quantities.

For full weight matrices, the weighting equals the inverse of the variance–covariance matrix of the observation errors.

GUIDELINE 6: ASSIGN WEIGHTS THAT REFLECT ERRORS

For any observation, errors result from many processes. Determining the statistic (the variance, standard deviation, or coefficient of variation) used to calculate the weight requires quantifying as many of the major error sources as possible. In this section, we first consider quantifying a single component of error, then many components of error, and finally errors for observations that are sums of or differences between measurements when the errors are independent, additive, and normally distributed. Chapter 3, Section 3.3.3 discussed situations in which these assumptions may not apply.

Quantify One Component of Error: An Observation Well Example The statistics used to calculate observation weights can often be determined using readily available information about likely errors and a simple statistical framework. For example, consider the common situation in groundwater modeling of error in the elevation of an observation well used to determine head measurements at the well.

The data on the well are as follows: the well elevation was determined by an altimeter and is thought to be accurate to within 3 ft. To estimate the variance of the error, this statement needs to be quantified. For example, the statement that "the probability is 95 percent that the true elevation is within 3 ft of the measured elevation" might apply. If, in addition, the errors are assumed to be normally distributed, a table of areas under the standard normal curve (Table D.1 in Appendix D) can be used to determine the desired statistics. This process is outlined in Table 11.3.

For some data and/or instrument types, error studies have been conducted. For example, for determining elevations from U.S. Geological Survey (USGS)

TABLE 11.3 Steps Needed to Determine a Standard Deviation that Can Be Used by MODFLOW-2000 and UCODE_2005 to Calculate a Weight

Step	Description	Example
1.	Quantify the statement about measurement accuracy; include a significance level.	"The probability is 95 percent that the true elevation is within 3 ft of the measured elevation." The significance level is 5 percent $(5 = 100 - 95)$
2.	Determine the critical value. For normally distributed errors, use areas under the standard normal curve (Table D.1) to obtain the critical value.	A significance level of 5 percent has a critical value of 1.96.
3.	Construct a confidence interval on the measured value, y_i, using the critical value and standard deviation of the error. Equate it to the confidence interval expressed in the statement developed in step 1.	Confidence interval $= y_i \pm 1.96 \times s_{y_i}$ $= y_i \pm 3$ ft Thus, $1.96 \times s_{y_i} = 3$ ft
4.	Solve for s_{y_i}.	$s_{y_i} = 1.53$.

topographic maps, the USGS (1980, p. 6) states that on these maps, "not more than ten percent of the elevations tested shall be in error more than one-half the contour interval." This statement indicates that a 90-percent confidence interval for this error equals plus and minus one-half the contour interval. Assuming that the error is normally distributed, a 90-percent interval is constructed by adding and subtracting 1.65 times the standard deviation of the measurement error. Thus, the standard deviation of the measurement error can be calculated as one-half the contour interval divided by 1.65, or (contour interval)/(2 × 1.65). The value of 1.65 was obtained from Table D.1 as described in Table 11.3.

Errors can also be evaluated by modeling the sampling process. For example, Schäfer et al. (2003) use what they call virtual aquifers to evaluate solute concentrations measured at wells in heterogeneous materials.

Many situations are not as definitive as the examples above. Difficulties in determining weighting are discussed in Section G6.2.

Accumulate All Error Components Generally, for any observation there are many sources of error. For example, possible errors for hydraulic heads in the simulation of a groundwater system include:

1. Error in measuring the water level in the well.
2. Error in determining the elevation of the well. (For drawdowns, this cancels out.)
3. Aspects of well construction. If we could drill 100 wells in the same place, different gravel packs, screen settings, grouting, and so on would produce variations in the measured water levels. Unfortunately, such repeated sampling is not practical, so these variations are not well characterized. Errors related to well construction are likely to be greatest in dynamic situations such as during an aquifer test.

There are other errors in placing the well in the context of the model:

4. Errors in placement horizontally.
5. Errors in placement vertically.

There are errors that can be classified as model errors in that they could be corrected with a finer grid or time step, but can be included in the weighting if they have a mean of zero. These include:

6. Incorrect placement of hydrogeologic units caused by grid size.
7. Unrepresented temporal variations in recharge, pumping, and so on (see Chapter 9, Section 9.1.2).
8. Unrepresented flow fields, such as typically local flow fields in a regional flow model (see Section G6.2).
9. Unrepresented flow fields, such as regional flow omitted from a site model.

Often the errors can be considered independent and normally distributed. In this situation, the variance of the sum of the errors equals the sum of the variances. That

GUIDELINE 6: ASSIGN WEIGHTS THAT REFLECT ERRORS   297

is, $\sigma_{total}^2 = \sigma_1^2 + \sigma_2^2 + \cdots$. Only variances are additive; standard deviations and coefficients of variation cannot be added. This is because the standard deviation (from which the coefficient of variation is calculated) is defined as the square root of the variance, and the square root of the sum of two quantities does not equal the sum of the square roots of the two quantities. An example of accounting for a number of different types of error for transient head observations in a groundwater model is presented in Chapter 15, Section 15.2.1.

Weights for Observations that Are Sums of or Differences Between Measurements Observations can be sums of or differences between measured values. For example, consider streamflow measurements between two streamflow gauging stations. In groundwater modeling, the difference between two flow measurements often is used as an observation in the regression. These observations are called streamflow gains or losses.

Consider a situation in which the upstream and downstream flow measurements are 3.0 ft^3/s and 2.5 ft^3/s, so that there is a 0.5 ft^3/s loss in streamflow between the two measurement sites, and in which the following assumptions apply:

1. The measurements are each thought to be accurate to within 5 percent (using the error analysis of Carter and Anderson, 1963).
2. There is a 90-percent probability that the first measurement is within 0.15 ft^3/s (5 percent) of the true value, and a 95-percent chance that the second measurement is within 0.125 ft^3/s (5 percent) of the true value.
3. The errors in the two measurements are independent and are normally distributed.

The procedure for calculating the coefficient of variation of the streamflow loss is as follows:

1. Calculate the standard deviation of the first measurement using the method described in Table 11.3.

$$1.65 \, s_{q_1} = 0.15 \, \text{ft}^3/\text{s}, \quad \text{so} \quad s_{q_1} = 0.091.$$

2. Calculate the standard deviation of the second measurement in the same manner.

$$1.96 \, s_{q_2} = 0.125 \, \text{ft}^3/\text{s}, \quad \text{so} \quad s_{q_2} = 0.064.$$

3. Square the standard deviations to calculate variances.

$$s_{q_1}^2 = 0.0083 \, (\text{ft}^3/\text{s})^2 \quad \text{and} \quad s_{q_2}^2 = 0.0041 \, (\text{ft}^3/\text{s})^2.$$

4. Calculate the variance of the 0.5-ft^3/s streamflow loss (the difference between the two flows) by adding the variances.

$$s_{q_1}^2 + s_{q_2}^2 = 0.0124 \, (\text{ft}^3/\text{s})^2.$$

5. Take the square root of the variance to obtain the standard deviation of measurement error.

$$[0.0124(\text{ft}^3/\text{s})^2]^{1/2} = 0.111 \text{ ft}^3/\text{s}.$$

6. Calculate the coefficient of variation of the loss by dividing the standard deviation by the loss.

$$c.v. = (0.111 \text{ ft}^3/\text{s})/(0.5 \text{ ft}^3/\text{s}) = 0.22, \quad \text{or 22 percent.}$$

In UCODE_2005 and MODFLOW-2000, a variance, standard deviation, or coefficient of variation can be specified by the user for each observation. The choice generally is based on achieving statistical values that are most meaningful to the modeler. For many types of flow observations, coefficients of variation are often most meaningful.

Determine Covariances for Weight Matrices Some circumstances clearly produce correlations between errors of different observations. For example, consider three streamflow measurements, $q1$, $q2$, and $q3$, along the length of a stream, and three associated measurement error variances, σ_1^2, σ_2^2, and σ_3^2. Gains or losses are calculated by subtracting each measurement from the next downstream measurement. For the three measurements, this results in two gain/loss observations, $q2 - q1$ and $q3 - q2$ and, from the preceding discussion, error variances $\sigma_1^2 + \sigma_2^2$ and $\sigma_2^2 + \sigma_1^2$. The errors in the two differences are not statistically independent, because the error in $q2$ is included in both differences. Hill (1992, p. 43) reported that in this circumstance the covariance between the two differences equals $-\sigma_2^2$. Christensen et al. (1998) extended this result to measurements along branching streams and indicate that the covariance equals -1 times the sum of the variances of the flows shared by any two gain/loss observations. Covariances can be included in UCODE_2005 or MODFLOW-2000. In some situations, inclusion of off-diagonal covariance terms in the weight matrix have had a negligible effect on estimated parameters (unpublished results by the first author of this book and S. Christensen, 1996, Aarhus University, oral communication). In others they have been important (Bentley, 1997).

It is not known how large the covariances need to be before a diagonal weight matrix produces significant errors in parameter estimates or measures of uncertainty. Additional work and definitive publications would be useful. In some situations, correlated errors can be accommodated by differencing, as discussed in Chapter 9, Section 9.1.2 in the context of temporal observations.

G6.2 Issues of Weighting in Nonlinear Regression

The following common issues are considered: difficulties in determining the weights, confusion about the term weighting, measurement error versus model error, the utility and difficulty of using exaggerated weights, the importance of weighting strategy in detecting model error and overfitting, and weighting system information on parameter values.

GUIDELINE 6: ASSIGN WEIGHTS THAT REFLECT ERRORS

Difficulties in Determining the Weights In practice, it is generally impossible to identify all errors that contribute to an observation. In addition, the variances, standard deviations, and coefficients of variation calculated using the methods discussed in this guideline are clearly approximate. Thus, determining proper weighting can seem problematic and has discouraged some from using regression methods.

Yet, it is rarely difficult to determine weighting that adequately represents errors for use in regression. If one poses different levels of potential error, almost always some can clearly be identified as realistic while others are not realistic. Indeed, posing a range of values generally reveals a believable range of error. Such evaluations can be used to create statements for step 1 of Table 11.3, and the statistics can then be determined using steps 1 through 4. While the resulting statistics are not rigidly defined, such an analysis generally is able to determine the weights well enough. This is because regression results generally are not sensitive to moderate variation in the weighting: nearly identical results are typically obtained given weighting within a range that reasonably represents the likely observation error. If the weighting is changed beyond reasonable ranges, large variations in regression results can occur, causing the regression to lose meaning and become arbitrary.

In applications of multiobjective optimization, which was discussed in Chapter 3, Section 3.2.3 and Chapter 5, Section 5.3, alternative weighting schemes are considered. In those methods, the weighting changes as multipliers on the different objective functions change. An example in which only four weightings are considered is presented by Ghandi et al. (2002a), who used head, concentration, and interwell flow data to calibrate a transport model of a groundwater recirculation system used for in situ bioremediation. As for most applications of multiobjective optimization, the weighting strategy differed from that suggested in this book in that none of the observation weights were based on likely errors, and they may have varied over a larger range than would have been supported by an analysis of errors. Even so, the four strategies only resulted in moderately different estimates of some parameters. The authors selected the final weighting strategy on the basis of its ability to produce a good overall fit to all three data types. Because no analysis of observation error is presented, the relationship between the weighting and the errors cannot be evaluated.

Confusion About Terminology Confusion about weighting occurs for many reasons. One reason is that very different definitions for the terms "weights" and "weighting" are used in different parameter-estimation methods. In this book we use these terms only to describe a term in the objective function (Eq. (3.1)). However, other authors have used these terms very differently. For example, Yeh et al. (1996) use terms "weights" and "weighting" to describe quantities that reflect the smoothness of the parameter field (through the spatial variance–covariance matrix of, typically, hydraulic conductivity) and the sensitivities. Other methods may have no formal mechanism for accommodating expected data errors, a role suggested for weights in this book. To avoid confusion about the role of weights in different parameter-estimation methods, careful reading and writing are important. This will help all modelers clearly understand the function of weights and weighting in any application of any method.

Measurement Error and Model Error Model errors are defined here as any errors that could be eliminated by changes in the model given greater computer capacity, more time, or more complete information about the groundwater flow system even if the information is not attainable given present technology. Model errors are caused by, for example, inaccurate interpolation of simulated equivalents to observations, inability of the model to represent some processes, fluctuations in properties that are smaller than the grid size, and parameterizations that limit the spatial or temporal variability of parameters. As noted in Chapter 1, parameterization is needed to attain a tractable problem, but it does produce model errors. Dealing with this conflict is the topic of Guideline 3.

Here we consider whether the observation errors accounted for by weighting should include only measurement errors, or whether some types of model error can be included as well. While this point can be, and is, argued extensively, a useful definition is:

Observation error is error related to any aspect of the observation not accounted for by the model considered, for which the expected value is zero.

Unambiguous types of measurement errors are those associated with the measuring device and the spatial location of the measurement. Ambiguous contributions include, for example, heads measured in wells that only partially penetrate the numerical layer to which they are assigned, or temporally averaged head measurements or single measurements that are clearly affected by transient effects used in a steady-state model. These are more ambiguous because the model could be modified to better accommodate the measurements. Despite such ambiguities, the above definition for observation error works well in practice, because it produces sufficiently accurate weighting, and, as mentioned above, the regression often is not highly sensitive to moderate changes in the weighting.

For example, in a groundwater model of the Madison aquifer in the northern Great Plains, USA, Cooley et al. (1986, p. 1764) anticipated that the small error with which the hydraulic head in shallow wells could be measured would produce accurate observations at these points, and thus assigned them large weights. During calibration it was determined that "the model fit no better at these points than elsewhere" (Cooley et al., 1986, p. 1772). Apparently, heads in the shallow wells were affected by shallow, local flow systems not represented by the regional-scale model, and this situation produced residuals that were as large as those associated with inaccurately measured heads in deep wells. Decreasing the weights (increasing the variances) for observations from shallow wells to account for the model not simulating the shallow flow dynamics produced better results.

If weights are determined based on observation errors that include measurement errors and possibly some model errors, the standard error of the regression is significantly greater than 1.0 (as determined using the methods described in Chapter 6,

GUIDELINE 6: ASSIGN WEIGHTS THAT REFLECT ERRORS

Section 6.3.2), and the fitted error statistics (see Section 6.3.2) are too large to be accounted for by measurement error, it is likely that model error is involved in the misfit. If the weighted residuals are randomly distributed, it is possible that unaccounted for model errors have zero means and a variance–covariance matrix that is proportional to the variance–covariance matrix of the observation errors. In this situation, the large value of the calculated error variance produces a variance–covariance matrix on the parameters of Eq. (7.1) that appropriately accounts for the unaccounted for model errors.

Using Large Weights It can be useful to assign large weights to selected observations or prior information, as discussed in Sections G4.2 and G4.3. Figure 5.4 and Exercise 5.1b showed how a regression could become better posed by increasing the weighting on an observation to place more emphasis on it than is warranted given likely observation errors. This is frequently done to establish the existence of a solution (Backus, 1988), especially for observations that provide unique information. Such observations may be identified from an understanding of the simulated processes, or because statistics such as scaled sensitivities or measures of leverage or influence are distinctive. Examples include observations that are a different measurement type or that are collected at a different location or time.

For example, in groundwater flow modeling there are typically many hydraulic-head observations but very few flow observations (such as streamflow gains and losses, or spring flows). There is a perception that the small number of special observations (here, flows) will not be properly accounted for in the regression, and thus there is often an inclination to assign larger weights than are consistent with likely errors in these observations. The concern is heightened if predictions of interest are closely related to the few special observations.

However, the possibility that keeping large weights throughout both model calibration and uncertainty analysis might diminish the accuracy of the model, predictions, and/or measures of uncertainty is suggested by the theoretical requirements of weighting and needs to be considered.

To investigate this issue, consider a simple problem in which linear regression is applied in a situation known to be characterized by a linear model. Figure 11.5 shows that of 10 observations only one is located in the range of relatively large x values for which predictions are of interest. The important question is whether the accuracy of the predictions can be improved by increasing the weighting of the special observation. It is apparent from Figure 11.5 that the answer is no, because the other data are clearly relevant to predictions at larger values of x, given that a linear model is valid. Increasing the weighting of the observation with large x would produce a model that closely matches the error of that measurement, but is likely to degrade the accuracy of the resulting calibrated model.

The one observation for large x in Figure 11.5 is analogous, for example, to the few flow observations in a groundwater system, because in both cases these observations have sensitivities that are special in some way. For this linear regression problem, the sensitivity of each data point with respect to the intercept parameter equals 1.0, and the sensitivity to the slope parameter equals the x value

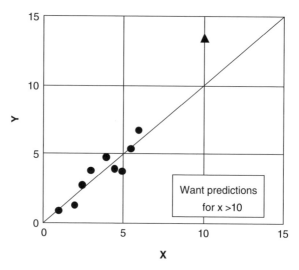

FIGURE 11.5 The true model of $y = 0.0 + 1.0x$, and possible observations to be used in linear regression. The data represented by dots are clustered and have x values that are distinctly different from that of the data point represented by the triangle.

of the observation (i.e., for the model $y = a + bx$, $\partial y/\partial a = 1.0$, and $\partial y/\partial b = x$). Thus, for the special observation, the sensitivity to the slope parameter is larger than that for the other observations. In groundwater systems, flow data provide special information, which is expressed by sensitivities that often reduce correlations among parameters, and thus produce a regression better able to uniquely estimate parameter values. Also, errors in flow data tend to have very different sources and magnitudes than errors in head data, as well as different units (e.g., m^3/day versus m).

To more closely examine this issue, consider the results of Exercise 5.1a and define a prediction of interest as the advective-transport distance toward the river from the center of the top layer after 10 years. The true predicted value is 1737 m. Objective-function surfaces for the parameter-estimation problem are shown in Figure 5.4. With hydraulic-head observations alone (Figure 5.4a), the objective-function surface is composed of parallel lines, and no minimum exists.

Addition of the flow observation using weighting that realistically represents the observation error produces the objective function in Figure 5.4b; imposing a large weight that assumes an unrealistically small observation error produces the objective function of Figure 5.4c. Increasing the weighting of this observation obviously produces a better-defined minimum and might be justified if the existence of a plausible solution is being explored, but the consequences need to be considered. As shown in Table 11.4, for this simple model the parameter estimates and the advective-transport prediction are very similar for the two different weightings. The regression with the large, unrealistic weight on the flow observation produced a slightly more accurate prediction. However, in general, this accuracy could either improve or deteriorate, depending on the actual error in the flow observation.

TABLE 11.4 Selected Regression Results Using Weighting of the Flow Observation Resulting from Reasonable (10 percent) and Unrealistically Small (1 percent) Coefficients of Variation

Flow Coefficient of Variation	Parameter Estimate		Prediction (Distance Traveled Toward River in 10 years) (m)	Confidence Interval[b] (m) on the Prediction	Interval Includes True Value of 1737 m?
	K_MULT	RCH_MULT			
10 percent	1.16	0.89	1017	71; 1964	Yes
1 percent	1.18	0.91	1036	940; 1131	No

[a]A smaller coefficient of variation produces a larger weight.
[b]Ninety-five percent linear confidence intervals constructed assuming a normal probability distribution.

The most significant consequence of using an unrealistic weight is related to the confidence intervals on the predictions (Table 11.4). With the unrealistically small coefficient of variation of 1 percent, the small prediction confidence interval does not reflect a realistic level of prediction uncertainty, as indicated by its omission of the true predicted value by a wide margin. In contrast, with a more reasonable coefficient of variation of 10 percent, the interval is more realistic and contains the true predicted value. This results from the unrealistic weight being included in the calculation of the parameter variance–covariance matrix (Eq. (7.1)) and the effects being propagated to the standard deviation on the prediction using Eq. (8.1).

In more complex situations, unrealistic weightings may produce different estimates and predictions compared to when realistic weights are used, but there is no assurance that the different values will be more accurate. In addition, the consequences for uncertainty analysis are likely to be similar to those shown here for the simple two-parameter model. Ultimately, exaggerated weighting cannot be expected to produce more accurate models; that goal can only be achieved by better data and/or better use of data. This book focuses on the latter.

Detecting Model Error and Overfitting Specifying weighting that equals the inverse of the variance–covariance matrix of the observation errors establishes a context for detecting model error and for identifying fits that arc too good (as shown in Figure 11.1b). This analysis uses common measures of model fit and is discussed in Chapter 6, Section 6.3.2 and considered later in Guideline 8. This analysis is useful and is often overlooked, so a summary of the analysis is included here.

If the model fit is consistent with the assigned weighting, then the calculated error variance and the standard error of regression will be close to 1.0. Larger values (common in practice) indicate that the model fits the data more poorly than is consistent with the weighting. For example, if the standard error is 5.0, the model fit is, on average, five times worse than is consistent with the expected observation error. Possible sources of the additional error are neglected measurement error or model error. If model error is suspected, but no bias is evident in the weighted residuals, the error may be accumulated from small contributions, and model predictions and measures of uncertainty may still be useful (Hill et al., 1998) (see Chapter 6, Section 6.3.2).

The calculated error variance and the standard error of regression also can be less than 1.0. This is not common in practice but may occur if too many model parameters are estimated. The value of the standard error might not increase as the number of parameters increases if prior information is added for each added parameter value. Small values of the standard error indicate that the model fits the observations better than expected based on the analysis of observation errors. Thus, for example, if the standard error is 0.1 and the confidence interval on the value (Eq. (6.2)) does not include 1.0, then the model fit is, on average, 10 times better than is consistent with the preliminary analysis of observation error. In this situation, the expected errors and the model fit should be closely examined for evidence that the model is fitting errors rather than system processes. Overfitting can be more easily identified if the observation errors have been carefully evaluated. This can be accomplished when weighting is defined as suggested here.

Defining weights on the basis of an analysis of errors encourages comparison of the weighting to theoretical ideals. If nonideal weighting is used to achieve regression results, the nonideal weighting can be compared to likely errors. For example, if an observation has a weighted residual that is distinctly larger than other weighted residuals in absolute value, reducing the weight (by increasing the variance, standard deviation, or coefficient of variation) can make the weighted residual more comparable to other weighted residuals. Indeed, robust regression automatically makes such adjustments to the weights (Huber, 1981). However, it is important to evaluate whether the final statistics believably represent the error. If manual adjustment or robust regression methods result in variances, standard deviations, or coefficients of variation that seem to represent an unrealistic level of error, evaluate the magnitude of the associated unweighted residual for indications of important model error or observation bias. Clearly, resolving these two problems is more likely to yield an accurate model than hiding the problem by reducing the weight (increasing the value of the statistic).

The example above and comments in Chapter 6, Section 6.3.2 suggest that the relatively simple idea of making the weights equal to one over the variance of the error or making the weight matrix equal to the inverse of the error variance–covariance matrix has proved to be very useful. It respects the statistical theory, provides a framework for identifying model error and/or measurement bias, and contributes to using the standard error of regression as a measure of model error. That is quite an accomplishment for a simple idea!

Weighting System Information on Parameter Values The discussion above focused on observations but is directly applicable to weighting prior information. Errors in system information often result from scale issues; see, for example, Beckie and Harvey (2002).

Weighting on prior information can be determined by, for example, constructing a 95-percent confidence interval on the basis of the likely range of parameter values, using independent field data or knowledge about hydrologic or geologic processes related to the quantities represented by the parameters. Two issues of special importance to prior information are the use of large weights and resulting "regularization,"

GUIDELINE 6: ASSIGN WEIGHTS THAT REFLECT ERRORS   305

and the use of log-transformed parameters. While observations can be log-transformed, as mentioned in Chapter 3, Section 3.3.3, it is not very common. Thus, the effects of log-transformations on weighting are discussed here in the context of system information.

If the weighting realistically represents the uncertainty, the system information on parameter values included in the regression is called prior information and fits into the framework of either classical statistics or Bayesian statistics (the latter being the framework from which the term "prior information" originates). Sometimes, however, larger weights (smaller statistics) are assigned to the system information to achieve a stable regression, in which case the term "regularization" needs to be used instead of prior information (Backus, 1988). Setting parameter values to constants that are not changed by the regression can be thought of as an extreme case of regularization. When regularization is used, confidence intervals on parameters and predictions tend to underestimate actual uncertainty, as demonstrated in Table 11.4. Thus, it is very important in practice to appropriately classify prior information and regularization.

Prior information and regularization can be imposed on individual parameter values, or on characteristics of a parameter distribution, such as smoothness, as discussed in Chapter 1, Section 1.3.2 and Section G5.1. Extreme examples of the latter are (1) requiring that the model input value be constant over a region, a volume, or a specific type of material wherever it exists, or (2) requiring a specific interpolation scheme.

The capability of defining many parameters is implemented in PEST as its regularization capability. An example is presented by Doherty (2003). PEST is programmed to allow the user to specify the desired fit to observations and then adjust the weighting to achieve that fit. As suggested by Doherty (2003), users need to take care that the model fit specified is not less than expected observation errors. In addition, the resulting weighting of the observations and regularization need to be checked to determine if the weights are supportable based on hydrogeologic data. If not, it is important to modify the weighting before proceeding with uncertainty analysis.

The second issue unique to prior information occurs when the associated parameter is log-transformed. In this situation, the statistic used to weight the prior information generally needs to relate to the log of the parameter value. The methods discussed above for quantifying errors are directly applicable, but an extra step often is needed because usually it is easier to establish a range of plausible values for native than for transformed values. Thus, if the prior estimate for a hydraulic conductivity is 1×10^{-5} m/s, and the true value is expected to fall between 1×10^{-6} and 1×10^{-4} m/s with a certainty of about 95 percent, a 95-percent confidence interval for the native value has approximate limits of 1×10^{-6} to 1×10^{-4} m/s. Taking the log (base 10) of these values produces limits of -6 and -4 about a prior estimate of -5. If it is assumed that the uncertainty in the hydraulic conductivity can be approximated by a log-normal distribution, the log-transformed value is normally distributed. The methods described above can be used to determine that the standard deviation relevant to the log-transformed parameter equals 0.51. This value would be specified as the statistic used to calculate the weight for the prior estimate.

GUIDELINE 7: ENCOURAGE CONVERGENCE BY MAKING THE MODEL MORE ACCURATE AND EVALUATING THE OBSERVATIONS

Nonlinear regression models of complex systems often do not converge despite using the ideas suggested in Guideline 3 to maintain a well-posed problem. The major reasons for convergence problems are insensitive parameters, nonlinearity of the forward model with respect to estimated parameters, and inconsistencies between the processes important to the observations and simulated processes caused by poor representation of the system by the model and/or misinterpretation of data. These causes are listed in Table 11.5. Parameter correlation is

TABLE 11.5 Possible Actions to Encourage Convergence and Obtain an Accurate Model

General Comments

Identify the parameters associated with the largest values of *max-calculated-change*. This information is provided by the computer codes used in this book, as shown in Figure 5.5, and possible actions are as follows.

The Three Main Problems that Plague Convergence and Possible Solutions[a]

Insensitivity

If *css* for any parameter is less than 1 percent of the largest *css*, consider ideas presented in Guideline 3. These include:
(a) Specify the parameter values with small *css*.[b]
(b) Check problematic parameters. Consider combining existing parameters or redesigning the parameterization. Consider the suggestions in Guideline 2 about creative use of system information.

Nonlinearity

Evaluate simulated results for parameter values from intermediate parameter-estimation iterations. Look for evidence of nonlinearity. Consider weighted residuals that are largest in absolute value, observations omitted because simulated equivalents could not be obtained, and whether parameter values are realistic. If forward model nonlinearities are problematic, consider using a linear approximation, as suggested in Guideline 1.

Inconsistencies

Check the representation of the parameters. Check dominant observations identified using *dss*, DFBETAS, and leverage statistics.
Evaluate observations, prior information, and their simulated equivalents.

(Continued)

TABLE 11.5 *Continued*

All

Consider reducing the amount by which parameters are allowed to change within one parameter-estimation iteration—called *max-allowed-change* in Chapter 5, Section 5.1.3 before Eq. (5.7). Alternatively, consider using the trust-region approach available in UCODE_2005 or other methods.

Diagnosing Problems that Plague Convergence

Regression Performance	Problem[c]
(1) In the first parameter-estimation iteration, values of sensitive parameters move far from their starting values.	Inconsistencies. If evaluation indicates no inconsistencies, move starting parameter values closer to those from the first iteration, after first checking model fit.
(2) *max-calculated-change* remains large in absolute value and is either consistently positive or consistently negative.	Insensitivity.
(3) *max-calculated-change* goes through a repeated sequence in which it is reduced in size over several iterations only to dramatically increase.	Nonlinearity and/or insensitivity.
(4) *max-calculated-change* oscillates between large positive and negative values (as in Figure 5.5).	Insensitivity of one or more parameters.

[a]The solution of obtaining more data on insensitive parameters is not listed. Potential data acquisition efforts often are most advantageously considered in the context of predictions, as discussed in Guidelines 11 and 12 in Chapter 13.
[b]If parameter values are specified to alleviate convergence problems, calculate their *css* and *pcc* in later regression runs. If possible, estimate the parameters using regression. Generally, specified parameters need to be included to assess uncertainty. See Chapters 7 and 8.
[c]The problems are listed with the performance to which they are most likely to apply. However, consider all problems and actions listed to address convergence problems.
Note: *css*, composite scaled sensitivity; *dss*, dimensionless scaled sensitivity; *pcc*, parameter correlation coefficient.

not likely to result in lack of convergence of a single regression, as shown in Exercise 5.1b and discussed in Guideline 3. Similar results are expected for local minima.

Convergence is usually improved as the model becomes a better representation of the system that produced the observations being matched by the regression. This means that, generally, the goal of achieving convergence with a valid regression and the goal of achieving a model that accurately represents major processes are identical.

Information available from regressions that fail to converge provide substantial insight. This insight can be obtained by careful consideration of dimensionless, one-percent, and composite scaled sensitivities; parameter correlation coefficients;

weighted and unweighted residuals; parameter values; parameter updates calculated by the regression; and other information from the regression. These items can be used to detect inaccuracies in model construction. Review of Tables 10.1 and 10.2 and the questions of Table 10.3 may help to suggest useful approaches.

In addition, evaluation of regression performance can be useful, as suggested in Exercise 5.2b. The *max-calculated-change* defined after Eq. (5.7) is the largest calculated fractional change, in absolute value, for any parameter in one parameter-estimation iteration and is reported for each iteration. Generally, *max-calculated-change* must be less than a user-defined criterion for the regression to converge (see Chapter 5, Section 5.1.3). When it does not diminish sufficiently, the regression is said to not converge. Possibilities include, but are not limited to: *max-calculated-change* remains large in absolute value and is either consistently positive or consistently negative; *max-calculated-change* goes through a repeated sequence in which it is reduced in size over several iterations only to dramatically increase; *max-calculated-change* oscillates between large positive and negative values (as in Figure 5.5).

These possibilities are listed in Table 11.5. They are related to the three likely causes of nonconvergence mentioned above in this guideline and in the first part of Table 11.5. Suggested causes and solutions are listed. There is no suggestion to change observation weighting, which is tempting but rarely helpful in this circumstance and can be very time-consuming. Also, when the regression performs in the four ways listed in the second part of Table 11.5, increasing the number of parameter-estimation iterations is rarely helpful for achieving convergence.

GUIDELINE 8: CONSIDER ALTERNATIVE MODELS

There is always more than one possible representation of natural systems, because there are different possible interpretations of the incomplete data about the systems. Guideline 8 encourages considering as many alternative models as possible and offers strategies for designing, organizing, and comparing them.

Formal parameter-estimation methods that produce optimal parameter values are essential to use when considering alternative models if results are to be at all definitive. If parameter values are determined using a clear process such as optimization, model fit and other model attributes can be compared without speculation about whether conclusions would be different if only this parameter value was a bit higher or that one a bit lower.

Commonly, to begin the modeling process, one model is constructed using Guidelines 1 through 7. During development of this model, model fit and parameter values are evaluated at various stages using Guidelines 9 and 10 in Chapter 12. In addition, predicted quantities are evaluated, and their relationships to model calibration issues are considered, as discussed in Guideline 13 in Chapter 14. This process is an example of how the guidelines are not always used sequentially. Guideline 8 is positioned to emphasize that alternative models are fundamental to the study of any natural system.

G8.1 Develop Alternative Models

Alternative model evaluation often indicates that many plausible models exist. Considered another way, the data are insufficient to further limit the possible alternatives. Development of alternative models can be motivated by many different circumstances, including different equally plausible interpretations of incomplete system information (Guideline 2) and difficulties with initial models of the system, such as problems with model fit or optimal parameter values (Guidelines 9 and 10 in Chapter 12).

Alternative models typically differ in their representation of the characteristics and properties of the simulated system and/or in their simulated processes. Commonly, they have the same set of observations, and this is required by some methods of analysis.

The mechanisms for developing alternative models fall into three categories: deterministic, stochastic, or a combination of these two. These approaches are discussed below.

Deterministic methods of developing alternative models generally use different conceptual models. For example, in groundwater systems it is common that different interpretations of geologic processes yield different hydrogeologic framework models. Different choices of included processes are also usually determined from a deterministic decision process: for example, including the effects of temperature or subsidence on groundwater flow. Deterministic development of alternative conceptual models often is facilitated for complex three-dimensional problems by using the data organization, visualization, and analysis tools discussed in Chapter 15 for the Death Valley regional groundwater flow system.

Stochastic methods for developing alternative models usually identify one aspect of the system that is expected to dominate simulated results of interest (e.g., predictions) and randomly generate model input realizations. Each realization is then used in the model to produce simulated results. The model input may be a single number; or it may be many numbers that define a spatial and/or temporal field of—using groundwater model examples—hydraulic conductivity and areal recharge. The model input might also be numbers that define the spatial distribution of a system property, but not the actual values. For example, Poeter and McKenna (1995) present an innovative method in which alternative models are developed using indicator kriging to generate different zonation arrays that are used in the model to define the hydraulic-conductivity distribution. Hydraulic-conductivity values for each zonation are then estimated by regression. This example is discussed in Guideline 14 in Chapter 14. The transition-probability method of Carle et al. (1998) and the indicator simulation method of Gomez-Hernandez (2006) also are designed to generate many realizations of a three-dimensional field.

For field systems, methods of developing alternative models that depend on both deterministic and stochastic contributions are likely to be very useful. This includes generating stochastic distributions within a deterministic structure. For example, in a groundwater model the alternative structures may be different interpretations of large-scale hydrogeologic units; stochastic methods might be used to generate alternative models of the interior variability of selected units and/or selected parts of the system.

G8.2 Discriminate Between Models

Models that are more likely to be accurate tend to have three attributes: lower values of overall fit statistics (Chapter 6 and Guideline 9); weighted residuals that are more randomly distributed (Chapter 6 and Guideline 9); and more realistic optimal parameter values (Chapter 7 and Guideline 10).

Often these criteria are used to identify a single most likely model and all subsequent simulations of predictions and other analyses are pursued with this one model. However, for most natural systems, one model generally is insufficient to represent the variety of defensible ideas about how the system works, and many alternative models should be evaluated. In this context, these criteria are used to identify strengths and weaknesses of the developed models.

The first attribute of more accurate models is a better match to observed data, as indicated by smaller values of the calculated error variance (Eq. (6.1)), the standard error of the regression (the square root of Eq. (6.1)), fitted error statistics (Chapter 6, Section 6.3.2), AIC_c and BIC statistics (Eqs. (6.3) and (6.4)), or the maximum likelihood criteria (Eq. (3.3)). These measures are printed by UCODE_2005 and MODFLOW-2000. The UCODE_2005 and MMA (Multi-Model Analysis; Poeter and Hill, in press) computer codes report additional statistics, such as Kashyap's measure (Medina and Carrera, 1996).

Figure 11.6 shows a graph of AIC_c and BIC statistics and the sum of squared, weighted residuals for five models of the Maggia Valley in southern Switzerland. The models differed in that the hydraulic-conductivity distribution was represented with between one and six parameters defined using geologic mapping of fluvial deposits. As is typical, the sum of squared, weighted residuals diminishes or is unchanged as parameters are added. The AIC_c and BIC statistics are smallest for the model with three hydraulic-conductivity parameters, which suggests that, of the models considered, this one is preferable.

The second attribute of better models is that weighted residuals (defined in Chapter 3, Section 3.4.3) are more randomly distributed. This attribute generally is determined using the graphs and related statistics discussed in Chapter 6, Section 6.4 and Guideline 9 in Chapter 12. Graphs of weighted residuals against weighted simulated values are shown for two models of the same system in Figure 11.7. The weighted simulated values have been adjusted because the coefficients of variation for the weighting are calculated using the observed values, as discussed in Chapter 6, Section 6.4.2. The weighted residuals from model CAL0 tend to be larger than those of CAL3, as indicated by the greater spread about the 0.0 weighted-residual line. In this example, the weighting on the streamflow gains and lake loss was modified within reasonable limits during the course of model development to achieve statistically consistent weighted residuals (Hill et al., 1998, Table 1). A consequence is that the spread of weighted residuals for flows in model CAL3 does not necessarily indicate a closer fit between simulated and observed flows, compared to model CAL0. However, the smaller spread for hydraulic heads in model CAL3

GUIDELINE 8: CONSIDER ALTERNATIVE MODELS

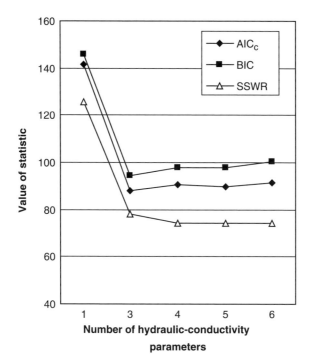

FIGURE 11.6 AIC_c and BIC statistics and the sum of squared, weighted residuals (SSWR), defined in Chapter 3, calculated for five models of the Maggia Valley in southern Switzerland. (From data presented by Foglia et al., in press).

does indicate a better fit, as also is evident in Figure 11.7. The two sets of weighted residuals of Figure 11.7 are both reasonably random, although the dominance of positive CAL0 residuals in Figure 11.7a for weighted simulated values between 15 and 30 may indicate some model bias.

In analyzing the distribution of weighted residuals when comparing alternative models, it also is important to consider additional types of figures that display the observations, simulated values, residuals, and weighted residuals, as discussed in Chapter 6, Section 6.4 and Guideline 9 in Chapter 12.

The third attribute of better models is that optimal parameter values tend to be more reasonable. Evaluating the optimal estimates and confidence intervals is discussed in Chapter 7, in Guideline 10 in Chapter 12, and in Guideline 14 in Chapter 14.

For the complex synthetic system considered by Hill et al. (1998), analyses of the optimal parameter estimates resulted in elimination of model CAL0 as a viable model. In the calibrated CAL0 model, one hydraulic-conductivity estimate was unreasonable, and the confidence interval on the parameter excluded all reasonable values. Analyses of optimal parameter estimates showed that all of the other alternative models were viable, and that none could be considered clearly better than the others on the basis of analyzing these estimates.

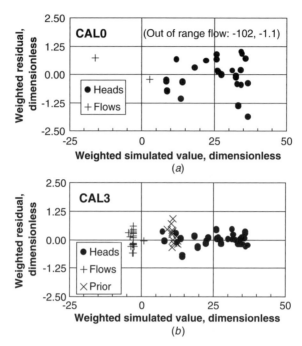

FIGURE 11.7 Weighted residuals versus weighted simulated values for models (*a*) CAL0 (with 34 heads and 3 flows) and (*b*) CAL3 (with 54 heads, 19 flows, and 16 prior) of Hill et al. (1998).

G8.3 Simulate Predictions with Alternative Models

There is general agreement that predictions need to be evaluated using alternative models, but there is disagreement about what alternative models should be included. Some suggest that all models developed should be included in any analysis of predictions (e.g., Burnham and Anderson, 2004; Poeter and Anderson, 2004), while others suggest a more selective approach. The argument for including all models is that those that do not fit the observations well, as indicated by large values of one or more of the measures of overall fit listed in Section G8.2, are given little credence in the analysis, and leaving them in allows all underlying conceptual models to be represented. The argument for a more selective approach is that results from clearly unreasonable models can be confusing to resource managers and the public.

The presentation of predictions from alternative models is important because usually this communicates the most important result of a typically substantial investment by a government, commercial, or nonprofit entity. Ideally, the presentation reveals the predictions, measures of prediction uncertainty, and possibly a separate indicator of model plausibility.

Measures of prediction uncertainty can be calculated using the methods described in Chapter 8, Sections 8.4 and 8.5. Quantifying prediction uncertainty using alternative models is discussed in Chapter 8, Section 8.6 and Guideline 14 in Chapter 14.

G8.4 Application

Here we discuss three alternative models presented by Tiedeman et al. (1997, 1998a). The models represent a regional groundwater flow system in fractured crystalline rock near Mirror Lake, New Hampshire. In each of the models, two model layers represent surficial glacial deposits, and three layers represent fractured crystalline bedrock. The three alternative models differ in their representation of the bedrock hydraulic conductivity, as shown in Figure 11.8. In model A, the bedrock hydraulic conductivity is homogeneous; in model B, it varies with depth; and in model D, it varies with land-surface elevation. Model C is not discussed here. The variations each have a hydrogeologic rationale. For example, consider weathering processes, where weathering of the fractured crystalline rock is expected to

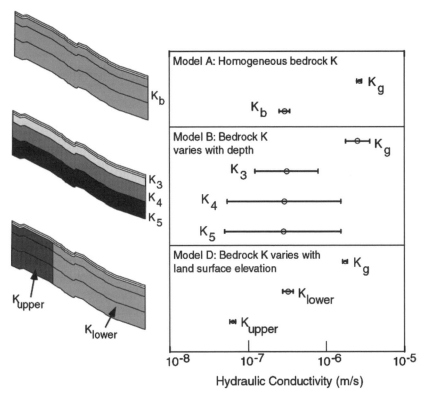

FIGURE 11.8 Representations of bedrock hydraulic conductivity along a hillside cross section (vertical exaggeration approximately 5:1), parameter estimates, and linear individual 95-percent confidence intervals for alternative models A, B, and D of a regional groundwater flow system in fractured crystalline rock near Mirror Lake, New Hampshire. K_g is the hydraulic conductivity of surficial glacial deposits; other K parameters are hydraulic conductivities of the bedrock, as shown on the left for each alternative model. (Adapted from Tiedeman et al., 1997, 1998a.)

increase hydraulic conductivity. Homogeneity (model A) suggests little weathering; model B suggests weathering concentrated at the surface and evenly distributed throughout the area, and model D suggests weathering concentrated in the lower elevations.

Each model was calibrated by nonlinear regression using three flow observations. The number of hydraulic-head observations was 90 for models A and B and 91 for model D. The additional head observation lies in the upper elevations of the system and displays a fairly shallow water level. Only one such observation is available, and it is not known if the water level detected at this well represents the regional water table. As a result, all models were tested with and without this observation; it is only included in the results shown here for model D. Models A and B, which lack any change in hydraulic conductivity with elevation, cannot simultaneously match this observation and head observations at lower elevations. The importance of this depends on the validity of the high-elevation head observation, which is unclear. The optimal hydraulic-conductivity estimates and linear confidence intervals produced by the three models are shown in Figure 11.8.

In model B, the parameter estimates for the hydraulic conductivity of each bedrock model layer (K_3, K_4, and K_5) are nearly the same as the estimate of homogeneous bedrock hydraulic conductivity (K_b) in model A. The confidence intervals for K_3, K_4, and K_5 are each significantly larger than that for K_b. These results suggest that the observations do not support the hypothesis that hydraulic conductivity varies with depth. The model fit to the observations is almost identical in models A and B (the standard error of regression, s, equals 3.0 in both models), which also supports the conclusion that model B is not an improved representation of the flow system compared to model A. Further discussion considers only models A and D.

In model D, the estimate of K_{lower} is about the same as that of K_b in model A, but the estimate of K_{upper} is substantially smaller, and the confidence intervals for K_{lower} and K_{upper} are relatively small and do not overlap, suggesting that the calibration observations support the hypothesis that conductivity varies with elevation. However, the standard error of regression for model D of 3.4 is somewhat larger than that for model A, primarily because model D produces a poorer match to the three flow observations.

Models A and D appear to represent conceptual models that are reasonably well supported by the observations. Additional hydraulic-head data at the upper elevations are needed to better delineate the regional water table. If collected, these data could then help discriminate between the two models. Because models A and D are both plausible, predictions of the regional water budget and of the three-dimensional groundwater basin are simulated for both models (Tiedeman et al., 1997, 1998a).

12

GUIDELINES 9 AND 10—MODEL TESTING

A basic attribute of nonlinear regression methods is that, given a well-posed problem, parameter values are calculated that produce the best fit between simulated and observed values. The model can then be evaluated without speculation about whether a different set of parameter values would produce a better model fit.

A primary purpose of evaluating model fit is to detect ways in which the model incorrectly represents the real system. This incorrect representation is commonly referred to as model error. Model error that causes systematic problems with model fit is denoted model bias. Two common problems are strong indicators of model error: (1) the model does a poor job of matching observations in that the lack of fit is large and/or the weighted residuals are not randomly distributed in time, in space, and/or relative to simulated values and (2) the optimized parameter values are unrealistic and confidence intervals on the optimized values do not include reasonable values. The fundamental premise is displayed in Figure 12.1. Model fit issues are discussed in Guideline 9; estimated parameter-value issues are discussed in Guideline 10.

Effective Groundwater Model Calibration: With Analysis of Data, Sensitivities, Predictions, and Uncertainty. By Mary C. Hill and Claire R. Tiedeman
Published 2007 by John Wiley & Sons, Inc.

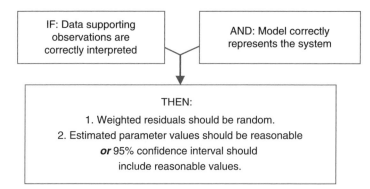

FIGURE 12.1 Premise underlying much of the analysis of model fit and estimated parameter values suggested in Guidelines 9 and 10. If 1 and/or 2 are not true, a better model can be obtained by reevaluating the observations and the model.

GUIDELINE 9: EVALUATE MODEL FIT

The match to observations can be evaluated using the methods described in Chapter 6. The evaluation generally involves the following: (1) determine model fit, including both overall fit and variation in fit among individual observations and (2) diagnose the cause of poor model fit. Evaluations of model fit have been presented in many publications, including Cooley et al. (1986), Yager (1993, 1996), D'Agnese et al. (1997), Tiedeman et al. (1998a,b), Hill et al. (1998), and other studies cited in Chapter 15.

G9.1 Determine Model Fit

Overall measures of model fit were discussed in Chapter 6, Section 6.3; graphical measures are discussed in Section 6.4. Here, we present additional example analyses of model fit.

Weighted residuals have the advantage of indicating model fit in the context of expected observation error (Guideline 6). Model misfit is often more useful when presented in this context. This is especially true if observation errors are proportional to the observed or simulated value and this value varies over many orders of magnitude. In such situations, unweighted residuals can be very misleading. Examples include flow observations in surface-water models and concentration observations in any type of transport model. On the other hand, weighted residuals can be confusing because they are dimensionless. Often it is useful to include maps and other figures of both weighted and unweighted residuals in reports. The discussion of these figures can then indicate whether any large unweighted residuals are actually less problematic than their magnitudes suggest, because of observation error. Figures constructed using unweighted and weighted residuals from a model of the Death Valley regional flow system are presented in Chapter 15.

GUIDELINE 9: EVALUATE MODEL FIT 317

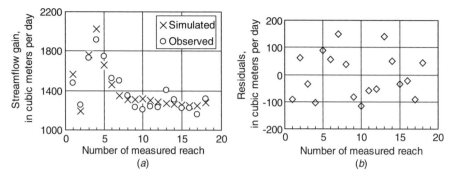

FIGURE 12.2 (*a*) Observed and simulated streamflow gains for model CAL3 of Hill et al. (1998). (*b*) Streamflow gain residuals, equal to the observed minus the simulated values.

Two graphs that illustrate model fit are presented in Figure 12.2. Figure 12.2*a* shows observed and simulated streamflow gains along the length of a river. Figure 12.2*b* shows the related residuals, which are a good indication of model fit if the observed gains are all similarly reliable. Although the two figures present identical information about model fit, each display is useful in a unique way. Figure 12.2*a* places the model fit in the context of the observed quantity. Figure 12.2*b* more clearly displays the variation of misfit with the number of the measured reach.

Figure 12.3 shows an example of weighted residuals displayed on a map of a groundwater model domain. This type of figure is effective for assessing the details of model fit and the spatial randomness of the weighted residuals. Figure 12.3 shows that the weighted residuals are generally small and appear to be randomly distributed in the southern part of the domain. However, in the northern part, weighted residuals are larger and clusters of residuals with similar signs are present, illustrating some bias in the model fit. In this model, the subsurface hydrogeology in the north was not as well characterized as that in the central and southern part of the region, which helped explain this bias (Sanford et al., 2004a).

Identifying trends (lack of randomness) by visual inspection is not always reliable and is made more difficult by the small sample size typical of many regression problems. Often it is useful to evaluate randomness using formal methods to avoid false identification of trends and to identify trends that are difficult to detect. One such method is the runs test, as discussed in Chapter 6, Section 6.4.4. The runs test statistics are calculated using Eq. (6.16) and (6.17).

MODFLOW-2000 and UCODE_2005 each calculate a runs statistic that evaluates the randomness of the weighted residuals with respect to the order in which the observations are listed in the model input files. This statistic can be used to quickly and roughly assess whether the spatial randomness of the weighted residuals is improving as changes are made in the model during the calibration process, which can be advantageous when it is time-consuming to produce maps of weighted residuals such as those shown in Figure 12.3. For example, if water-level observations are listed in the model input file in order from north to south, and initially

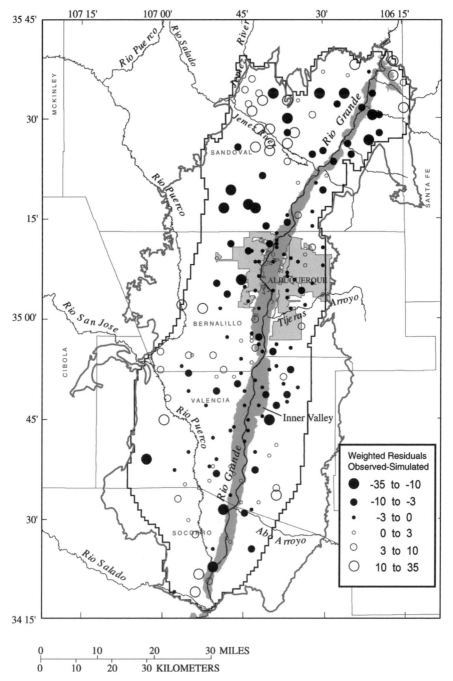

FIGURE 12.3 Distribution of weighted hydraulic-head residuals in a model of steady-state predevelopment groundwater flow in the Middle Rio Grande Basin, New Mexico. (From Sanford et al., 2004a.)

GUIDELINE 9: EVALUATE MODEL FIT   319

the model fit is such that simulated water levels are consistently too large in the north and too small in the south, then the runs statistic will indicate too few runs (a large negative value). As the model is refined and additional regression runs are made, the runs statistic can be evaluated rather than producing a new map after every regression run in which some model aspect has been modified. A runs statistic that becomes smaller in absolute value (closer to 0.0) indicates that the weighted residuals are becoming more randomly distributed.

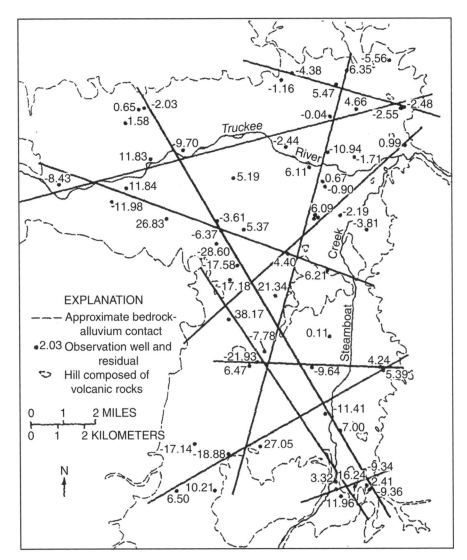

FIGURE 12.4 Hydraulic-head residuals from a model of the Truckee River Basin, Nevada, with lines used to conduct runs tests. The lines are located in the center of swaths for which the runs statistic is calculated. (From Cooley, 1979.)

The runs statistic also can be used to assess the spatial randomness of weighted residuals plotted on the model domain as illustrated in Figure 12.4. In this study, Cooley (1979) used runs tests to evaluate randomness for residuals distributed within a specified distance of selected transects. The results indicated that all the runs along each transect could have occurred by chance. MODFLOW-2000 and UCODE_2005 do not calculate runs statistics to evaluate the residuals in this manner, so this type of analysis requires the modeler to use a custom spreadsheet or code.

The fit of a calibrated model also can be tested using simulated equivalents for observations that either were not included in the model calibration process or are collected after model calibration. Such testing commonly is referred to as validation or a postaudit and is discussed in Guideline 13 in Chapter 14.

G9.2 Examine Fit for Existing Observations Important to the Purpose of the Model

It is important to closely examine the model fit for observations important to the purpose of the model. If the purpose is related to some aspect of model construction, the statistics listed in Table 10.2 that connect observations and parameters can be used to identify these observations. If the purpose is to predict unmeasured quantities, the *opr* statistic that connects observations and predictions can be used. The *opr* statistic is presented in Guideline 12 in Chapter 13 in the context of guiding additional field work. An example of using this statistic is presented in Chapter 15, Section 15.2.1.

G9.3 Diagnose the Cause of Poor Model Fit

Detailed evaluations of weighted residuals, such as those shown in Figures 12.2–12.4, can be used to diagnose the cause of poor model fit. Obvious locations of potential problems include areas in which the model fit is poor and/or biased. However, in models of natural systems, simulated conditions are generally sensitive to both local and distant aspects of model construction. Thus, discovering the cause of problems with model fit often requires considering problems located not only where the misfit occurs, but in a potentially large surrounding volume and, for transient models, at earlier simulated times. For example, discharge to springs in regional groundwater models can be influenced by hydrogeologic and hydrologic conditions at great distances upgradient and downgradient from the spring location. Water needs to supply the spring and some system characteristic or dynamic needs to make the water flow to the spring instead of to downstream locations. Similarly, recharge in high-elevation regions of a model can affect hydraulic heads and discharge at distances far downgradient.

In some cases, aspects of model construction make it impossible for the regression to match a given observation. Dimensionless scaled sensitivities (*dss*) (Chapter 4, Section 4.3.3) and leverage and influence statistics (Chapter 7, Section 7.3) can help reveal this problem. If the values of these statistics for an observation are near zero for all parameters, then the simulated equivalent is insensitive to changes in all parameters. This suggests that model construction precludes a good fit to the observation and can reveal problems with model construction. In groundwater modeling, an extreme example occurs when an observed head is located in a cell defined as constant head, and the value of the constant head is not being estimated by the regression. For a less extreme example of this problem, consider a geologic feature with very low hydraulic conductivity that has been interpreted as discontinuous, based on geologic data. However, differences in hydraulic head across the feature indicate that it is continuous and forms a substantial barrier to groundwater flow. If the model construction is not changed, the model fit to these head observations will be poor.

In many types of models, it can be difficult to address problems with model construction because of the inaccessibility of the true system and/or the expense of data collection. These limitations to investigating the true system may preclude modifying the model to alleviate all problems with poor model fit. In this case, the modeler needs to carefully evaluate the implications for the model predictions of the poor model fit.

An additional potential type of model error involves omission of processes that are important to simulating the observed values, or including processes that actually do not occur in the true system. This type of error can strongly affect the quality of the model fit to the observed values. Thus, when diagnosing the cause of poor fit, it is important to assess whether the model includes the relevant and important processes thought to occur in the true system. Determining the appropriate processes is especially important in transport models, where there are typically a large number of potential transport mechanisms that affect simulated concentrations, as discussed in Chapter 9, Section 9.2.

If the model fit is unsatisfactory, five aspects of the model or calibration effort can be investigated and possibly changed as described below. The magnitude of the changes can range from correcting data entry errors, to adding simulated processes, to completely reevaluating some or all facets of the conceptual model.

Parameter Definition Parameter definition can be modified, for example, by adding, omitting, dividing, or combining parameters. As always, the final parameterization needs to be consistent with all known information about the system; for example, in groundwater problems, hydrogeologic information needs to be respected. See the methods described in Chapters 4 and 7 and Guidelines 2 and 3 in Chapter 11.

Simulated Equivalents of the Observations Problems with the calculation of simulated equivalents to observations may become apparent as model calibration proceeds.

For example, consider a groundwater system with an observation from a well in which the screen spans several layers of the corresponding groundwater model. The simulated equivalent of the observed head is typically calculated as a weighted average of the simulated hydraulic heads in the model layers spanned by the screen. It is not always straightforward to define the contribution from each layer and the appropriate contribution may change as the model changes during calibration. If there is a poor fit to this type of observation, definition of the simulated equivalent may need to be modified. Alternatively, the methods included in, for example, the Multi-Node Well (MNW) Package (Halford and Hanson, 2002) of MODFLOW could be adapted to calculate the contributions from each model layer.

Problems with calculation of simulated equivalent also can occur when the defined elevations for head-dependent boundaries of a discretized model are determined from digital elevation maps (DEMs). Within the area of the model-grid cell the appropriate value might be the average elevation, the lowest elevation, or, depending on the resolution of the DEM, an elevation that is even somewhat lower than the lowest elevation. The latter circumstance can occur when there are narrowly incised rivers or springs ensuing from small depressions. Problems are generally indicated by too little simulated flow to the head-dependent boundaries, unexpected flow from the boundaries to the groundwater system, or too much flow from the boundaries into the groundwater system. Inspection of areal photographs and/or field work at selected representative sites often is needed to determine appropriate elevations to use in the model.

Other Aspects of Model Construction Model construction can be modified, for example, by correcting input data, changing the representation of boundary conditions and parameterization, and changing the processes simulated. In groundwater models, a surface-water body represented by a constant-head boundary might be changed to a head-dependent boundary, the pumping rates at wells might be updated based on new information, or temperature variations might be included explicitly or implicitly. Most of these changes are no different from modifications a modeler would consider as part of any model calibration effort.

Observations Affected by Processes that Are Not Simulated It is sometimes necessary to remove observations from the regression; however, this should be done only after careful consideration. For example, in groundwater systems, wells that intersect perched water do not directly reflect the dynamics of a regional flow system. Including measurements from such wells as observations in the calibration of a regional model is likely to produce fallacious results, and thus, these observations will typically be omitted from the regression.

Perched wells are one example of observations affected by systematic errors caused by omission from the model of a process important to the observation. Unlike random measurement error, systematic error cannot easily be accommodated by weighting and can be impossible to separate from the effects of simulated processes. An example is given by Pavelko (2004) for aquifer compaction and expansion data recorded by an extensometer in Las Vegas, Nevada. Extreme heating in the extensometer shed caused thermal effects on the extensometer, which were recorded as apparent diurnal fluctuations in aquifer deformation. When these data were employed as observations in regression runs used to calibrate a model of aquifer deformation, this resulted in poor model fit, unreasonable parameter estimates, and problems with regression convergence (M. Pavelko, U.S. Geological Survey, written communication, 2004). To correct this problem, the calibration observations were defined as aquifer deformation over time periods of 25 days or longer. This definition of the observations minimized the influence of the thermal effects.

Weighting Errors   Errors in the weighting of the observations or prior information also are possible. However, use caution when considering changes to the weighting as an approach to resolving problems with model fit. It is easy to invest a great deal of time modifying weights and running regressions, with little consequent gain in model accuracy or in understanding the system dynamics. This does not mean that weighting never needs to be modified. Adherence to the principles described in Guideline 6 in Chapter 11 will help keep the effort spent determining and modifying the weight matrix consistent with its importance to the purpose of the model.

GUIDELINE 10: EVALUATE OPTIMIZED PARAMETER VALUES

Evaluating optimized parameter values involves five steps. (1) Quantify parameter-value uncertainty. (2) Detect model error by comparing the estimates and their linear and nonlinear confidence intervals to reasonable ranges determined from field data. (3) Diagnose the cause of unreasonable parameter values. (4) Identify observations important to the parameter estimates. (5) Determine whether fewer parameters are likely to produce as good a fit or if additional parameters can be supported by the available observations. These issues are discussed in the following five sections.

G10.1   Quantify Parameter-Value Uncertainty

Parameter-value uncertainty can be quantified using parameter confidence intervals, which are an integral part of the analyses discussed in Sections G10.2 to G10.5. Calculation of parameter confidence intervals is presented in Chapter 7.

The relative uncertainty of parameters can be important to the evaluations of Sections G10.4 and G10.5. Confidence intervals can be directly compared for parameters with the same units, such as in Figure 12.5. To compare the uncertainty of parameters with different units, such as hydraulic conductivity and recharge, confidence intervals can be expressed in terms of percent of estimated value, as shown in Figure 7.7, 7.8, and 9.19. Alternatively, parameter coefficients of variation (Eq. (7.4)) can be used.

G10.2   Use Parameter Estimates to Detect Model Error

The use of optimized parameter values to detect model bias was presented in Chapter 7, Section 7.6. This simple test can be an unexpectedly powerful indicator of model error, even given the wide ranges of reasonable values for many characteristics of natural systems. For example, in groundwater systems hydraulic

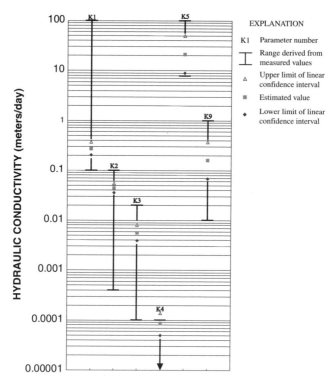

FIGURE 12.5 Optimized hydraulic-conductivity values, 95-percent linear confidence intervals, and the range of hydraulic-conductivity values derived from field and laboratory data. (From D'Agnese et al., 1997.)

conductivity can vary by many orders of magnitude in a single field site. Use of reasonable ranges to detect model error has been demonstrated using synthetic numerical test cases by Poeter and McKenna (1995), Anderman et al. (1996), Poeter and Hill (1996), Barlebo et al. (1998), and Hill et al. (1998). Field studies that have found this test to be useful include those by D'Agnese et al. (1997, 1999), Tiedeman et al. (1998b), McAda (1999), Faunt et al. (2004), and Gannett and Lite (2004). Relevant results from Barlebo et al. (1998) are presented in Chapter 15.

A graphical comparison of estimated hydraulic conductivities and ranges of expected values is presented in Figure 12.5 for the Death Valley regional flow system study of D'Agnese et al. (1997, 1999), which is discussed further in Chapter 15. Two features of Figure 12.5 deserve discussion: the large reasonable ranges and the small linear confidence intervals on the estimates.

The reasonable ranges in this example are large, but a number of conceptual models were rejected because optimized parameter values were outside these ranges. Thus, even in this circumstance with large ranges of expected values, requiring reasonable optimized parameter values produced an important constraint on model development.

GUIDELINE 10: EVALUATE OPTIMIZED PARAMETER VALUES

Examination of the confidence intervals of Figure 12.5 could lead to the conclusion that the intervals are too small to realistically represent the uncertainty in the estimate. This judgment, however, needs to be made in the context of the meaning of the parameter estimates and confidence intervals. In this example, and in many situations, the defined parameters result from simplifying assumptions. The most relevant assumption in this example is that a few parameters and simple functions are used to represent the very complicated hydraulic-conductivity distribution that exists in the true flow system. In the Death Valley model, the simple functions are a zonation scheme in which the hydraulic conductivity is set to the same parameter value at all locations where rocks with certain characteristics occur. The simple functions also could involve using a few defined parameters to implement an interpolation scheme. Definition of a few parameter values to represent hydraulic conductivity throughout a model is very useful if broadly defined variations in hydraulic conductivity dominate system dynamics. The resulting estimated parameters represent effective or average values. The confidence intervals for these parameters represent the uncertainty in these effective or average values. In contrast, the reasonable ranges often represent the breadth of local values.

Confidence intervals on average (mean) values depend on the standard deviation of the original population, and on the sample size used to calculate the estimated average. Because the population statistics often are unknown, the sample standard deviation is commonly used. To demonstrate the importance of these dependencies, consider a simple example using a generated population of 300 normally distributed random numbers. Figure 12.6 shows the mean and range of the 300 numbers, as well as the mean and the confidence interval on the mean for different sample sizes drawn from the population. This simple example illustrates that even with very few samples, the confidence interval for the average is significantly smaller than the range of the population.

In Figure 12.5, the ranges of hydraulic conductivities are derived from measured field and laboratory values. Each of the six ranges is analogous to the population range in Figure 12.6. The six parameter estimates shown in Figure 12.5 for the Death Valley regional flow model are analogous to the three means of Figure 12.6, with one important difference. In Figure 12.5, the parameter estimates are derived through nonlinear regression. Thus, most of the data used to estimate the effective hydraulic-conductivity values are measurements of other quantities (hydraulic heads and spring flows). In contrast, the means of Figure 12.6 are calculated from samples taken from the population for which the range is plotted on the left side of the figure. Figure 12.6 reveals something important about Figure 12.5. That is, the wide ranges and much narrower confidence intervals such as those shown in Figure 12.5 are to be expected given that the confidence intervals are on the expected value.

Using independent information on the parameters to identify model error, as suggested here and in Chapter 7, Section 7.6, is an alternative to using the information on the parameters to define prior information or to impose limits on estimated parameter values, which are discussed in Chapter 5, Section 5.5 and in Guideline 5 in Chapter 11. As noted there, unreasonable optimized parameter

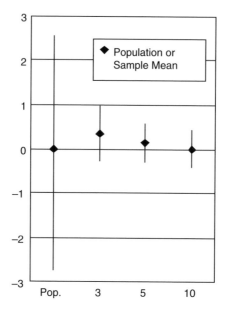

FIGURE 12.6 Population range and mean, and confidence intervals and means for different sample sizes. Means are calculated as the arithmetic average value (Table 7.1) The population range is noted by the bar labeled "Pop." The other bars are labeled with the sample size used (3, 5, and 10), and the lengths of the bars display the associated confidence interval on the mean, calculated as the mean $\pm 2s/n^{1/2}$, where s is the sample standard deviation (Table 7.1) and n is the sample size (Ott, 1993, pp. 201–202).

values can be disconcerting but can be important indicators of problems with model construction, the observations, or both.

G10.3 Diagnose the Cause of Unreasonable Optimal Parameter Estimates

If the analysis of parameter estimates and confidence intervals reveals imprecise and/or unreasonable parameter estimates, investigation of the issues discussed in Section G9.3 can help reveal the cause. Here, two strategies for diagnosing unreasonable estimates are discussed: influence statistics and inconsistencies between true and simulated processes. When attempting to diagnose why the regression is producing unreasonable parameter estimates, it is important to keep in mind that error in other model attributes (those not associated with the parameters with unreasonable estimates) also might be contributing to the problem.

Yager (1998) used *DFBETAS* influence statistics to help identify model error. In the model of regional groundwater flow through fractured dolomite depicted in Figure 11.2, an unreasonably large optimal value of horizontal anisotropy was estimated by nonlinear regression. The *DFBETAS* statistics were used to

GUIDELINE 10: EVALUATE OPTIMIZED PARAMETER VALUES   327

identify a set of influential head observations that were important to the anisotropy parameter and were located at the edge of a rural recharge zone that was adjacent to an urban recharge zone. The recharge rate in the rural zone was expected to be substantially lower than that in the urban zone. Evaluation of the calibrated model revealed that to optimize the overall model fit, the unrealistically large anisotropy value was estimated because it compensated for a recharge rate that was too low to provide a good fit to the influential heads. To resolve this problem, Yager (1998) modified the position of the boundary between the recharge zones, which was not clearly defined by field data, so that the influential observations were in the urban recharge zone. With this modification to model construction, the regression produced a realistic optimal estimate of horizontal anisotropy and of other model parameters and provided a good fit to the influential heads.

Investigating whether processes important to observed values are simulated in the model, as discussed in Guideline 9 with regard to diagnosing poor model fit, also can lead to identification of model error when diagnosing unrealistic optimized parameter values. For example, if observed solute concentrations in fractured rock are strongly controlled by advection, dispersion, and matrix diffusion, transport simulations using only advection and dispersion can produce unreasonable estimates of dispersion parameters. Matrix diffusion needs to be simulated to obtain reasonable estimates. In the example from Pavelko (2004) described in Guideline 9, unreasonable parameter estimates, as well as poor model fit, resulted when observations were defined in a way that emphasized diurnal signals in the data caused by heating of the extensometer.

G10.4 Identify Observations Important to the Parameter Estimates

The statistics presented in this book that can be used to identify observations important to parameters are listed in Table 10.2.

Statistics that can be used to identify observations that are important to individual parameter estimates include dimensionless scaled sensitivities (*dss*, Chapter 4, Section 4.3.3) and *DFBETAS* (Chapter 7, Section 7.3.2). The *dss* are fit-independent, but this attribute is not as important for Guideline 10 applied to a model that is substantially calibrated, as it is for Guideline 3 applied to a newly constructed model. *DFBETAS* has the advantage of representing the effects of both sensitivity and parameter correlation, so it is usually a better choice for the evaluations conducted as part of Guideline 10. Using *dss* to identify observations important to one parameter estimate is illustrated in Figure 8.1c. *DFBETAS* statistics can be presented similarly.

The statistics that can be used to identify observations that are important to a set of parameters include leverage (Chapter 4, Sections 4.3.6 and 4.4.3 and Chapter 7, Section 7.3.1) and Cook's *D* (Chapter 7, Section 7.3.2) Using Cook's *D* to identify observations important to all parameter values is illustrated in Chapter 15, Section 15.2.1. The statistics differ in that the leverage statistics are fit-independent, so that they do not account for model fit to observations. Cook's *D* is a measure of influence that accounts for model fit to observations. The choice of statistic, therefore, is likely to be based on whether the model is mature enough that inclusion of model fit is preferred.

G10.5 Reduce or Increase the Number of Parameters

If individual linear confidence intervals (Chapter 7, Section 7.5.1) for two or more parameters overlap, it may imply that the true parameter values are similar, even if the estimated values are different. An example is presented in Figure 11.8 and discussed in Section G8.4. In model B, the overlapping 95-percent linear individual confidence intervals and similar estimates suggest that the true hydraulic conductivities may be similar and it may be possible to assign the same hydraulic conductivity to model layers 3, 4, and 5 without a significant deterioration in model fit. Indeed, this was achieved using model A. If model fit significantly deteriorates, the parameters probably should not be combined. For nonlinear models, linear intervals are approximate. Nonlinear intervals (Chapter 7, Section 7.5.1) can be considered, but each limit of each interval requires as much execution time as a regression. It is often more effective to use linear intervals to identify likely parameter combinations that can then be tested using a single regression.

Analysis with the confidence intervals is analogous to performing a standard two-tailed, or two-sided, hypothesis test (Davis, 2002, pp. 61–64; Helsel and Hirsch, 2002, p. 104) in which the hypothesis for model B is that the hydraulic conductivity of the bedrock is uniform with depth. If the test results show that this hypothesis cannot be rejected, then it may be possible to define one hydraulic-conductivity parameter that applies to layers 3, 4, and 5 (model A).

At any stage of model calibration, composite scaled sensitivities can be analyzed as described in Guideline 3 (Chapter 11) to determine if the available data are likely to support additional detail in representing the system characteristics associated with the defined parameters. Parameters with composite scaled sensitivities that are significantly larger than 1.0 and large compared to css values for other parameters might be divided in ways that are consistent with other data, such as geologic and hydrogeologic data in groundwater problems. The new set of defined parameters could then be evaluated using the methods of Guideline 3, and regression pursued if warranted.

New parameters can also be added and estimated using, for example, the representer, super parameter, or constrained minimization method. See Chapter 1, Section 1.3.2 for references.

13

GUIDELINES 11 AND 12— POTENTIAL NEW DATA

In most natural systems, collecting meaningful data is expensive and time-consuming. Thus, it is important to collect data most beneficial to the modeling objectives. These objectives commonly include (1) better understanding of the processes and properties governing system dynamics and/or (2) simulating predictions of future conditions. Of course, (2) depends on (1), but identifying predictions that are of primary importance can be used to focus data collection efforts.

Models are powerful tools for guiding additional field data collection, as suggested in Chapter 8, Section 8.3. This effort is best achieved using the fit-independent statistics listed in Table 10.2. Fit-independence is important because the value of the potential new data is unknown. If fit-dependent statistics are used, they need to be evaluated for a reasonable range of potential observed values. The statistics can be used to identify observations important to parameters, parameters important to predictions, and observations important to predictions. Knowledge of the important parameters and observations can then be used to guide the data collection effort, as discussed in Guidelines 11 and 12.

To be useful for the task of identifying important potential new data, a model needs to represent the system with a reasonable level of accuracy. It can be difficult to determine when a model is sufficiently accurate, but at the very least, obvious errors in the system representation and in simulated equivalents to observations need to be resolved. Strategies to resolve these problems using analyses of model fit and optimal parameter estimates are presented in Guidelines 9 and 10 in Chapter 12.

Effective Groundwater Model Calibration: With Analysis of Data, Sensitivities, Predictions, and Uncertainty. By Mary C. Hill and Claire R. Tiedeman
Published 2007 by John Wiley & Sons, Inc.

In this chapter, Guideline 11 discusses identifying new data to improve the parameter estimates and distribution. Guideline 12 discusses identifying new data to improve model predictions. The analyses of model uncertainty discussed in Chapter 14 also often motivate and provide guidance for new data collection efforts. Modelers are encouraged to be creative in how they use the methods discussed in the guidelines.

For the approaches discussed here, results need to be considered carefully because they inherit all the simplifications and approximations in the model. To determine how to proceed with data collection, generally it is wise to use model results in combination with other information, such as existing observation data, existing knowledge of the system characteristics and observations, and information about past and future stresses, such as pumping in groundwater systems. That is, we do not suggest depending only on the model-generated results that are the primary focus of this discussion.

GUIDELINE 11: IDENTIFY NEW DATA TO IMPROVE SIMULATED PROCESSES, FEATURES, AND PROPERTIES

Sometimes models are constructed primarily to better understand the processes, features, and properties that govern system dynamics. New data can serve four roles in improving the representation of these entitites in models. Data can be used to support: (1) modifying system processes; (2) modifying the geometry of system features, including parameter structures such as zonation or interpolation; (3) defining system property values that relate directly to parameter values; and (4) new observations that provide indirect data about system information. Guideline 11 focuses on the fourth role and on methods for identifying useful new observations. Commonly, data related to observations are much easier and less expensive to collect than is system information. First, roles (1)–(3) are briefly discussed.

Collecting new data to support modification of processes or system features often is motivated by poor model fit or unreasonable parameter estimates. The first step typically is to evaluate the likely importance of the process or feature conceptually. Next, the process or feature is modified in the simulation model, and methods in Guidelines 8–10 can be used to test model improvement. If supported by these analyses, field data can be collected to further characterize the process or feature, and to improve its representation in the model, as discussed in Guideline 2.

Obtaining data that relate directly to parameter values can be difficult. System properties commonly are measured at scales that differ from those to which model parameters apply, as discussed in Chapter 9, Section 9.2.3 in relation to transport models. To the extent that the measurements do apply to parameter values, they can be used as reasonable ranges, prior information, or specified values, as discussed in Guidelines 2 and 5.

Data that support additional observations can be evaluated with the model using fit-independent statistics. We discuss methods that use sensitivities to evaluate potential new observation types and locations for the information they provide

TABLE 13.1 Dimensionless Scaled Sensitivities (*dss*) and Leverage for Two Potential Observations from Exercise 8.1c, and Composite Scaled Sensitivities (*css*) Calculated for the Existing Observations

	Dimensionless Scaled Sensitivities[a] (*dss*) for Parameter[b]:				
	HK_2	VK_CB	K_RB	RCH_2	Leverage
Potential head observation	−3.5	8.0×10^{-3}	−0.105	54.8	0.988
Potential flow observation	-3.2×10^{-5}	1.1×10^{-6}	-0.349×10^{-5}	−4.50	0.491
css for existing observations	3.1	0.22	0.20	25.3	

[a] For four of six parameters.
[b] *Parameter labels*: HK_2, hydraulic conductivity of model layer 2; VK_CB, vertical hydraulic conductivity of the confining unit; K_RB, vertical hydraulic conductivity of the riverbed; RCH_2, recharge rate away from the river.

about the model parameters. These methods include dimensionless and composite scaled sensitivities (*dss*, *css*), parameter correlation coefficients (*pcc*), and leverage statistics (Chapter 4, Sections 4.3.6 and 4.4.3, and Chapter 7, Section 7.5.2). For *dss*, *css*, and leverage statistics, the anticipated accuracy of the potential observations also can be considered because observation weights are included in the calculation. Weights for potential observations can be determined using the same strategies as for existing observations, discussed in Guideline 6.

Table 13.1 shows the *dss* calculated in Exercise 8.1c for two potential observations in the simple steady-state model with pumping, and the *css* calculated using only existing observations obtained before pumping began. In Exercise 8.1c, the potential observations were evaluated with respect to their contribution to reducing prediction uncertainty. Here they are evaluated with respect to their contribution to improving parameter estimates.

The *dss* in Table 13.1 are shown for three model parameters for which the existing observation data provide relatively little information (HK_2, VK_CB, and K_RB), as indicated by the *css*. The *dss* also are shown for parameter RCH_2, for which the existing observations provide ample information. The *dss* suggest that the potential head observation, which is located in the top model layer far upgradient from the river, is more important to all four model parameters than is the potential flow observation, which is the discharge along the entire length of the river.

In analyzing the *dss* in the context of the importance of potential observations to improving parameter estimates, it is important to assess them relative to the *css* calculated for the existing observations. To be helpful for improving a parameter estimate, the absolute value of the *dss* for a potential observation needs to be roughly of the same or greater magnitude than the *css* for the parameter. Comparison of the *dss* and *css* values in Table 13.1 suggests that the potential head observation is likely to improve the estimates of HK_2, K_RB, and RCH_2 but would contribute little toward estimating VK_CB. The potential flow observation is only likely to help improve the estimate of RCH_2. When evaluating the *dss* to determine the value of a potential observation, there is an additional consideration. As discussed in Chapter 4, Section 4.3.4, parameters with *css* less than 1.0 are more likely than those with larger *css* to be poorly estimated, and to cause regression convergence problems. Thus, potential observations that are likely to increase the *css* of a parameter to greater than 1.0 are of special interest. Potential observations that are likely to increase the *css* to a value less than 1.0 are not likely to improve the estimate of that parameter. By this analysis, the potential head observation is not likely to improve the estimate of K_RB.

Parameter correlation coefficients (*pcc*) also need to be considered when evaluating potential new observations. Potential observations that provide little information as indicated by the *dss* might be very important to improving the parameter estimates, if they help to reduce parameter correlations. This can be tested by comparing *pcc* calculated only with the existing observations to those calculated with the existing and potential observations. The latter calculation uses the parameter variance–covariance matrix with potential observations, discussed in Chapter 7, Sections 7.2.1 and 7.2.5. The results of this comparison are given in Exercise 8.1c for the example presented above and show that addition of the potential head observation reduces the absolute value of several correlations that are very large when only the existing observations are included. Addition of the flow observation as well further reduces the correlations, indicating that it is more important to improving the parameter estimates than is indicated by the *dss* alone, but how much more is not clear.

Leverage statistics also can be used to evaluate the potential effect of one or more observations on a set of parameter estimates. The actual effect is measured by influence statistics, which depend on the observed value, and so are not useful for evaluating potential observation data. In addition to the effects measured by *dss*, leverage statistics reflect the ability of the potential observation to reduce parameter correlations. The leverage statistics for the example are listed in Table 13.1. Leverage statistics can range from 0.0 to 1.0, so the potential head observation has extremely high leverage and the potential flow observation has moderate leverage. A disadvantage of leverage statistics is that they do not indicate the particular parameter(s) to which a potential observation is most important. In this example, the leverage statistic suggests that, overall, the head observation is likely to contribute more information than the flow observation, which is consistent with the analyses of the *dss* and *pcc*. Final decisions about data collection often also depend on which parameters are important to predictions, which is the topic of Guideline 12.

It can also be useful to plot the *dss* for potential observations in relation to independent variables such as time and location. The graph of *dss* versus time shown in Figure 13.1 indicates the relative importance of potential drawdown observations during pumpage. For parameters HK_1, HK_2, and Q_1&2, the sensitivity increases with time, indicating drawdown observations later in time provide the most information about these parameters. In contrast, drawdown at an intermediate time is most likely to improve the estimates of the storage coefficient parameters.

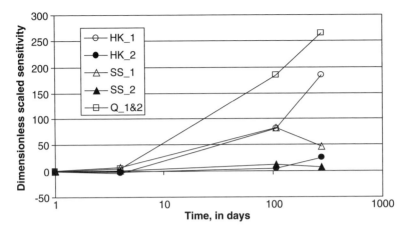

FIGURE 13.1 Dimensionless scaled sensitivities plotted in relation to time for an existing head observation (at time = 0 days, with no pumpage) and potential drawdown observations (at time >0 days, with constant pumpage) from well 2 of Exercise 9.6. Model parameter HK_1 represents the hydraulic conductivity in model layer 1, HK_2 is used to calculate the hydraulic conductivity of model layer 2, SS_1 and SS_2 are storage coefficients of the top and bottom layers, respectively, and Q_1&2 is the pumping rate in wells 1 and 2.

Additional uses of scaled sensitivities are discussed in Chapter 14 under Guideline 14 and in Chapter 4, Section 4.3.3. Using *dss* in this manner is similar to how sensitivity measures were used by Knopman and Voss (1988).

Maps of one-percent scaled sensitivities for hydraulic heads, such as those shown in Figure 4.4, are an additional tool for identifying areas and depths where hydraulic heads are important to one or more parameters but there are no existing observations. However, there are important limitations to the use of these maps, as discussed in Chapter 4, Section 4.3.7.

When evaluating potential new observations using *dss*, *css*, *pcc*, and leverage, model nonlinearity can produce misleading results. This is illustrated using the example from Anderman et al. (1996), discussed in Chapter 11, Section G3.1. Although in this example existing observations are considered, the results of analyzing the *css* and *pcc* would be identical if these observations were potential data. This example also provides an example of how nonlinearity can affect sensitivity analysis. Using initial parameter values, the advective-transport path entered a lake near the source instead of continuing a greater distance within the groundwater system. The longer path is more probable given the concentration data. The unrealistic short advective-travel path resulted in an underestimate of the importance of the advective-transport data when evaluated using the *css* and *pcc* calculated for the initial parameter values.

This situation demonstrates the importance of calculating statistics for multiple sets of parameter values, which also is discussed in Chapter 4, Section 4.4 and portrayed in Figure 4.2. If the statistics change considerably when calculated at a different, reasonable set of parameter values, then they may not be reliable indicators of the worth of the potential data to the model calibration. Reasonable parameter values are those that both respect the system information and produce a reasonable fit to observations. If the simulations produce statistics that support very different data acquisition efforts, the improved understanding of the system obtained from the analysis may still be helpful in making decisions about how to proceed with data acquisition and model development efforts.

GUIDELINE 12: IDENTIFY NEW DATA TO IMPROVE PREDICTIONS

Often models are developed primarily for predictive purposes. In this case, a high priority for field data collection is to improve predictions by increasing their accuracy or reducing their uncertainty. As noted in Guideline 11, new data can serve four roles in improving the representation of model processes, features, and properties governing system dynamics. Data can be used to (1) modify the processes included; (2) modify the geometry of system features, including the structure of parameterizations such as zonation or interpolation; (3) define system property values that relate directly to parameter values; (4) support additional observations. In the context of improving predictions, we expand the fourth role to include improving existing observations with poorly characterized attributes. The first role is discussed briefly below, (2) and (3) are considered in Section G12.1, and (4) is considered in Section G12.2.

Considering additional processes in the context of predictions follows the steps described for Guideline 11, except the effect on predictions also is considered. Once the process is included in the model, the methods used are identical to those described in Section G12.1.

G12.1 Potential New Data to Improve Features and Properties Governing System Dynamics

The most common method for identifying system features that are important to predicted values is to simulate the predictions using alternative conceptual models of the system in which selected features are added, removed, or modified. The process is similar to the sampling methods described in Chapter 14, Section G14.2, except that here the simulations are used to identify potential new data instead of evaluating uncertainty.

Methods presented in Chapter 8, Section 8.2 also can be used to guide collection of data important to the predictions. These include (1) combined use of composite and prediction scaled sensitivities (css and pss) and parameter correlation coefficients (pcc), as illustrated in Figure 8.2, and (2) the parameter–prediction (ppr) statistic. These methods focus on identifying parameters that are most important to the predictions. The results of these methods can be used to guide collection of data about the values of parameters associated with system features. Field activities to obtain this type of data in groundwater systems include, for example, hydraulic tests for estimating transmissivity and storativity values.

The results from the *css–pss–pcc* and *ppr* methods also can be used to guide collection of system information related to the representation of model features. By this approach, it is assumed that there is a link between model parameter importance and system feature importance. That is, it is assumed that information about system features is important to predictions if parameters related to the features are identified as important to predictions. The parameters identified as most important to the model predictions may not always correspond to the features of model construction that are most important to the model predictions, but it is expected that there will often be such a correspondence. In groundwater systems, such features might include the geometry and internal variability of a hydrogeologic unit associated with a hydraulic-conductivity parameter identified as important. Field activities might include geologic and geophysical investigation and interpretation of the extent and thickness of the hydrogeologic unit.

Tiedeman et al. (2003) discuss in more detail issues related to using the *pss* and *ppr* statistics to guide field data collection and provide an example of their application, which is summarized in Chapter 15, Section 15.2.1.

G12.2 Potential New Data to Support Observations

New data can be used to improve existing observations or to obtain new observations. It can be beneficial to improve existing observations if they are shown to be important to predictions and there is a resolvable deficiency in the observations. For example in a groundwater study, wells for which existing head observations are shown to be important to predictions might be the focus of downhole methods to better understand the condition of well screens where corrosion is suspected. In a surface-water study, high streamflow at a site may be important to predictions, but the flow might be derived from a stage measurement and a rating curve extrapolated beyond streamflow measurements. The new data might involve better delineating the local topography and vegetation and using the methods of Kean and Smith (2005) to improve the streamflow observation.

There are two primary tools for evaluating potential new observation data for improving the predictions, both of which are presented in Chapter 8, Section 8.3. The first involves using *dss*, *css*, *pss*, and *pcc* together, and the second is the observation–prediction (*opr*) statistic.

Using *dss–css–pss–pcc* involves first using *pss* to identify parameters that are important to a prediction, then using *css* to identify whether any of these parameters are not well supported by the existing observation data. Then, the methods discussed in Guideline 11 can be used to identify potential new observations likely to provide information about the identified parameters. Finally, *pcc* can be used to evaluate if the potential new observations help reduce parameter correlations that are problematic for predictions. This process has the advantage that each of the separate statistics is conceptually easy to understand and to convey to others. Disadvantages include that it can be cumbersome to display and evaluate the four different measures and associated graphs and it does not reflect the importance of parameter correlations to predictions.

The *opr* statistic of Chapter 8, Section 8.3.2 addresses the disadvantages of the *dss–css–pss–pcc* method by integrating the effects of both sensitivity and correlation. It can be used to evaluate an existing monitoring network to identify locations and types of data that are most advantageous to continue measuring under anticipated future scenarios. It also can be used to identify potential new observation types and locations that would be most beneficial to add to a monitoring network. The primary disadvantage is that *opr* may be more difficult to understand. Tiedeman et al. (2004) provide an example of applying the *opr* statistic to evaluate a hydraulic-head monitoring network associated with the Death Valley regional groundwater flow system. This application is summarized in Chapter 15, Section 15.2.1.

14

GUIDELINES 13 AND 14—PREDICTION UNCERTAINTY

An advantage of using optimization for model development and calibration is that optimization provides methods for evaluating and quantifying prediction uncertainty. Both deterministic and statistical methods can be used. Guideline 13 discusses using regression and postaudits, which we classify as deterministic methods. Guideline 14 discusses inferential statistics and Monte Carlo methods, which we classify as statistical methods.

GUIDELINE 13: EVALUATE PREDICTION UNCERTAINTY AND ACCURACY USING DETERMINISTIC METHODS

Deterministic methods are useful for evaluating and understanding prediction error. Here we consider two methods. The authors have discussed the first method with a number of people, including John Doherty (Watermark Consulting, Corinda, Australia, oral communication, 2002), but we are not aware that it has appeared in any previous publication.

G13.1 Use Regression to Determine Whether Predicted Values Are Contradicted by the Calibrated Model

In some circumstances the regression can be a useful tool for evaluating predictions and their uncertainty. For example, consider a model in which the simulated

---

Effective Groundwater Model Calibration: With Analysis of Data, Sensitivities, Predictions, and Uncertainty. By Mary C. Hill and Claire R. Tiedeman
Published 2007 by John Wiley & Sons, Inc.

concentration of a contaminant at a location is always below the drinking water standard, but resource managers question whether a small change in the parameter values could result in the simulated concentration exceeding the standard. This question can be addressed using linear and nonlinear confidence intervals, but the answer can sometimes be conveyed more clearly by revealing the parameter values or conditions that would be required to produce a specific simulated prediction or set of predictions. This can be accomplished using MODFLOW-2000, UCODE_2005, PEST, or other inverse models as follows.

1. Add the predicted value to the regression as an "observation" using a large weight (small statistic). In the example given above, this value would be a concentration that exceeds the drinking water standard.
2. Perform regression.

The conclusion is that the predicted value is contradicted by the observations and the calibrated model if (a) the predicted value cannot be matched by the regression, (b) it can be matched but the parameter values required produce a poor match to the calibration observations, or (c) the parameter values required to achieve the match are unreasonable. To the extent that the model represents the relevant aspects of the system, this suggests that the predicted value is unlikely to occur. This result also can be communicated using nonlinear confidence intervals on the predicted value, and possibly linear intervals.

The conclusion is that the predicted value is not contradicted by the observations and the calibrated model, and the concerns of the resource manager are substantiated, if the predicted value is matched without producing a poor match to the observations or unreasonable parameter values. This result can be communicated using the results of the regression and linear or nonlinear confidence intervals.

G13.2 Use Omitted Data and Postaudits

Model accuracy can be evaluated by comparing simulated predictions with existing data intentionally omitted from model calibration or new data. Here we concentrate on situations when the omitted or new data are related to predictions. Sometimes these tests are called model validation, but we agree with concerns expressed by, for example, Konikow and Bredehoeft (1992) and Bredehoeft and Konikow (1993), that this terminology is misleading. Different tests lead to different levels of confidence in the model, and saying each "validates" the model ignores that important distinction. Tests against new data are sometimes called postaudits.

These tests are meaningful when the new data represent stress conditions or aspects of the system that differ from those represented in the data used for model calibration. For example, consider a model calibrated using two cycles of tidally induced fluctuations. It is less meaningful to test the model using another cycle with the same amplitude and phase and measured at the same locations than to assess the ability of the model to reproduce a cycle with a different amplitude or system response to some other type of stress entirely, such as the imposition of pumpage. As another example, consider a groundwater model calibrated with hydraulic-head and flow data. A meaningful test might be to assess its ability to predict (1) heads and/or flows collected when pumpage had increased significantly or (2) another process such as transport.

Here we present a few published examples of using new data to test model predictive capabilities. The first two examples use new data collected under conditions similar to the calibration conditions.

Van Loon and Troch (2002) calibrated a suite of distributed hydrological models with varying temporal and spatial resolutions, using sets of soil moisture observations with different temporal and spatial densities. They then tested the ability of the calibrated models to predict a subset of soil moisture data collected later in time under similar hydrologic conditions. They concluded that prediction accuracy did not necessarily increase as model resolution increased.

Saiers et al. (2004) examined the dependence of prediction accuracy on the types of observations used to calibrate a groundwater flow and transport model. The observation sets consisted of heads; heads and flow; or heads, flows, and concentrations. The predictions were heads and flows measured at different times under similar conditions. The authors found that, for predicting heads, use of all three calibration observation sets performed equally well. For predicting flows, use of head observations alone did a poor job, and use of concentration observations did not produce increased prediction accuracy compared to use of only head and flow data. The latter conclusion resulted because the information about flow that the concentration observations provided was similar to that provided by the flow observations.

Most published postaudits of regional-scale groundwater flow and transport models have found that actual system responses differ from responses predicted by the model. For example, see the postaudits presented by Konikow and Person (1985), Alley and Emery (1986), Konikow (1986), Reichard and Meadows (1992), Hanson (1996), and Stewart and Langevin (1999), which involved regional-scale models with prediction times several years after the model calibration period. Results of these postaudits were used to gain considerable insight into the simulated groundwater systems, even though the predictions were incorrect to some degree. This insight was used to help detect model error and to identify data needs and changes to the conceptual model that could help reduce this model error. Andersen and Lu (2003) present a study in which remediation results are used for a postaudit analysis, which helped reveal error in the initial model. However, capture zones simulated with an updated model were similar to those with the initial model, indicating that the initial model was useful for designing remedial strategies despite the error.

GUIDELINE 14: QUANTIFY PREDICTION UNCERTAINTY USING STATISTICAL METHODS

Guideline 14 suggests two methods for quantifying prediction uncertainty: inferential statistics and random sampling (Monte Carlo) methods. For both these methods, the mechanism for communicating uncertainty is often some type of interval around the prediction.

The prediction uncertainty that can be quantified most readily by both inferential statistics and Monte Carlo methods is that produced by uncertainty in the defined parameters. Indeed, as noted in Chapter 8, Section 8.5.2, the two types of methods project parameter uncertainty onto predictions in ways that produce identical results in some circumstances. If the parameters do not represent all aspects of the model that may be incorrect, then the uncertainty represented by these methods tends to underestimate the actual uncertainty. If, however, defined parameters represent many aspects of the system, and other aspects of the model accurately represent system characteristics, then these methods can capture a substantial amount of the prediction uncertainty. This implies that one approach for better characterization of uncertainty is to represent more aspects of the system using defined parameters. This is a largely unexplored approach.

Predictions tend to be less accurate as they differ more from observations and as prediction conditions differ more from calibration conditions. The example by Saiers et al. (2004) presented in Section G13.1 showed that a groundwater model calibrated using heads produced poor predictions of flow. It is not clear how much measures of uncertainty can account for the differences between predictions and calibrations. Certainly if the predictions are affected by processes not represented in the model, the uncertainty calculated using any of the methods discussed here would be too small. There is no clear solution to this problem, but it is important to be aware of its possible existence in many types of models.

Inferential statistics, Monte Carlo methods, and other methods of uncertainty analysis, such as those presented by Sun (1994), are based on the assumption that the model accurately represents the real system. In truth, all models are simplifications of real systems, and the accuracy of the uncertainty analysis is in question. This accuracy is very difficult to evaluate definitively. Christensen and Cooley (1999) compared nonlinear prediction intervals with measured heads and flows and found good correspondence between the expected and realized significance level of the intervals. If model fit to data indicates model bias, theory suggests that the calculated intervals do not reflect all aspects of system uncertainty, and thus they might be best thought of as indicating the minimum amount of uncertainty. That is, actual uncertainty might be larger than indicated by the confidence intervals. If prediction intervals are dominated by the measurement error term, they are less likely to be prone to error. Unfortunately, in many circumstances the confidence intervals are of greater interest because they reflect model uncertainty most clearly. Cooley (1997, 2004) provides additional analysis of nonlinear confidence intervals.

Inferential statistics and Monte Carlo methods also can be used together. For example, Monte Carlo simulations based on alternative models could each calculate linear or nonlinear confidence intervals based on inferential statistics. Model uncertainty might then be represented by the range of predictions represented by the full set of confidence intervals. Such ideas are promising and are just beginning to be considered in the literature, as noted in Chapter 8, Section 8.6.

Clearly, prediction uncertainty is an area where there is much to be done and ongoing improvements in computer technology make advances more accessible than ever before. Here, we comment on using the methods presented in Chapter 8.

G14.1 Inferential Statistics

The most common and useful inferential statistics for quantifying prediction uncertainty are confidence and prediction intervals, which can be constructed using the methods described in Chapter 8, Section 8.4. Instead of reporting a single predicted value, a predicted value and a confidence or prediction interval are reported.

Given the different types of intervals discussed in Section 8.4—confidence and prediction, individual and simultaneous, and linear and nonlinear—confusion can arise as to when to use them. The following three points are provided for guidance.

1. Use *prediction intervals* to compare measured equivalents to predictions.
2. Use *simultaneous intervals* for multiple or vague predictions.
3. Suggested steps: calculate *linear intervals* and *test model linearity*. If the model is nonlinear, calculate a few *nonlinear intervals*. If needed, calculate more *nonlinear intervals*.

As noted in Chapter 8, Christensen and Cooley (1999) show that in nonlinear problems, nonlinear confidence intervals can be very different from linear intervals for some quantities, and can be very similar for others. It appears that linear confidence intervals are useful as a general indication of uncertainty in many circumstances, but, if at all possible given computer resources, some nonlinear intervals need to be calculated if the model is nonlinear. Brooks et al. (1994) calculated nonlinear confidence intervals for drawdowns. Keating et al. (2003) present nonlinear confidence intervals calculated on boundary fluxes predicted by a groundwater flow model. Besides quantifying the uncertainty, inferential statistics on predictions have been used to include risk assessment in design criteria by Tiedeman and Gorelick (1993).

Predictions and their confidence intervals need to be calculated for all reasonably accurate models to evaluate how different sets of observations and conceptual models are likely to affect both the simulated predictions and their likely uncertainty. Indeed, it can be useful to include at least linear confidence intervals when calculating predictions for each model calibration run.

Christensen et al. (1998) examined how nonlinear confidence intervals on predictions of streamflow gain varied with different observation sets used for calibrating a groundwater model. The observation sets included hydraulic heads and from zero to 18 streamflow gains. As expected, the confidence intervals were large for the model calibrated using only head data, but some intervals also were large even when the full set of streamflow gains was used for calibration.

G14.2 Monte Carlo Methods

Monte Carlo methods were described in Chapter 8, Section 8.5.

Computer technology and processor speed have greatly improved in recent years, making it much more feasible to conduct Monte Carlo analyses for models of natural systems. Parallel processing capabilities are advantageous to Monte Carlo studies in which the different runs are independent, because individual simulations easily can be distributed to different computer processors. These parallel processing capabilities can be achieved cost-effectively by using clusters of networked personal computers employing, for example, the parallel processing capabilities of the JUPITER API (Banta et al., 2006).

Monte Carlo methods in groundwater modeling have been used to assess the uncertainty of contributing areas to wells. For example, Evers and Lerner (1998) identified a zone of confidence defined as the area that is common to all contributing areas predicted by models that provide a reasonable fit to the calibration data. They also identified a zone of uncertainty, defined as the total area covered by all reasonable contributing areas. Starn et al. (2000) varied parameter values in a three-dimensional model using the variance–covariance matrix produced using regression, and simulated contributing areas using the generated parameter sets. Several Monte Carlo evaluations of capture zones that consider small-scale variations in hydraulic conductivity in two-dimensional systems have been published; for example, van Leeuwen et al. (2000), Feyen et al. (2001, 2003), and Stauffer et al. (2004). The latter was briefly described in Chapter 8, Section 8.5.2. Additional references are cited in the listed works.

Monte Carlo methods also have been integrated with regression to quantify model prediction uncertainty. Examples in groundwater modeling include Poeter and McKenna (1995) and McKenna and Poeter (1995). The Poeter and McKenna (1995) model was briefly described in Guideline 8 (Chapter 11) and provides an example of the six elements of a Monte Carlo analysis presented in Chapter 8, Section 8.5.1 for a groundwater problem. The work includes Monte Carlo runs conducted using three sets of information on the hydraulic-conductivity field, including (a) only hydrogeologic information (measurements of hydraulic conductivity), (b) hydrogeologic and geophysical information, and (c) hydrogeologic and geophysical information as well as hydraulic-head and streamflow gain and loss data integrated using nonlinear regression.

In all cases the goal was to quantify the uncertainty of concentration at a well. The six elements for the analyses were as follows:

1. The model input changed was the zonation used to represent the hydrogeology of the aquifer material.
2. The realizations were generated using indicator kriging.
3. Each Monte Carlo run for (a) and (b) consisted of a forward model simulation. For (c) each run consisted of an inverse model simulation to obtain the best-fit parameter values for the generated zonation. The observations were hydraulic heads and streamflow gains and losses. The concentration at a well was simulated for each Monte Carlo run. The system was simulated using MODFLOW, MODFLOWP, and MT3D.

4. Four hundred runs were conducted. The number of runs was determined largely by computational limitations.
5. For each Monte Carlo run, the following were saved: the zonation, estimated parameter values, information about parameter-estimation convergence, the standard error of the regression, and the predicted concentration.
6. Final results were analyzed by plotting histograms of the predicted concentrations. For (*a*) and (*b*), results for all 400 runs were plotted. For (*c*), results were omitted from the Monte Carlo analysis if one of the following conditions occurred: (i) the best-fit parameter values were unrealistic in that they were

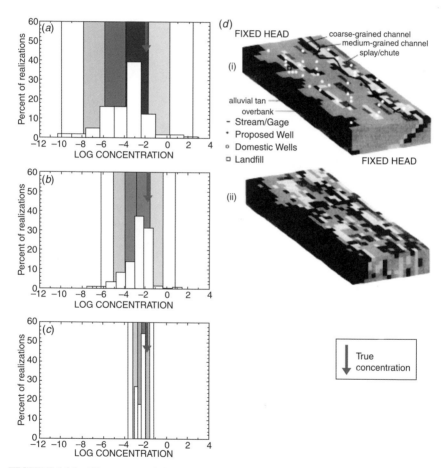

FIGURE 14.1 Histograms of simulated concentrations from models calibrated using three sets of data: (*a*) only hydrogeologic information (measurements of hydraulic conductivity), (*b*) hydrogeologic and geophysical information, and (*c*) hydrogeologic and geophysical information as well as hydraulic-head and streamflow gain and loss data integrated using nonlinear regression. (*d*) The true system and imposed boundary conditions, and a generated zonation. (From Poeter and McKenna, 1995.)

not in order from largest to smallest when such relations could be determined from system information, (ii) the best-fit parameter values were substantially different than expected, (iii) the model fit was significantly worse than for other models, or (iv) the regression did not converge. Ten realizations remained after these conditions were considered.

In this synthetic test case, the true solution was known so that performance of the different methods of characterizing the system could be definitively tested. Results are shown in Figure 14.1. Using nonlinear regression produced much more accurate predictions than were attained by (a) and (b). This is because nonlinear regression allowed conditioning to observations and comparison of estimated parameter values with realistic ranges and rankings based on system information. The dramatic improvement in the predictions produced by models screened using these criteria indicates that their application is likely to be useful for identifying more accurate models.

15

USING AND TESTING THE METHODS AND GUIDELINES

Nonlinear regression is a useful, imperfect tool for model calibration, and its application is not always straightforward. In many situations, forward and inverse model computer execution times are a concern. This chapter first discusses this issue.

It can be helpful to examine other, similar examples when designing, conducting, and reporting a modeling study using nonlinear regression. The nonlinear regression methods, diagnostic and inferential statistics, and guidelines described in this book have been used successfully in many field applications and tested using many synthetic test cases. This chapter lists some of these studies and their references and provides information to help readers identify examples with selected characteristics. The chapter concludes with results from two field sites.

15.1 EXECUTION TIME ISSUES

Computer execution time is often a problem when calibrating models. Whether regression is used or not, model calibration requires many model runs to investigate the interactions between the data, simulated processes, and parameter values. Thus, effective model calibration needs to address how model construction affects execution time. Guideline 1, with its suggestion to start with a relatively simple model of the groundwater system and build complexity as warranted by the system information and by the available data, is relevant to the issue of minimizing execution time. Starting with a simple model often results in shorter execution times and an

---

Effective Groundwater Model Calibration: With Analysis of Data, Sensitivities, Predictions, and Uncertainty. By Mary C. Hill and Claire R. Tiedeman
Published 2007 by John Wiley & Sons, Inc.

opportunity to understand simulated characteristics that might be obscured by a more complicated model.

Without parallelization (Poeter et al., 2005, pp. 131–133), execution times for regression simulations can be estimated using execution times for forward simulations (e.g., a simulation that solves for hydraulic heads in a groundwater flow model) as

$$T_i = (2 \times NP) \times [T_f \times (1 + NP)] \qquad (15.1)$$

where T_i is the execution time for the regression (inverse) solution, T_f is the execution time for the forward solution, and NP is the number of parameters to be estimated by regression. Parallelization can be used to reduce T_i (Poeter et al., 2005; Doherty, 2005).

This assumes that the number of parameter-estimation iterations approximately equals twice the number of parameters ($2 \times NP$), which is typical. The $(1 + NP)$ term accounts for one forward simulation and one simulation to calculate sensitivities for each of the NP parameters. For the forward or backward perturbation sensitivities commonly used by UCODE_2005 and PEST, the NP sensitivity simulations solve a slight variation of the forward problem (Poeter et al., 2005, Chapter 3). In MODFLOW-2000, these simulations solve sensitivity equations that result from taking the derivative of the forward equation with respect to the parameter (Hill et al., 2000, p. 67). In both cases, each of the sensitivity simulations requires approximately the same amount of execution time as a forward simulation. The longer execution times typically experienced with UCODE_2005 or PEST are caused by the effort required to coordinate the different model runs, not a difference in computational effort required to solve sensitivities.

Equation (15.1) applies when sensitivities are calculated using the sensitivity-equation method and using forward or backward differencing. Faster execution time can be achieved for some problems with the sensitivity-equation method using direct solvers because a single matrix decomposition can be used to solve for heads and sensitivities for all parameters (Hill et al., 2000, p. 70). More execution time is needed by UCODE_2005 and PEST when sensitivities are calculated by central differencing.

Inverse model execution times that exceed about 15 hours (an overnight simulation) commonly occur when the forward execution time exceeds 30 minutes. The model spatial and temporal resolution that this execution time allows depends on the speed of the computer, the number of computer processors, and the characteristics of the simulated system.

Simple changes in the simulation can dramatically improve execution times. For example, the initial hydraulic-conductivity structure of the groundwater flow model described by D'Agnese et al. (1997, 1999) was characterized by values in neighboring finite-difference cells that differed by more than five orders of magnitude in many parts of the model. Introducing single cells of moderate hydraulic conductivity between the high and low valued cells in most of the model resulted in about a sixfold decrease in execution time, with little effect on simulated results. The contrasts were preserved where they were important.

Replacing nonlinear forward problems with linear approximations as much as possible, as suggested in Chapter 11, Section G1.2, also can dramatically reduce execution time without substantially diminishing model accuracy.

15.2 FIELD APPLICATIONS AND SYNTHETIC TEST CASES

Selected references describing field applications using some or all of the methods and guidelines described in this book are listed in Table 15.1. Similar approaches were used by Cooley (1977, 1979, 1983a), Gailey et al. (1991), Tiedeman and Gorelick (1993), Yager (1993), Kuiper (1994), Olsthoorn (1995), Christensen (1997), and Christensen et al. (1998). Selected aspects of the field applications for the Death Valley region, USA, and for Grindsted Landfill, Denmark, are described later in this chapter.

Table 15.2 lists models of synthetic numerical systems to which the methods and guidelines have been applied. These test cases provide the opportunity to conclusively evaluate the accuracy of models calibrated using the methods described in this book because all aspects of the synthetic numerical systems are known. Hill et al. (1998) used a complex hypothetical groundwater model to test many of the methods, and to develop and test the guidelines. Poeter and McKenna (1995) present a synthetic groundwater transport model that evaluates stochastically generated parameter zonations using nonlinear regression methods. In both studies, better models, as gauged using the ideas discussed in this book, produced more accurate predictions. Poeter and Hill (1996, 1997) demonstrate many of the ideas presented using simple examples. Selected characteristics of these and other examples are listed in Table 15.2.

Three studies listed in Table 15.1 or 15.2 did not include optimization of parameter values. Scheibe and Chien (2003) use a very extensive set of hydrogeologic data to investigate the predictive (uncalibrated) use of stratigraphy and hydraulic-conductivity data from three sources: borehole flowmeter data normalized to slug tests, cross-borehole radar tomography, and ground-penetrating radar. Hill and Østerby (2003) investigate the accuracy of parameter correlation coefficients. Shoemaker (2004) investigates the importance of observations and parameters in density-dependent groundwater flow using a sensitivity method that follows the suggestions provided in this book.

15.2.1 The Death Valley Regional Flow System, California and Nevada, USA

The Death Valley regional flow system (DVRFS) is located in southern California and Nevada, USA, west of Las Vegas, Nevada (Figure 15.1). The groundwater system has been studied extensively because of the potential effects of activities at the Nevada Test Site, where underground nuclear testing has been conducted, and Yucca Mountain, where high-level radioactive waste is proposed to be stored in thick unsaturated volcanic strata located above the water table (D'Agnese et al., 1997). In addition, several communities within the DVRFS pump groundwater for a variety of uses, such as domestic and agricultural water supply. These activities have raised concerns about the transport paths and times of potential contaminants from the Nevada Test Site, and the effects of these potential contaminants at possible groundwater discharge locations, such as Death Valley National Park and Ash Meadows, and pumping wells in many locations. The DVRFS model was built to

TABLE 15.1 Selected Field Investigations of Groundwater Systems that Demonstrate All or Some of the Guidelines

Reference	Location[a]	System[b]	Observations[c]	Parameters[d]	Predictions[c,e]
Cooley et al. (1986)	Madison aquifer, WY, MT, ND, SD, NE (USA) {Chapter 11, G6.2}	2D SS gw flow	Heads	**7-10HK, 8Q**[f] **(spring flows), CH**[f]	None
Brooks et al. (1994)	Birmingham, England	2D TR gw flow	Heads	**S, HK, KHD, 6RCH**	Drawdown. NI
McKenna and Poeter (1995)	Golden, CO (USA)	3D SS gw flow	Heads	8HK	None
Anderman et al. (1996), Anderman and Hill (1998)	Cape Cod, MA (USA) {Chapter 11, Section G3.1}	2D SS gw flow	Heads, flow, ADV	**HK, RCH, 2Q, KHD**	None
Barlebo et al. (1996, 2004)	MADE[g] Site, Columbus, MS (USA)	3D SS gw flow, transp	Heads, conc	**8HK, 8VK, 2RCH, DISP, POR**	None
Holtschlag et al. (1996)	Tri-County Region near Lansing, MI (USA)	3D SS gw flow	Heads, flows	HK, VK, KHD, RCH	None
Yager (1996)	Niagara Falls, NY (USA) {Chapter 11, Section G2.4}	3D SS gw flow in fractured rock. Ten-layer model	Heads, flows	**2HK, 3VK, 2RCH,** 3HK, VK, 3KHD, RCH	None
Yager (2002)	Niagara Falls, NY (USA)	3D SS gw flow, reactive transp in fractured rock. Five-layer model.	Heads, flows, conc	**2HK, VK, RCH, 3KD,** Por, Disp, Retardation factor	None
Tiedeman et al. (1997, 1998a)	Mirror Lake, NH (USA) {Chapter 11, Section G8.4}	3D SS gw flow	Heads, flows	**2-4HK, RCH,** 2HK, KHD	Groundwater basin
D'Agnese et al. (1997, 1999); Hill et al. (2000); Tiedeman et al.	Death Valley region, CA, NV (USA) {Chapter 11, Section G3.1; Chapter 12, Section G10.2;	3D SS gw flow. Three-layer model.	Heads, spring flows	**6HK, VANI, 2RCH,** 3HK, VANI, 2RCH, 5CHD, ETM, 2Q	ADV

D'Agnese et al. (2002)	Death Valley region, CA, NV (USA)	3D SS gw flow. 15-layer model.		**11HK, 3VANI, 3RCH,** 2HFB, 22HFB, 98HK, 11VANI, 10RCH, 57KHD	None
Faunt et al. (2004)	Death Valley region, CA, NV (USA) {15.2.1}	3D SS&TR gw flow. 16-layer model.		**2RCH, 3HFB, 40HK, 4KHD, 6KDEP,** 2VANI, 17HK, 3KHD, 6RCH, 7S, 6HFB, 4VANI, 4KDEP	None
Tiedeman et al. (1998b)	Albuquerque, NM (USA)	3D TR gw flow	Heads, flow	**6HK, 3RCH,** ET, VANI, 6HK, 2S, Q, 7RCH, 2KHD	None
McAda (1999)	Albuquerque, NM (USA)	3D TR gw flow	Drawdown	**HK, VANI, 2S, 4Q,** 4HK, 2KHD,	Induced river recharge
Sanford et al. (2004a,b)	Albuquerque, NM (USA) {Chapter 11, Section G9.1}	3D SS gw flow, advective transport	Heads, age dates, hydrochemical zones	**20HK, 12VANI, 2KHD, RCH**	None
Barlebo et al. (1998)	Grindsted Landfill, Denmark {15.2.2}	3D SS gw flow, transp	Heads, conc	**2HK, 3VK,** RCH	None
Christensen and Cooley (1999)	Zealand, Denmark {8.4.3}	2D SS gw flow	Heads	**9HK, 2VK**	Heads. NI
Lebbe (1999)[h]	French–Belgian border	2D TR gw flow, density-dependent transp	Heads, electrical resistivity, salinity	**HK, VK, Disp, Por, ERS**	None
Eberts and George (2000)	Basins and Arches system, IN, OH, MI (USA)	3D SS gw flow	Heads, flows	**5HK, 4RCH, 2VK, KHD,** 2HK, 3RCH, 2VK	None

(*Continued*)

TABLE 15.1 Continued

Reference	Location[a]	System[b]	Observations[c]	Parameters[d]	Predictions[c,e]
Hunt et al. (2000)	Middle Gene see Lake, WI (USA)	2D SS gw flow	Heads, flows	**HK, RCH, KHD**	Change in lake stage
Hunt and Steuer (2000)	Frederick Springs, Dane County, WI (USA)	3D SS gw flow	Heads, flows	**2HK**, 6HK, 4CHD	Spring capture zone
Yager et al. (2001)	Genesee valley, NY (USA)	3D SS&TR gw flow, subsidence	Heads, flows, subsidence	**2S, 5HK, 2VK, 3RCH, 3S, 2HK, 2VK**	None
Kelson et al. (2002)	Forest and Langland Counties, WI (USA)	2D SS gw flow	Heads, flows	**3HK, 1RCH**, 2CHD	Mine flow, streamflow reduction
Pint et al. (2003); Hunt et al. (2006)	Vilas County, north central WI (USA)	3D SS&TR gw flow	Heads, flows, lake-water depth, ADV	**5HK, 2RCH, 4KHD, 2S**	None
Hoffmann et al. (2003)	Antelope Valley, CA (USA)	3D TR gw flow, subsidence	Heads, subsidence	**Six time constants**. Local variations handled separately	None
Scheibe and Chien (2003)	Oyster, VA (USA) Extensive hydrogeologic data {Chapter 11, Section G1.2}	3D SS gw flow, transp, LGR	Heads, conc	Six distributions of HK: three deterministic and three stochastic (no optimization)	None

Reference	Site	Model	Observations	Parameters	Predictions
Keating et al. (2003)	Las Alamos, NM (USA)	3D SS&TR gw flow, transp, LGR	Heads, flows	**3RCH, 26HK, 2S**	Site model: flux to stream and transp. NI
Doherty (2003)	Unidentified	2D SS gw flow	Heads	92HK[f]	None
Scott et al. (2003)	Uvas Creek, CA (USA)	1D transient sw flow, transp	Conc	**Disp, lateral inflow, cross-sectional areas of river and storage zone, transient storage, adsorption rates**	None
Gannett and Lite (2004)	Upper Deschutes Basin, OR (USA)	3D SS&TR gw flow	Heads, flows	14HK, 2VK, 1VANI, 7KHD, 1RCH, 3HK, 4VK, 2-4S	Effects of pumpage, streamflow reduction

[a] Section numbers in curly brackets refer to section in *this* book in which results are illustrated.
[b] 2D, two dimensional; 3D, three dimensional; SS, steady state; TR, transient; transp, transport; gw, groundwater; sw, surface water; LGR, local grid refinement used in model construction.
[c] Flows, streamflow gain or loss, groundwater inflow to a lake, or other; conc, concentrations; ADV, advective transport.
[d] Approximate number of different types of defined parameters estimated by nonlinear regression are in bold type. Disp, dispersivity; ERK, ERS, parameters relating electrical resistivity to hydraulic conductivity and salinity, respectively; ETM, maximum evapotranspiration; HK, horizontal hydraulic conductivity or transmissivity; KD, reaction rate; KDEP, depth decay of horizontal hydraulic conductivity; KHD, hydraulic conductivity or conductance of a head-dependent boundary: HFB, conductance of a horizontal-flow barrier; S, storage coefficient, specific storage, or specific yield; Por, effective porosity; Q, flow at a boundary or well(s); RCH, areal recharge; VANI, vertical anisotropy; VK, vertical hydraulic conductivity. For many studies, the log of HK, KHD, S, VANI, and/or VK are estimated instead of the native value.
[e] NI, Nonlinear confidence intervals are presented.
[f] Prior information or regularization is used.
[g] MADE, MAcroDispersion Experiment.
[h] Also includes a synthetic test case.

TABLE 15.2 Selected Investigations of Synthetic Groundwater Systems that Demonstrate All or Some of the Guidelines

Reference[a]	Purpose and System Complexity	2D/3D[b]	Calibrated System[b]	Observations[c]	Parameters[d]	Predictions[c]
Poeter and McKenna (1995) {Chapter 11, Section G8.1; Chapter 14, Section G14.2}	Test alternate conceptual models. Complex.	3D	SS flow	Heads, river gain	5HK	Conc
Poeter and Hill (1996)	Test impact of alternative constructed models. Complex.	2D	SS flow	Heads, river gain	4HK, 1RCH	None
Poeter and Hill (1997) {4.4.2}	Demonstrate guidelines. simple.	3D	SS flow	Heads, river gain	2HK, Q	
Hill et al. (1998) {Chapter 11, Section G8.2; Chapter 12, Section G9.1}	Test guidelines. Complex.	3D	SS flow	Heads, river gains	1-19HK, RCH, 1-3KHD, VANI, 0-4HK	Drawdown, river gain
Hill and Østerby (2003)	Test parameter correlation coefficients. Simple.	3D	SS flow	Heads	2HK, VANI, KHD, RCH	None
Dam and Christensen (2003)	Demonstrate new method. Simple.	2D	SS flow	Heads	1, 16, or 49 HK, ERK	None
Shoemaker (2004)	Sensitivity analysis. Simple.	2D	SS flow, density-dependent transp	Heads, salinities, flows	HK, VK, Disp, Por, RCH, flow from inland	None
Moore and Doherty (2005, 2006)	Test pilot points, Simple.	2D	SS flow	Heads	1-104HK[f]	ADV

Note: Please see the footnotes for Table 15.1.

15.2 FIELD APPLICATIONS AND SYNTHETIC TEST CASES

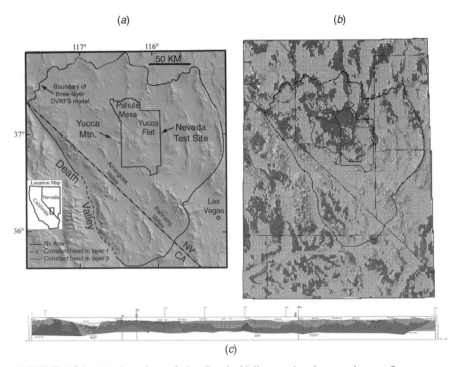

FIGURE 15.1 (*a*) Location of the Death Valley regional groundwater flow system (DVRFS) and the boundary of the three-layer flow model, (*b*) a surficial geologic map, and (*c*) a hydrogeologic cross section. (Adapted from D'Agnese et al., 1997, and Belcher, 2004.)

(1) provide a regional evaluation of the effects of pumpage, (2) provide boundary conditions for local-scale models at selected sites, and (3) provide information about regional-scale transport. Predictions related to (3) are discussed in the section below on Guideline 12.

Several groundwater flow models of the DVRFS have been developed: a steady-state, three-layer model (D'Agnese et al., 1997, 1999), a steady-state, 15-layer model (D'Agnese et al., 2002), and a transient, 16-layer model (Faunt et al., 2004). Examples from the three-layer model and the transient model are used in this section to provide examples of applying Guidelines 2, 5, 6, 9, 10, and 12. Examples from the DVRFS three-layer model also are included in the discussion of Guidelines 3 and 10 in Chapters 11 and 12 (Figures 11.3 and 12.5).

Evaluate System Information (Guideline 2) Evaluating and organizing data as needed for Guideline 2 ("Use a Broad Range of System Information to Constrain the Problem") can require a simple or complicated database and visualization methods and software, depending on the characteristics of the system and the model being constructed. The DVRFS has very complex geology characterized by faulted, fractured, and deformed rocks with a wide range of compositions, ages, and depositional histories. A complex database and visualization system was used for integrating the hydrologic and geologic information and incorporating it into a groundwater flow model. This involved developing a series of conceptual, digital, and numerical models that build upon one another:

   I. Hydrogeologic framework conceptual model
  II. Hydrogeologic framework digital model
 III. Groundwater system conceptual model
  IV. Groundwater system numerical model

The following discussion, modified from D'Agnese et al. (1999), describes the approach and specific tools used in developing each of these models. This process was iterative. For example, testing of initial groundwater system numerical models revealed problems with model fit to the observation data, which led to revision of the hydrogeologic framework conceptual and digital models, and ultimately to improvement of the groundwater models.

Figure 15.2 shows the steps and software products used to evaluate system information and develop the models listed above. Simpler systems, such as groundwater flow in less complex geologic settings, often only use a subset of the methods and products employed for this study. Characterizing the DVRFS hydrogeology and groundwater flow system required integrating extensive regional-scale data. These data included point hydraulic-head data, geologic maps and cross sections, vegetation maps, surface-water maps, spring data, meteorological data, and remote-sensing imagery. The data were converted into consistent digital formats using various traditional two-dimensional geographic information system (GIS) products, as described by Faunt et al. (1993). Fully three-dimensional geoscientific information system (GSIS) products also were needed because the hydrogeology is not well represented by a sequence of layers. GSIS refers to digital data management and modeling systems designed to handle a wide variety of three-dimensional data types (Raper, 1989). Use of GIS and GSIS products to integrate the system data allowed ease of data manipulation and significantly helped development of the conceptual, digital, and numerical models.

A three-dimensional hydrogeologic framework conceptual model (model I above) is a set of ideas and hypotheses about the dynamics and processes in the hydrogeologic system that are considered important to the observed and predicted quantities. Typically, conceptual hydrogeologic model development involves the following steps:

1. Describe the geometry, general composition, and hydraulic properties of the materials that control groundwater flow.
2. Characterize surface and subsurface hydrologic conditions that affect groundwater movement.
3. Evaluate hypotheses about the system to develop a conceptual model for simulation.

For the DVRFS, some aspects of the hydrogeologic framework conceptual model follow directly from data that are organized and visualized using the products listed in Figure 15.2. Other aspects are hypotheses that need to be constantly tested

15.2 FIELD APPLICATIONS AND SYNTHETIC TEST CASES

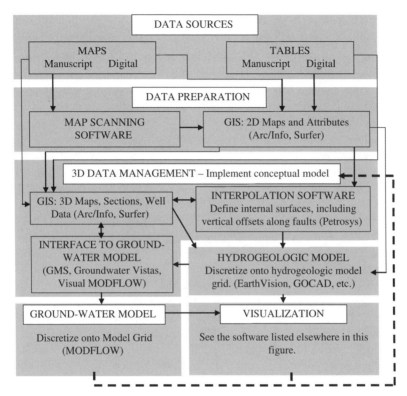

FIGURE 15.2 Flowchart showing logical movement of data used to investigate the Death Valley regional groundwater flow system. (Adapted from D'Agnese et al., 1997, 1999.) Dark dashed lines indicate the groundwater model results that affect the conceptual model. Software program names are listed as examples in parentheses; their listing does not imply endorsement by the authors, the publisher, or the U.S. Geological Survey. (Web sites accessed June 7, 2006 for the listed software are www.esri.com, www.petrosys.com.au, www.goldensoftware.com, www.dgi.com/earthvision, www.ems-i.com, www.groundwater-vistas.com, and www.visual-modflow.com, www.earthdecision.com.)

against the data and that lose credence if they are contradicted by the data. Thus, data organization and visualization are important when developing conceptual models.

The DVRFS hydrogeologic framework digital model (model II above) is a digital representation of the conceptual hydrogeologic framework. Construction of this model began by assembling digital elevation models (DEMs), hydrogeologic maps and sections, and lithologic well logs. DEMs and hydrogeologic maps were manipulated by standard GIS techniques. The merging of these four primary data types to form a single coherent three-dimensional hydrogeologic framework digital model required more specialized, fully three-dimensional GSIS software products. Construction of this model involved five steps:

1. Combine DEM data with hydrogeologic maps to provide a set of points representing the outcrops of hydrogeologic units.

2. Locate hydrogeologic sections and well logs properly in three-dimensional coordinate space to define locations of hydrogeologic units in the subsurface.
3. Interpolate surface and subsurface data to define the tops of hydrogeologic units, incorporating any offsets along major faults.
4. Integrate hydrogeologic unit surfaces using appropriate stratigraphic principles to represent natural stratigraphic and structural relations accurately. This modifies the interpolations of step 3.
5. Analyze hydrologic and hydraulic property values to define parameter starting values and reasonable ranges for the groundwater system models.

Assigning hydraulic properties to the framework digital model also required fully three-dimensional GSIS capabilities. Following development of the groundwater conceptual and numerical models, parameter values and other aspects of the framework model were revised as observations were incorporated into the evaluation through model calibration. In this way, hydraulic data and model IV were used to test hypotheses about the hydrogeology represented in models I and II.

To develop a groundwater system conceptual model (model III above), the hydrogeologic characterization was combined with the hydrologic components of the system. For the DVRFS, these components included groundwater recharge through infiltration of precipitation, and groundwater discharge through evapotranspiration (ET), spring flow, and pumpage. Maps describing these components of the groundwater flow system were developed using remote sensing and GIS techniques (D'Agnese et al., 1996).

Groundwater recharge estimates were developed from data related to varying soil-moisture conditions (including elevation, slope aspect, parent material, and vegetation) using the empirical methods described by D'Agnese et al. (1996). GIS methods were used to produce maps describing recharge potential on a relative scale. The recharge potential maps were used to describe groundwater infiltration as a percentage of annual precipitation.

For groundwater discharge, a variety of different data sets were used. Multispectral satellite data were evaluated to produce a vegetation map, which was combined with ancillary data sets in a GIS to delineate areas of ET, including wetland, shrubby phreatophyte, and wet playa areas. Estimated water consumption rates for these land surface types were then applied to approximate total likely ET discharge from the areas. Spring flow and pumpage were incorporated by developing point-based GIS maps. For springs, this map contained spring location, elevation, and measured discharge rates. Likewise, water-use records were used to develop a spatially distributed water-extraction map describing long-term average withdrawals.

Once completed, the three-dimensional data sets describing the hydrologic system were integrated and compared to develop groundwater system conceptual model configurations which included:

1. The three-dimensional hydrogeologic framework, including a complete definition of the geology throughout the volume simulated, and identification of aspects for which the data supported multiple interpretations.

15.2 FIELD APPLICATIONS AND SYNTHETIC TEST CASES

2. A description of system boundary conditions.
3. Estimates of the likely average values of hydraulic properties of the hydrogeologic units.
4. Estimates of groundwater flow at sources and sinks.
5. Hypotheses about regional and subregional flow paths and the global water budget.

Groundwater system numerical models (model IV above) were then developed and used with the hydraulic head and spring-flow observation data to test the alternative conceptual models. This effort produced the following:

1. System interpretations that were feasible given the entire available database, including all system information as well as the head and flow observations.
2. The location and type of additional data that would be needed to reduce the uncertainty of values simulated by the flow model.
3. The location of likely physical boundaries within the flow system.

The feasibility of multiple conceptual models for the hydrogeologic framework were evaluated. This evaluation showed that one of the alternatives considered was clearly the most likely representation of the system. This model is described in D'Agnese et al. (1997, 1999) and is used in this chapter to demonstrate selected guidelines.

Include Many kinds of Observations (Guideline 4) Water levels stored in the U.S. Geological Survey Ground-Water Site Inventory (GWSI) database (http://waterdata.usgs.gov/nwis/gwsi) were used to compute hydraulic-head and head-change observations for calibrating the DVRFS transient model. Water levels were retrieved from GWSI and stored in a companion database. In this database each water level was flagged to indicate the general condition at the time of measurement. The flags identified whether the water level represented regional or local, steady-state or transient, or some other general or specific condition. Only measurements flagged as regional steady-state and regional transient were used to calculate observations. The model has one prepumping steady-state stress period followed by an 86-year period (1913–1998) with pumping in annual stress periods. Regional steady-state measurements represent prepumped, equilibrium conditions, and regional transient measurements reflect water levels thought to be affected by pumping from the regional flow system. Preliminary model-calibration efforts included only the steady-state stress period and used observations derived as the average of all of the regional steady-state water levels at each location. When calibrating the final model composed of the initial steady-state and subsequent transient stress periods, the regional steady-state water levels were used to calculate annual average head observations that were specified in the stress period for the applicable year.

For wells with measurements that were thought to be affected by pumping, an initial head and subsequent temporal changes in head were used as observations defined for the transient stress periods. Temporal changes in head were calculated within the Observation Process of MODFLOW-2000 as the difference between the initial head and subsequent heads. The initial head observation at each well was the mean of the heads measured during the first year of record. The subsequent heads were calculated as annual means in later years. This approach maintained sufficient resolution of temporal changes in head and corresponded with the annual stress periods used for the transient model.

Assign Weights that Reflect Errors (*Guideline 6*) The following methodology was developed to determine weights for the transient 16-layer model of the DVRFS (Faunt et al., 2004, pp. 279–283; San Juan et al., 2004, pp. 128–131). This example illustrates typical concerns that need to be addressed when determining weights. Each modeled system poses its own challenges, and different approaches may be needed for other applications.

Calculation of Weights The weights for the observations of steady-state head and transient change in head were calculated by MODFLOW-2000 from standard deviations (see Guideline 6), which represent errors associated with the water-level measurements used to compute each observation. Individual weights are calculated from the sum of the variances associated with each of the individual error sources:

$$1/(sd_1 + sd_2 + sd_3 + \cdots)^2$$

where sd is the standard deviation of the error source represented by the subscript. In the analysis below all errors are assumed to be normally distributed so that the methods described in Chapter 11, Section G6.1 can be used to calculate error standard deviation. In these calculations, the critical value of 1.96 has been rounded to 2.

Steady-State Head Observations The sources of greatest error in determining the head observation errors are listed and the method for determining the standard deviations that represent these errors are described. Some are clearly measurement error; some have at least a component of model error.

ALTITUDE UNCERTAINTY Altitude uncertainty for each well was calculated using the altitude accuracy code given in the GWSI. The code expresses this potential error as a range distributed symmetrically about the measurement that relates directly to the method used to determine the altitude. This range varies from ± 0.03 m for high-precision methods (spirit level surveys and differential GPS) to ± 25 m for altitudes derived from maps with large contour intervals typical of areas with steep terrain. Assuming that $\pm$ the accuracy code defines a 95-percent confidence interval, the standard deviation can be computed as

$$sd = (\text{altitude accuracy code})/2$$

The standard deviation for this error ranges from 0.002 to 13 m.

15.2 FIELD APPLICATIONS AND SYNTHETIC TEST CASES

LOCATION UNCERTAINTY   This uncertainty is calculated from the hydraulic gradient and a distance representing the uncertainty in the assigned well coordinates. The hydraulic gradient is estimated from the generalized potentiometric-surface map presented by D'Agnese et al. (1997). The distance is determined from the coordinate accuracy code given in GWSI. This code, defined in seconds, ranges from about 0.1 to 100 seconds for wells within the DVRFS. A second represents a distance of about 30 m within this area, and thus, the accuracy expressed as a distance ranges between about 3 and 3000 m. The large values typically were associated with old well locations that were not updated; most occurred in areas with flat gradients so that the location error did not result in large observation error. Hydraulic gradients estimated at well locations for which observations are computed range from near zero to about 0.12. Assuming that a 95-percent confidence interval could be constructed by adding and subtracting the product of the hydraulic gradient and the coordinate accuracy code, and that the errors are approximately normally distributed, the standard deviation can be computed as

$$sd = (\text{hydraulic gradient} \times \text{spatial distance})/2$$

The standard deviation for this type of error ranges from near zero to 200 m. The large values occur for the few instances in which the well location is poorly defined in high-gradient areas. Despite the resulting small weight, the observation is included in the analysis to demonstrate the fit of the model to the historic data.

HYDROGEOLOGIC BOUNDARY POSITION/LOCATION ERRORS   These errors are caused by inaccuracies in the geometric representation of hydrogeologic units within the model grid. The error is likely to increase with coarser horizontal and vertical model-grid resolution. It probably increases with depth because knowledge of geologic units and structures decreases with depth. The error range is calculated as the product of the hydraulic gradient and nodal width multiplied by a depth function. The depth function was determined based on knowledge of the area and data. It increases linearly with the depth of the uppermost opening and varies from a value of about two for shallow wells to three for very deep wells. Assuming that the calculated range is a 95-percent confidence interval, the standard deviation can be computed as

$$[(\text{hydraulic gradient} \times \text{nodal width}) \times ((\text{opening depth}/\text{model thickness}) + 2)]/4$$

Model thickness is about 3000 m. The standard deviation for this error ranges from near zero to about 110 m.

MEASUREMENT ERRORS   The largest measurement error associated with any of the devices used to measure water levels in the DVRFS is about 0.1 percent of the depth to water in the well. Assuming that $\pm$ the measurement error defines a 95-percent confidence interval, the standard deviation was computed as

$$sd = (\text{observation depth} \times 0.001)/2$$

The standard deviation for this error ranges from near zero to about 0.4 m.

NONSIMULATED TRANSIENT STRESS ERRORS   Nonsimulated transient errors result mainly from seasonal water-level fluctuations. The calculation of uncertainty associated with these fluctuations requires a sufficient number and distribution of measurements within the year. This requirement is evaluated using a value calculated as the total number of measurements divided by the number of years during which measurements were taken. If this value is less than 7, then the range used to calculate the standard deviation is set to a number approximating the maximum seasonal range within the water-level database. The maximum range, as determined from measurements within the DVRFS database, is about 15 ft for open intervals within 50 ft of land surface, and about 5 ft for open intervals greater than 50 ft below land surface. If this value is 7 or greater, then the range is calculated directly from the measurements used to compute the observation. Assuming that the observation represents the mean, and that the range represents 95-percent confidence, the standard deviation can be calculated as

$$(\text{range}/4)$$

For shallow wells, the maximum standard deviation for wells having an open interval within 50 ft of land surface is about 3.75 ft, and for deeper wells is about 1.25 ft. Because the head observations are averages of heads measured over time, it is probable that some of the nonsimulated transient stress errors cancelled out as part of the averaging process so that the standard deviations calculated were too large. However, the measurement times generally were not evenly distributed throughout the years. In wells for which measurements were consistently in the same season, the error would not have been averaged out, while in wells for which measurements were distributed seasonally the errors would have been averaged out. Accounting for the time of year would have required considerably more effort and this error was a small enough component of the total error that the effects of averaging were ignored in determining the standard deviations.

After converting to variances, adding, and taking the square root, the standard deviations calculated for head observations used in the steady-state stress simulation range from 1 to 215 m. Thus, weights range from about 2.2×10^{-5} to $1\,\text{m}^{-2}$.

Transient Head-Change Observations   For initial heads at locations thought to be affected by pumpage, weights were calculated as for the steady-state observations except that only one year of measurements were used. In the model, the initial head was applied at the end of that year.

As mentioned in Chapter 11, Section G6.1, the differencing used to obtain head-change observation cancels errors that are constant. Here, errors that pertain to well altitude and location are considered to be constant enough in time that

15.2 FIELD APPLICATIONS AND SYNTHETIC TEST CASES

they are effectively cancelled for head-change observations. While the same dynamics that cause errors in unstressed conditions could cause errors in stressed conditions, the effect is more difficult to characterize, and in the DVRFS model another source of error is expected to dominate. Indeed the other source of error is even expected to dominate the effects for nonsimulated transient effects.

The error of concern is that associated with uncertainties in the amount and location of the pumpage, which are largely estimated based on crop type and electricity records. Personnel involved in estimating annual withdrawals for groundwater users within the DVRFS believe their estimates to be within about 30 percent of the actual value. Because the relation between pumping and head change is approximately linear, this component of the head-change error also is expected to be about 30 percent. To account for errors in pumpage location, a total error of 40 percent was proposed. However, strict use of this percentage was problematic because it produced unrealistically large and small weights for small and large head changes, respectively. Therefore, a function was developed to relate the standard deviation to the head change in a manner that did not produce problematic weights (San Juan et al., 2004, Eq. (4)), to yield:

$$sd_{hc} = 4 + [0.8 \times \log(hc^{obs}/40)] \quad \text{for } hc^{obs} > 1.0$$

$$sd_{hc} = 1 \quad \text{for } hc^{obs} \leq 1.0$$

where sd_{hc} is the standard deviation used to weight observed head change; log denotes the natural log of the value in parentheses; and hc^{obs} is the head-change observation, measured in meters for the Death Valley regional model.

Examples of head-change values, standard deviations, and coefficients of variation are as follows:

Head Change	Standard Deviation	Coefficient of Variation
0.5	1.00	2.00
0.6	1.00	1.67
0.7	1.00	1.43
0.8	1.00	1.25
0.9	1.00	1.11
1.0	1.00	1.00
5.0	2.34	0.47
6.0	2.48	0.41
10.0	2.89	0.29
20.0	3.45	0.17

Thus, a head change of about 6 has a coefficient of variation of about 0.40. Head-change values smaller than 6 have coefficients of variation that become larger as the head change becomes smaller, reflecting that additional errors are significant for small head changes, and, for changes less than 1.0, it is even doubtful that the direction of the head change is correct. Head-change values larger than 6 have coefficients of variation that become smaller as the head change becomes larger, reflecting that large head-changes result from stresses that often are better defined, such as pumpage from water-supply wells.

Investigate Model Fit to Observations (***Guideline 9***) Following calibration of the steady-state, three-layer DVRFS model, model fit was evaluated using Guideline 9 and the methods in Chapter 6. Figure 15.3 shows graphs of weighted residuals for the calibrated model. Figure 15.3a reveals that the weighted residuals are not entirely randomly distributed. For heads, the positive weighted residuals are more extreme and outnumber the negative weighted residuals. Also, if normality is assumed, it is expected that only 3 out of 1000 (or 1 to 2 of the 516 weighted residuals) would lie outside plus or minus three times the standard errors of the regression, s (here, $3s = 14.1$). These problems with the distribution of residuals suggest that some model bias exists. However, investigation of the observations associated with the largest weighted head residuals revealed that they probably were affected by perched conditions that are not simulated by the regional saturated-zone model, so the model bias is likely to be less than indicated by Figure 15.3a.

The weighted simulated values that are plotted in Figure 15.3a for the flow observations, which are weighted using coefficients of variation, have been modified using Eq. (6.7) and then are scaled so that they lie in a range compatible with the range of weighted simulated values for heads.

Figure 15.3b shows weighted observed values versus weighted simulated values. Here, unmodified weighted simulated values for all observations are plotted, so that the weighted residuals, which are represented as deviations from the line with a

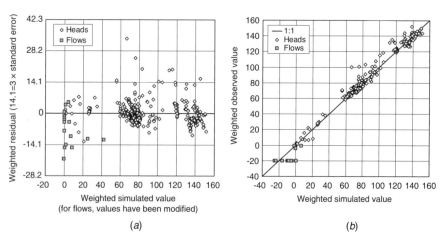

FIGURE 15.3 Graphs of (a) weighted residuals and (b) weighted observed values versus weighted simulated values for the three-layer Death Valley regional flow system (DVRFS) model of D'Agnese et al. (1997). Model misfit is more clearly displayed by graphs of weighted residuals.

15.2 FIELD APPLICATIONS AND SYNTHETIC TEST CASES

slope of 1.0, are equal to the weighted residuals shown in Figure 15.3a. In Figure 15.3b, the model misfit is much less obvious than in Figure 15.3a, and the details of the misfit cannot be as easily discerned. This is consistent with the remarks made in Chapter 6, Guideline 9, and Table 10.3 that suggest this type of graph is not as useful as the graph shown in Figure 15.3a for revealing model bias.

Figure 15.4 shows maps with weighted and unweighted head residuals. There are a large number of positive weighted and unweighted residuals in the northwest part the model domain. The poorer fit to the observed heads is partly because the quality of the data used for model construction in these areas is generally worse than that of data used for other areas. Also, in the region extending from Amargosa Valley southeast to Pahrump Valley (see Figure 15.1), residuals tend to be negative in the northeast part of this band and positive in the southwest part. This systematic trend is believed to be caused in part by the coarse vertical discretization in the model, which precludes accurate representation of the vertical hydraulic conductivity.

Some head observations with relatively large (in absolute value) unweighted residuals in Figure 15.4a have much smaller weighted residuals in Figure 15.4b. For example, in the mountainous area just east of Death Valley, several observations have large positive unweighted residuals, but most have weighted residuals that are much smaller in magnitude. This indicates that the match is as expected, given the quality of the head data. Most of these observations are in areas of steep hydraulic gradient, where observation location errors can be very large.

Figure 15.5 shows maps with weighted and unweighted spring-flow residuals. Most flow residuals are negative. In MODFLOW groundwater discharge is negative,

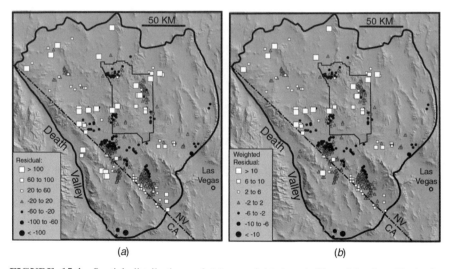

FIGURE 15.4 Spatial distributions of (a) unweighted and (b) weighted residuals for hydraulic-head observations in model layer 1 of the steady-state, three-layer DVRFS model. (Adapted from D'Agnese et al., 1997, Figures 48 and 49.)

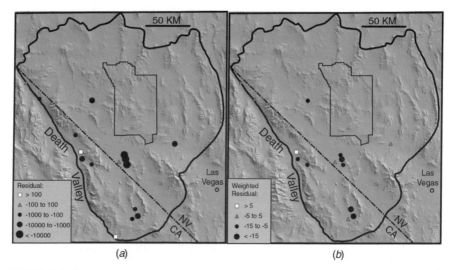

FIGURE 15.5 Spatial distributions of (*a*) unweighted and (*b*) weighted residuals for spring-flow observations of the three-layer DVRFS model. (Adapted from D'Agnese et al., 1997, Figures 54 and 55.)

so the dominance of negative residuals indicates that the simulated discharge at the springs is generally less than that observed. D'Agnese et al. (1997) concluded from this result that more detailed evaluation was needed of the conductance parameters that control the spring discharge.

Three flow observations with very large unweighted residuals, in absolute value (Figure 15.5*a*), have much more moderate weighted residuals (Figure 15.5*b*). At these locations, spring discharge is larger than elsewhere in the domain, and the weighted residuals provide the more meaningful result that, based on percent difference in flow, the residuals at these locations are similar to those at other locations. The flows are weighted using coefficients of variation.

Observations important to predictions can be identified using the *opr* statistic, as illustrated in the section for Guideline 12 below.

Determine Observations that Dominate Parameter Estimation (Guideline 10) The Cook's D values calculated for the 501 hydraulic-head observations of the three-layer model are shown in Figure 15.6. The critical value of Cook's D (see Chapter 7, Section 7.3.2) is $4/517 = 0.0077$, because there are 501 hydraulic-head observations and 16 flow observations for the steady-state DVRFS model calibration. Sixty-six of the head observations and 13 of the flow observations exceed this critical value. The head observations with large Cook's D are spread fairly evenly over most of the areas that contain head observations, indicating that observations from one part of the model domain do not dominate the regression. The large percentage of flow observations with large Cook's D indicates the extreme importance of flow observations in the calibration of this model.

15.2 FIELD APPLICATIONS AND SYNTHETIC TEST CASES

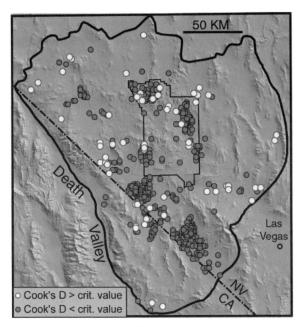

FIGURE 15.6 Cook's D values for the 501 head observations in the three-layer DVRFS model of D'Agnese et al. (1997, 1999). Sixty-six have Cook's D values larger than the critical value. Cook's D identifies observations important to estimated parameter values.

Identify New Data to Improve Predictions (Guideline 12) First, the predictions are defined. Then prediction scaled sensitivities (pss) and parameter–prediction statistics (ppr) are used to identify parameters important to predictions and to infer advantageous data to collect for improving simulated properties governing system dynamics. Finally, observation–prediction statistics (opr) are used to identify existing and potential new observations important to predictions and to infer advantageous field data to collect to support observations.

Predictions In the DVRFS, the predictions of interest involve potential transport of hypothetical contaminants. Accurate simulation of this transport is plagued by a number of problems, including the fractured nature of the subsurface rocks and the regional scale of the model. In a regional model, it is impossible to represent accurately processes such as dispersion and retardation or small features that can be important to transport. A useful approach is to consider only the transport processes appropriate to the scale of the model. Thus, advective transport is considered, which is the transport that would occur if the solute did not disperse and encountered no reactions with the surrounding rocks. It is simply the transport produced by bulk flow in the subsurface system. It can be considered the first building block of transport, upon which other complexities are added. Calculation of advective transport over large distances and times is consistent with the scale of a regional model, because regional conditions generally influence this component of transport more than they influence other aspects of transport.

Advective transport was simulated using the Advective-Transport Observation (ADV) Package of MODFLOW-2000 (Anderman and Hill, 2001). The ADV Package uses particle-tracking methods nearly identical to those in MODPATH (Pollock, 1994) to determine advective-travel paths. To compute the particle trajectory, particle displacement is decomposed into displacements in the three spatial dimensions of the DVRFS model: north–south (N–S), east–west (E–W), and vertical. This analysis of the directional components of transport allows parameter and observation importance to be evaluated on the basis of the information they provide for each direction of transport. The predicted transport paths from several locations on the Nevada Test Site are shown in Figure 15.7.

Using pss and ppr Prediction scaled sensitivities (*pss*, Eq. (8.2c)) and the parameter–prediction (*ppr*) statistic (Eq. (8.8)) were used to evaluate the importance of all 23 defined parameters (Tiedeman et al., 2003).

The *pss* (Figure 15.8a) show that parameters K5 (very high hydraulic conductivity) and K1 (high hydraulic conductivity) rank as the two most important parameters to the

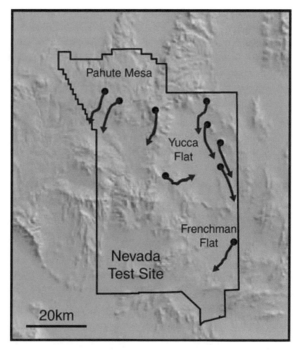

FIGURE 15.7 Selected predicted advective-transport paths from locations on the Nevada Test Site. The paths are simulated using the steady-state, three-layer model of D'Agnese et al. (1997). The paths shown start slightly below the water table. The vertical component of advective transport is small compared to the lateral component and is not shown. (Adapted from Tiedeman et al., 2004.)

15.2 FIELD APPLICATIONS AND SYNTHETIC TEST CASES

E–W and N–S components of advective transport. The results indicate that increasing the value of K5 by 1 percent of its standard deviation would change the distance traveled in the E–W direction by about 0.5 percent and the distance traveled in the N–S direction by about 0.7 percent. The *ppr* statistics for individual parameters (Figure 15.8*b*) show that K1 and Rch3 (relatively high recharge) are the two most important parameters to advective transport in the E–W and N–S directions. Reducing the standard deviation of K1 by 10 percent reduces the uncertainty of both the E–W and the N–S transport components by about 3.5 percent.

Figure 15.8 shows that some of the same parameters are identified by both the *pss* and *ppr* methods as important to the prediction of interest, but it also shows that the parameters identified as important can be different. For example, the *pss* indicate that predicted advective transport in the E–W and N–S directions is relatively insensitive

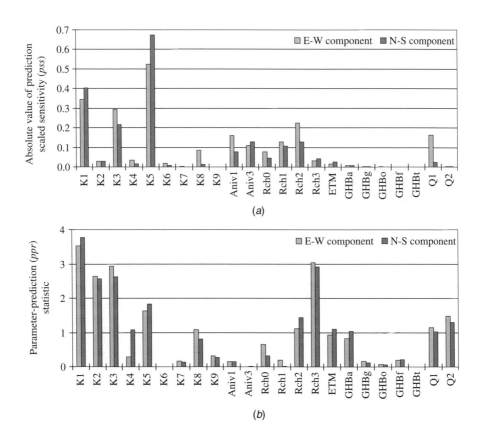

FIGURE 15.8 Prediction scaled sensitivities (*pss*) and parameter–prediction statistic (*ppr*) calculated to evaluate the importance of all 23 DVRFS model parameters to a predicted advective-transport path. (*a*) Absolute value of *pss*, defined as the percent change in predicted value produced by a 1 percent change in the standard deviation of a parameter (Eq. (8.2c)). (*b*) The *ppr* statistic, calculated as the percent decrease in prediction standard deviation produced by a 10 percent reduction in the standard deviation of a parameter (Eq. (8.8)). (From Tiedeman et al., 2003.)

to Rch3. However, the *ppr* indicate that Rch3 is important to advective transport in these directions. These differences occur because *ppr* accounts for parameter correlations, while *pss* does not. There are high correlations between Rch3 and K1 (0.96) and Rch3 and K3 (0.94). The *pss* show that transport in the E–W and N–S directions is sensitive to K1 and K3 (Figure 15.8a). Because of the correlations, specifying improved information on Rch3 improves the estimation of K3, which reduces the prediction uncertainty. This results in a relatively large value of *ppr*.

The effects of parameter correlations on *ppr* results have an important consequence for the cost-effectiveness of future data collection. For example, to improve predicted advective transport at this site, field data could be collected about Rch3 instead of about K1 or K3. Collecting data about a recharge rate or the geographic extent of a recharge zone is likely to be less expensive than collecting subsurface information about a hydraulic-conductivity value or hydrogeologic unit.

Using opr The observation–prediction (*opr*) statistic (Eq. (8.11)) was used to identify existing and potential observations important to the advective-transport predictions.

The existing 501 hydraulic-head observations were evaluated using the $opr_{(-1)}$ statistic averaged over a set of advective-transport paths, including those shown in Figure 15.7 (Hill et al., 2001; Tiedeman et al., 2004). Figure 15.9 shows the sets of 100 observations that are most and least important to the advective-transport

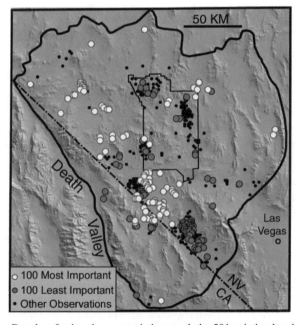

FIGURE 15.9 Results of using the *opr* statistic to rank the 501 existing head observations in the steady-state, three-layer DVFRS model of D'Agnese et al. (1997, 1999) by their importance to advective-transport predictions. (Adapted from Hill et al., 2001.)

15.2 FIELD APPLICATIONS AND SYNTHETIC TEST CASES

predictions. The advective paths are on or near the Nevada Test Site, in the center of the modeled region. These results clearly show that many important observations are located far from the predictions of interest, which reflects the distributed nature of most groundwater systems. The identification of important and unimportant observations can be used in Guideline 9 to focus evaluations of model fit, and as part of Guideline 12 to identify wells for which field work might be warranted, for example, to reduce errors in well altitude or screened depth.

The results shown in Figure 15.9 were obtained by evaluating each observation individually. While the results might be used to suggest that there is no need to investigate further the 100 least important observations, a concern is whether these 100 observations are important if considered together. This can be evaluated by calculating an opr statistic for which the entire group of 100 least important observations is omitted. This yields an $opr_{(-100)}$ statistic equal to only 0.6 percent, indicating that the 100 observations that are least important on an individual basis also are not important when considered as a group.

The $opr_{(+1)}$ statistics shown in Figure 15.10 for potential new hydraulic-head observations located anywhere in the uppermost layer of the Death Valley regional flow model. The very large values in some areas suggest locations where additional head observations would be extremely advantageous.

The opr values result from the parameterization using zonation and the simulated flow system dynamics. For example, in the area in the southern part of the model just

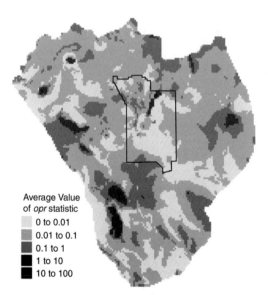

FIGURE 15.10 Results of using the $opr_{(+1)}$ statistic to rank potential new observation locations in the DVRFS by their importance to advective-transport predictions. (Adapted from Tiedeman et al., 2004.) High values identify areas where new head observations would be most advantageous, based on the steady-state, three-layer model described by D'Agnese et al. (1997, 1999).

east of Death Valley, *opr* is large because locally, steep gradients make the simulated hydraulic heads very sensitive to the hydraulic conductivity. The rocks here are thought to be hydraulically similar to rocks close to or along the advective-travel paths and are assigned the same hydraulic-conductivity parameter. If the rocks at the two locations are actually similar hydraulically, the steep gradients in this area provide a great opportunity to determine the hydraulic conductivity more accurately than is possible given the relatively flat gradients in the area of the advective-transport paths. However, if the hydraulic properties of these different rocks are actually not similar, then some observations ranked as important to the predictions may actually not be as valuable as indicated by the *opr* statistics. For a more detailed discussion of issues related to *opr* statistic results, see Tiedeman et al. (2004).

15.2.2 Grindsted Landfill, Denmark

The Grindsted landfill (Figure 15.11), located in the western part of Denmark, was the focus of a European Union study. Barlebo et al. (1998) characterized the underlying groundwater system using 100 head and 210 chloride concentration observations and an inverse groundwater flow and transport model developed using an early version of the finite-element model WATFLOW/WTC (Molson and Frind, 2002).

A map and cross section for the system are shown in Figures 15.11 and 15.12, respectively; the layers shown extend over most of the simulated area and were used to represent the system in all of the models developed. The model grid and other site features are shown in Figure 15.13. The concentrations were measured along the four transects labeled A, B, C, and D in Figure 15.13. This study investigated the possibility that transport occurred through the clay/silt layer

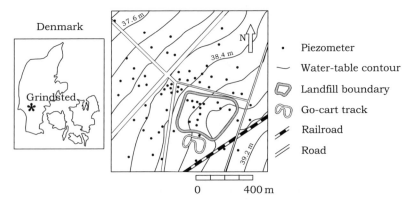

FIGURE 15.11 The Grindsted landfill, with local water-table contours and piezometer locations. Datum is sea level, contour interval is 0.2 m. Model area includes the landfill and extends to the northeast. (From Barlebo et al., 1998.)

15.2 FIELD APPLICATIONS AND SYNTHETIC TEST CASES

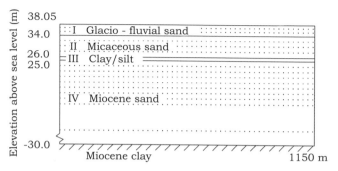

FIGURE 15.12 Northwest–southeast cross section of modeled area in the vicinity of the Grindsted landfill. Dots represent nodes in the model. (From Barlebo et al., 1998.)

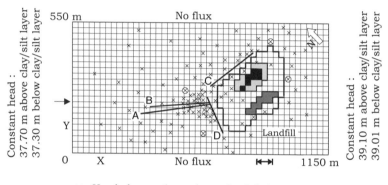

× Head observations above clay/silt layer
○ Head observations below clay/silt layer
A,B,C,D : Lines of wells
■ Source 2500 mg/L
▨ Source 800 mg/L
☐ Source 100 mg/L
→ Location of two-dimensional cross section
↔ Location of cells used in the two-dimensional model to represent the source

FIGURE 15.13 Plan view of modeled area in the vicinity of the Grindsted landfill, with the location of the cross section shown in Figure 15.12 identified by an arrow. The constant-head cells extend along each end of the model and the head imposed changes with depth as shown. (From Barlebo et al., 1998.)

shown in Figure 15.12. Most heads and all concentrations were measured using wells screened only above the silt layer.

Use a Broad Range of System Information (Guideline 2) and Include Many Kinds of Data as Observations (Guideline 4) Barlebo et al. (1998) focus on using the data set to investigate two issues: (1) the accuracy of simulated concentrations using a model calibrated with hydraulic-head observations and no concentration observations, and (2) the advantages and disadvantages of using a two-dimensional cross-sectional model instead of a three-dimensional model when the contaminant source is at the land surface. First, a three-dimensional model was developed using all available system information and was calibrated using the head and concentration observations. This was expected to be the most accurate model, and the plume simulated using this model is shown in Figure 15.14a. Then, the two issues of interest were considered. The absence of concentration data for calibrating a model used to predict transport was investigated by estimating hydraulic parameters for the three-dimensional model using only the head observations. The parameters needed only for simulating transport were assigned values equal to those estimated for the first model. The resulting plume is shown in Figure 15.14b. Use of a two-dimensional cross-sectional model (Figure 15.13) produced the plume shown in Figure 15.14c.

Two main conclusions were drawn from these analyses. First, the simulated plumes in Figure 15.14a,b are dramatically different, with the first indicating transport through the clay/silt layer and the second indicating transport to much greater distances from the source, but with the plume remaining above the clay/silt layer. This demonstrates the inaccurate transport behavior that is predicted using a model calibrated only with hydraulic-head observations. Second, the major benefit of two-dimensional cross-sectional modeling was that execution times were reduced

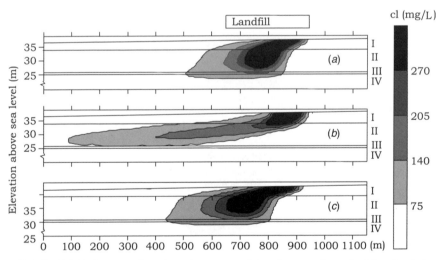

FIGURE 15.14 Cross section along the center of the plume at the end of the simulation period (1993) in models of the Grindsted landfill, Denmark. (a) Using a three-dimensional model with parameter values that produce the best fit to heads and concentrations. (b) Using a three-dimensional model with parameter values that produce the best fit to heads only. (c) Using a two-dimensional model with parameter values that produce the best fit to heads and concentrations. (From Barlebo et al., 1998.)

15.2 FIELD APPLICATIONS AND SYNTHETIC TEST CASES

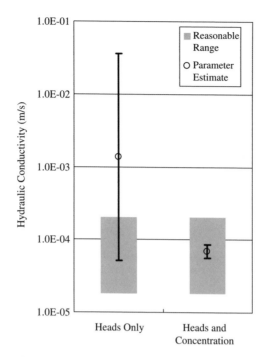

FIGURE 15.15 Estimates of parameter khII, in meters per second, from a groundwater model of the Grindsted landfill calibrated with only head observations and with head and concentration observations. Linear individual 95-percent confidence intervals (black bars) and the parameter reasonable range (gray boxes) also are shown. (From data reported by Barlebo et al., 1998.)

by a factor of 7. Advantages of the three-dimensional simulation were primarily that accurate representation of the source was possible (it had to be calibrated in the two-dimensional model), and that all hydraulic-head and concentration observations could be used. The simulated plumes are similar, though the three-dimensional plume matches the measured concentrations more closely.

Compare Estimated Parameters and Confidence Intervals to Reasonable Ranges (Guideline 10) During calibration of this model, estimated parameter values were compared to reasonable ranges. Figure 15.15 demonstrates why these comparisons also need to consider confidence intervals on the estimated parameters. The parameter value estimated using only head observations is unreasonable, but because of its large confidence interval, it is unclear whether the unreasonable estimate indicates errors in model construction. When concentration observations are included in the regression, the reasonable parameter estimate and small confidence interval indicate clearly that the problem was not model error. For the parameter estimate obtained when only heads were used, the information provided by the large confidence interval helped prevent an unnecessary search for model errors.

APPENDIX A

OBJECTIVE FUNCTION ISSUES

This appendix derives the maximum-likelihood objective function and shows that in most circumstances it reduces to the least-squares objective function used in MODFLOW-2000 and UCODE_2005. Because of the relationship between these objective functions, regression results from MODFLOW-2000 and UCODE_2005 can be interpreted in the context of either a maximum-likelihood or least-squares objective function. The practical differences between the two are as follows:

1. The maximum-likelihood objective function requires the assumption of normally distributed true errors, which is not required by the least-squares objective function except when constructing parametric confidence and prediction intervals.
2. The value of the objective function is different, even though the part involved in estimating optimal parameter values is the same. Values of both objective functions are printed by MODFLOW-2000 and UCODE_2005.

This appendix also discusses an issue related to the weighting used in the objective function—the assumptions required for diagonal weighting to be correct.

Effective Groundwater Model Calibration: *With Analysis of Data, Sensitivities, Predictions, and Uncertainty.* By Mary C. Hill and Claire R. Tiedeman
Published 2007 by John Wiley & Sons, Inc.

A.1 DERIVATION OF THE MAXIMUM-LIKELIHOOD OBJECTIVE FUNCTION

The maximum-likelihood objective function is developed by considering the random nature of y, the observations. This random nature results from conceptualizing observation error as random. If Y is the vector of jointly distributed random variables of which y is a realization, the joint probability distribution function (pdf), $f_Y(y)$, depends on the true model and true parameter values. For the purpose of estimating parameters for a given assumed model, consider the joint pdf conditioned on a particular set of parameter values, $f_Y(y|b)$. This joint pdf can be thought of as the probability that different sets of possible observations would occur given the parameter values b. In parameter estimation, the elements of y are known and we would like to estimate b. A reasonable requirement of the estimates in b is that they maximize the probability of obtaining the observations, y. This requirement is imposed by defining the objective function using the likelihood function, $\ell(b|y)$, which is defined as

$$\ell(b \mid y) = f_Y(y|b) \qquad (A.1)$$

If the true errors are from a joint, normal distribution, the likelihood function equals (Brockwell and Davis, 1987, p. 247)

$$\ell(b|y) = (1/2\pi)^{ND/2} |V(\varepsilon)|^{-1/2} \exp\{-(\tfrac{1}{2})e^T (V(\varepsilon))^{-1} e\} \qquad (A.2)$$

where, as in Eq. (3.2), e is a vector of residuals calculated as

$$e = y - y'$$

y' is a function of b and ND is the number of observations.

Replacing $V(\varepsilon)$ using Eq. (C.21) (see Appendix C), taking the natural log, and multiplying by (-2) produces the maximum-likelihood objective function:

$$S'(b) = -2\ln(\ell(b|y)) = ND\ln 2\pi - \ln|(1/\sigma^2)\omega| + e^T\{(1/\sigma^2)\omega\}e \qquad (A.3)$$

Multiplication by a negative number converts the maximization problem to a minimization problem; the objective is to determine the parameter estimates that minimize Eq. (A.3). To include prior estimates of the parameters, e and ω are augmented as shown in Eqs. (B.1) and (B.2), ND is replaced by $ND + NPR$, and the determinant of Eq. (A.3) is expanded so that Eq. (A.3) can be expressed as

$$S'(b) = (ND + NPR)\ln 2\pi + (ND + NPR)\ln \sigma^2$$
$$- \ln|\omega_d| - \ln|\omega_p| + e^T\{(1/\sigma^2)\omega\}e, \qquad (A.4)$$

where ω_d and ω_p are the sections of the weight matrix applicable to dependent variable observations and prior estimates of the parameters, respectively. Equation (A.4) is the maximum-likelihood objective function. Noting that $\ln 1.0 = 0.0$ and $(-\ln|\omega_d| - \ln|\omega_p|) = -\ln|\omega|$ as long as no observation errors are correlated with errors in prior information, Eq. (A.4) is equivalent to Eq. (3.3) when $\sigma^2 = 1.0$.

A.2 RELATION OF THE MAXIMUM-LIKELIHOOD AND LEAST-SQUARES OBJECTIVE FUNCTIONS

For any assumed model, set of observations, and defined weight matrix used in the parameter-estimation procedure, ND, σ^2, and ω do not change during a regression. Eliminating terms of Eq. (A.4) that do not depend on b and multiplying by σ^2 yields

$$S(b) = e^T \omega e \quad (A.5)$$

Thus, for the optimization process, the maximum-likelihood objective function equals the least-squares objective function (Eq. (3.2)). Burnham and Anderson (2002, p. 12) provide a different derivation that results in $S(b) = ND(\log(e^T \omega e))$, but for the purposes of this section the point is the same because both objective functions would result in the same estimated parameter values.

The development of Eq. (A.5) from the maximum-likelihood objective function requires that the true errors be from a joint, normal distribution, a condition not required when the equation is derived in other ways.

A.3 ASSUMPTIONS REQUIRED FOR DIAGONAL WEIGHTING TO BE CORRECT

Here we focus on errors in transient groundwater flow problems, though the concepts are generally applicable. The weighting is assumed to equal the inverse of the variance–covariance matrix of errors in observations and prior information, as suggested in Chapter 3, Section 3.4.2. Here the errors are referred to as true errors.

In transient groundwater flow problems, various dependent variables might be observed at many locations, and they might be observed at many times. The most general variance–covariance matrix of the true errors includes the following correlations:

1. Correlations between errors in observations made at different locations at the same time.
2. Correlations between errors in observations made at the same location at different times.
3. Correlations between errors in observations made at different locations and different times.

A.3 ASSUMPTIONS REQUIRED FOR DIAGONAL WEIGHTING TO BE CORRECT

Such a matrix is laborious to estimate.

Fortunately, assumptions about the true errors that are realistic for many circumstances can be made to simplify this variance–covariance matrix. Seven assumptions that result in a simple diagonal weight matrix are listed below and are further defined and discussed in this section. They are also discussed elsewhere in the book.

1. *All errors represented in the weighting have a mean of zero.* Required for a valid regression.
2. *Errors of different kinds of dependent variables are uncorrelated.* Commonly realistic for measurement errors.
3. *Errors of observations at different locations are uncorrelated.* Commonly realistic for measurement errors.
4. *Time-dependent deterministic components of the error are small and can be ignored.* Commonly realistic for measurement errors.
5. *At each observation location, total error for any observation* $= \varepsilon_1 + \varepsilon_2 + \varepsilon_3$. ε_1 errors are constant over time; ε_2 errors are temporally correlated, but not completely correlated like ε_1 errors; and ε_3 errors are temporally uncorrelated. Probably realistic.
6. *Errors are normally distributed.* Commonly realistic. Transformations may be needed in some circumstances.
7. *Either ε_1 or ε_2 or ε_3 errors dominate.* Transformations (generally differencing) can be applied to attain a diagonal matrix. Commonly questionable.

The set of assumptions presented here is not unique. They are presented so that one set of assumptions leading to a diagonal weight matrix can be thoroughly analyzed. The role of both measurement error and model error is included in the discussion, because Chapter 11, Section G6.2 suggests that both can play a role in determining weighting. Chapter 15, Section 15.2.1 provides an example of determining weighting.

1. *All errors represented in the weighting have a mean of zero.*

The sum of the errors accounted for by the weighting equals the ε of Eq. (C.1). The first assumption is discussed in Chapter 3, Section 3.3.2 and is required for a valid regression. Error with means other than zero cannot be accommodated correctly using the weight matrix.

2. *Errors of different kinds of dependent variables are uncorrelated.*

A similar assumption applied to observations and prior information produced the weight matrix of Eq. (B.1). The second assumption indicates, for example, that errors in observed hydraulic heads are independent of errors in observed streamflow gains and losses. This assumption is realistic for measurement errors; for example, it is unlikely that errors incurred when observing hydraulic heads are related to errors incurred when observing streamflow. This assumption, however, might not be realistic for model errors; for example, simulated hydraulic heads and adjacent simulated streamflows generally would be affected by the same deficiencies in the model, and associated model errors could be correlated.

3. *Errors of observations at different locations are uncorrelated.*

The third assumption is that errors in observations at different locations are uncorrelated. As with the second assumption, the third assumption is realistic for most measurement errors, but might not be realistic for model errors, especially at locations that are close to one another (Carrera, 1984, pp. 37–39). One exception occurs for measurement errors of head-dependent boundary gains and losses when one flow measurement is used to calculate more than one gain or loss, as discussed in Section G6.1 of Chapter 11.

If valid, the result of the second and third assumptions is that all correlations are eliminated except for the temporal correlations at each location.

4. *Time-dependent deterministic components of the error are small and can be ignored.*

The fourth assumption is that time-dependent deterministic components of the error (Brockwell and Davis, 1987, p. 15) are small and can be ignored. Such components are more typical of model error than measurement error, so the fourth assumption probably is realistic for measurement errors.

5. *At each observation location, total error for any observation* $= \varepsilon_1 + \varepsilon_2 + \varepsilon_3$.

ε_1 are errors that are constant over time; ε_2 are errors that are temporally correlated, but not completely correlated like ε_1; and ε_3 are errors that are temporally uncorrelated.

The fifth assumption is that the remaining correlations of the true errors, which are temporal, can be categorized by thinking of the error associated with each observation at a single time and location as the sum of three statistically independent types of errors:

$$\varepsilon = \varepsilon_1 + \varepsilon_2 + \varepsilon_3 \qquad (A.6)$$

where ε_1 is constant for all time for each observation location, so the temporal correlation coefficients between this error and the errors associated with observations at other times at this location equal 1.0; ε_2 varies with time and has correlation coefficients between this error and the errors associated with observations at other times at this location between 0.0 and 1.0, exclusive; and ε_3 varies

A.3 ASSUMPTIONS REQUIRED FOR DIAGONAL WEIGHTING TO BE CORRECT

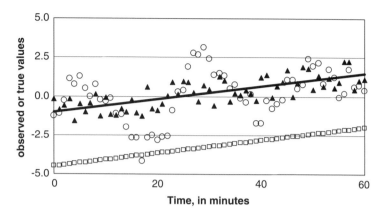

FIGURE A.1 True values represented by a solid line and observed values for which the errors are ε_1 errors (□), ε_2 errors (○), and ε_3 errors (▲). The ε_1 error equals -2.5; the ε_2 errors are from an autoregressive moving average process of order 7 based on a uniform distribution with a range of -1 to 1; the ε_3 errors are uniformly distributed with a range of -1 to 1.

independently over time and has correlation coefficients between this error and the errors associated with observations at other times at this location equal to zero.

Examples of observations with these types of errors are shown in Figure A.1. Considering their definitions, the assumed independence of ε_1, ε_2, and ε_3 is realistic, and this method of characterizing errors probably is valid for all groundwater flow models. The autocorrelated temporal errors considered by Sadeghipour and Yeh (1984), Carrera and Neuman (1986, p. 203), and Watson et al. (1990a, b) would be classified as ε_2, as defined above. Lu et al. (1988, p. 675) introduced a constant error similar to ε_1 to describe spatial correlations of errors in an estimated transmissivity field.

Processes that are likely to produce the three error types are different for measurement errors and model errors. For measurement errors, ε_1 might be the error in the measured elevation of the well; ε_2 might be an autocorrelated error in the recording device; and ε_3 might be random, uncorrelated inaccuracy in the recording device. For model errors, some examples are as follows: ε_1 errors might be produced because constant pumpage at a well near the observation location is not accounted for in the model; ε_2 errors might be produced because the parameterization of transmissivity is not realistic; and ε_3 errors might be a valid way to account for unrepresented processes with a mean affect of zero.

When written for a series of observations over time at an arbitrary observation location ℓ, Eq. (A.6) becomes

$$\boldsymbol{\varepsilon}_\ell = \boldsymbol{\varepsilon}_{1\ell} + \boldsymbol{\varepsilon}_{2\ell} + \boldsymbol{\varepsilon}_{3\ell} \tag{A.7}$$

where the length of each vector equals the number of temporal observations at location ℓ.

Here we discuss the characteristics of the variance–covariance matrix of each type of error.

$V(\varepsilon_{1\ell})$ is a matrix with all elements equal to the variance of the error. Thus, $V(\varepsilon_{1\ell}) = \mathbf{1}\sigma_{1\ell}^2$, where $\mathbf{1}$ is a matrix of ones and $\sigma_{1\ell}^2$ is the error variance associated with observation location ℓ. $V(\varepsilon_{1\ell})$ can be simplified if changes in the dependent variable over time are used in the regression instead of the actual values. For example, if hydraulic head is observed multiple times at one location, the first observation would be the first hydraulic head observation, and subsequent observations would be changes from that first hydraulic head defined as the temporal difference (see Chapter 9, Section 9.1.2). Because ε_1 errors are constant over time, they are included only in the first observation; the subtraction eliminates these errors from subsequent observations. Thus, by using temporal differencing, $V(\varepsilon_{1\ell})$ has one nonzero variance equal to $\sigma_{1\ell}^2$ associated with the first observation at observation location ℓ, and all other variances and covariances in the matrix equal zero.

$V(\varepsilon_{2\ell})$ depends on the correlation between the errors. If the correlation can be expressed by an autoregressive process (Brockwell and Davis, 1987, p. 79) and the observations are made at equally spaced times, differencing can be used to produce observations with independent errors and, therefore, a diagonal variance–covariance matrix. Sadeghipour and Yeh (1984) use differencing of a one-step autoregressive process and apply it to parameter estimation in a groundwater flow problem. The concern discussed in Chapter 9, Section 9.1.2 about differencing resulting in observations and associated sensitivities close to zero could be relevant in this situation and should be considered when designing the observations.

$V(\varepsilon_{3\ell})$ is a diagonal matrix because the ε_3 errors are uncorrelated. If, in addition, these errors are thought to have the same variance, $V(\varepsilon_{3\ell}) = I\sigma_{3\ell}^2$, where I is the identity matrix and $\sigma_{3\ell}^2$ is the error variance. When differencing is applied as discussed in the last two paragraphs as a way to create diagonal weighting for ε_1 and ε_2 errors, the weighting for $\varepsilon_{3\ell}$ errors goes from being diagonal to having off-diagonal terms. The differences have variances equal to the sum of the variances of the subtracted quantities and covariances are produced as discussed for streamflow measurements in Section G6.1 of Chapter 11.

6. *Errors are normally distributed.*

The sixth assumption is that the sources of error are numerous and varied enough that, by the central-limit theorem (Benjamin and Cornell, 1970, pp. 251–253), the joint probability distribution function (pdf) of the errors of Eq. (A.7) are normal, so

$$\varepsilon_3 \sim N[\mathbf{0}, V(\varepsilon_1)] = N[\mathbf{0}, V(\varepsilon_{1\ell}) + V(\varepsilon_{2\ell}) + V(\varepsilon_{3\ell})] \qquad (A.8)$$

where $\sim$ means "distributed as," and the three variance–covariance matrices can be added because of the assumed normality and independence of $\varepsilon_{1\ell}$, $\varepsilon_{2\ell}$, and $\varepsilon_{3\ell}$.

By using the expressions for $V(\varepsilon_{1\ell})$ and $V(\varepsilon_{3\ell})$ described above (assuming all ε_3 errors have the same variance), the joint pdf can be expressed as

$$\varepsilon_1 \sim N[\mathbf{0}, \mathbf{1}\sigma_{1\ell}^2 + V(\varepsilon_{2\ell}) + I\sigma_{3\ell}^2] \tag{A.9}$$

The sixth assumption probably is realistic for all groundwater flow problems.

Differencing of the observations result in $\mathbf{1}\sigma_{1\ell}^2$ being simplified further, but $I\sigma_{3\ell}^2$ becomes more complicated, as discussed above. If $\sigma_{3\ell}^2$ is small compared to $\sigma_{1\ell}^2$, an approximate diagonal matrix with the first variance equal to $\sigma_{1\ell}^2$ and variances for subsequent observations equal to $2\sigma_{3\ell}^2$ could be used. If ε_2 errors are important and first-order autoregressive, and the times between observations are constant, differencing methods can be used to make $V(\varepsilon_{2\ell})$ diagonal. However, $V(\varepsilon_{1\ell})$ and $V(\varepsilon_{3\ell})$ would then become more complicated.

7. *Either ε_1 or ε_2 or ε_3 errors dominate.*

The seventh assumption is that either ε_1 or ε_2 or ε_3 errors dominate, and if ε_2 errors dominate, the errors are autoregressive and the time between observations is constant. Although, as discussed earlier, it could be argued that ε_2 errors generally are a small part of the total error, the assumption that one type of error dominates is questionable under most circumstances.

Summary If all seven assumptions are valid, no flow observations are used to calculate more than one head-dependent boundary gain or loss, and the appropriate transformations are used to address the dominant ε_1 and ε_2 error terms, the variance–covariance matrix and, therefore, the weight matrix is diagonal. This structure is advantageous computationally and because the effect of a diagonal weight matrix on parameter estimation is easy for users to understand.

In practice, the seventh assumption is most likely to be violated. Violation of the seventh assumption means that more than one classification of error is significant and off-diagonal terms in the weight matrix are needed.

Little work has been done to determine the effect of using a diagonal weight matrix when the errors are actually correlated. Of concern is the effect on estimated parameter values and computed measures of uncertainty. The work that has been done suggests that the effects are small in typical circumstances, as mentioned in Chapter 11, Section G6.1.

A.4 REFERENCES

Benjamin JR, Cornell CA (1970). *Probability, Statistics, and Decision for Civil Engineers.* New York: McGraw Hill.

Burnham KP, Anderson DR (2002). *Model Selection and Multimodel Inference: A Practical Information-Theoretic Approach.* New York: Springer-Verlag.

Brockwell PJ, Davis RA (1987). *Time Series, Theory and Methods*. New York: Springer-Verlag.

Carrera J (1984). Estimation of parameters under transient and steady-state equations [unpublished dissertation]. Tucson: University of Arizona, Department of Hydrology and Water Resources.

Carrera J, Neuman SP (1986). Estimation of aquifer parameters under transient and steady-state conditions. *Water Resources Research* **22**(2):199–242.

Lu AH, Schittroth F, Yeh WW-G (1988). Sequential estimation of aquifer parameters. *Water Resources Research* **24**(5):670–682.

Sadeghipour J, Yeh WW-G (1984). Parameter identification of groundwater aquifer models—a generalized least squares approach. *Water Resources Research* **20**(7):971–979.

Watson TA, Gatens JM III, Lee JW, Rahim Z (1990a). An analytical model for history matching naturally fractured reservoir production data. *Society of Petroleum Engineers: Reservoir Engineering* **August**:384–388.

Watson TA, Lane HS, Gatens JM III (1990b). History matching with cumulative production data. *Society of Petroleum Engineers: Reservoir Engineering*, **January**:96–100.

APPENDIX B

CALCULATION DETAILS OF THE MODIFIED GAUSS–NEWTON METHOD

Three aspects of the calculations needed for the nonlinear regression methods described in this work require more detailed explanation. These include a more detailed description of vectors and matrices of equations in Chapters 3–8, presentation of an optional addition to Eq. (5.6a), and calculation of the damping parameter and convergence of Eq. (5.6b).

B.1 VECTORS AND MATRICES FOR NONLINEAR REGRESSION

The primary vectors and matrices of concern in nonlinear regression are the measured values of vector y, the simulated values of vector $y'(b)$, the sensitivities of matrix X, the weights of matrix ω, the residuals of vector e (equal to $y - y'(b)$), and the true errors of vector ε. These vectors and matrices, including terms for both the observations and prior information used in the regression, are as follows. Except for ε, these vectors and matrices are used in equations in Chapters 3–8 of this book. Vector ε is included here because it appears in Eq. (3.4) and Appendixes A and C.

Effective Groundwater Model Calibration: With Analysis of Data, Sensitivities, Predictions, and Uncertainty. By Mary C. Hill and Claire R. Tiedeman
Published 2007 by John Wiley & Sons, Inc.

A few common relationships are displayed using vector notation.

$$y = \begin{bmatrix} y_1 \\ y_2 \\ \vdots \\ y_{ND} \\ P_1 \\ P_2 \\ \vdots \\ P_{NPR} \end{bmatrix}, \quad X = \begin{bmatrix} x_{1,1} & x_{1,2} & \cdots & x_{1,NP} \\ x_{2,1} & x_{2,2} & \cdots & x_{2,NP} \\ \vdots & & & \\ x_{ND,1} & x_{ND,2} & \cdots & x_{ND,NP} \\ a_{1,1} & a_{1,2} & \cdots & a_{1,NP} \\ & & \vdots & \\ a_{2,1} & a_{2,2} & \cdots & a_{2,NP} \\ a_{NPR,1} & a_{NPR,2} & \cdots & a_{NPR,NP} \end{bmatrix}, \quad \omega = \begin{bmatrix} W & 0 \\ 0 & U \end{bmatrix} \quad (B.1)$$

W is the weighting for the observations; U is the weight matrix for the prior information, and it is assumed that the true errors in the observations are independent of the true errors in the prior information.

$$y'(b) = \begin{bmatrix} y'_1 \\ y'_2 \\ \vdots \\ y'_{ND} \\ P'_1 \\ P'_2 \\ \vdots \\ P'_{NPR} \end{bmatrix}, \quad y - y'(b) = e = \begin{bmatrix} e_1 \\ e_2 \\ \vdots \\ e_{ND} \\ u_1 \\ u_2 \\ \vdots \\ u_{NPR} \end{bmatrix}, \quad \varepsilon = \begin{bmatrix} \varepsilon_1 \\ \varepsilon_2 \\ \vdots \\ \varepsilon_{ND} \\ v_1 \\ v_2 \\ \vdots \\ v_{NPR} \end{bmatrix} \quad (B.2)$$

B.2 QUASI-NEWTON UPDATING OF THE NORMAL EQUATIONS

For problems with large residuals and a large degree of nonlinearity, Dennis et al. (1981) suggest substituting $X_r^T \omega X_r + R_r$ for $X_r^T \omega X_r$ in Eq. (5.6a) at selected iterations, where R_r is an estimate of the difference between $X_r^T \omega X_r$ and the Hessian matrix, $[\partial^2 S(b)/\partial b^2]$, and is calculated by quasi-Newton updating as (Dennis et al., 1981)

$$R_r = 0, \quad \text{for } r = 0$$

$$R_r = tR_{r-1} + \frac{u \, \Delta g_r^T + \Delta g_r \, u^T}{\rho_{r-1} d_{r-1}^T \Delta g_r} - \frac{\rho_{r-1} d_{r-1}^T u \, \Delta g_r \Delta g_r^T}{(\rho_{r-1} d_{r-1}^T \Delta g_r)^2}, \quad \text{for } r > 0 \quad (B.3)$$

where

$$\Delta g_r = g_r - g_{r-1}; \quad g_r = -X_r^T \omega e_r$$

$$e_r = [y - y'(b)]; \quad u = (X_r - X_{r-1})^T \omega e_r - tR_{r-1}\rho_{r-1}d_{r-1}$$

$$t = \min\left\{\left|\frac{\rho_{r-1}(d_{r-1}^T)(X_r - X_{r-1})^T \omega e_r}{\rho_{r-1}(d_{r-1}^T)R_{r-1}\rho_{r-1}(d_{r-1})}\right|; 1.0\right\}$$

and all other variables are defined after Eq. (3.2) and (5.6). R_r is calculated starting at $r = 1$ but is only included in Eq. (5.6a) in later iterations. Performance of the method depends on when R_r is included. Cooley and Hill (1992) found that it is most advantageous to include R_r after the sum of squared, weighted residuals no longer changes very much at each parameter-estimation iteration. When R_r is included in Eq. (5.6a), the elements of the diagonal scaling matrix, C, are calculated as $[(X_r^T \omega X_r + R_r)_{ii}]^{-1/2}$.

In UCODE_2005 and MODFLOW-2000, if quasi-Newton updating is used, R_r is included for all iterations after one of two criteria are satisfied: (1) the sum of squared, weighted residuals decreases by less than a user-defined percentage over two iterations or (2) after a user-specified number of iterations. The more elaborate criteria for inclusion of R_r suggested by Dennis et al. (1981) require additional model simulations. Given that many problems have lengthy execution times and considering the modest expected benefit demonstrated by Cooley and Hill (1992), the more elaborate criteria seem impractical and are not included.

B.3 CALCULATING THE DAMPING PARAMETER

For problems with one or more log-transformed parameters, requiring the absolute value of Eq. (5.7) to be less than *max-allowed-change* for any parameter-estimation iteration, and requiring Eq. (5.9) to be satisfied to achieve convergence can produce inconsistent results. The following example illustrates the problem as manifested when applying *max-allowed-change* which is labeled MAX-CHANGE here.

If the estimated parameter is $b_i = \log K$, where K is hydraulic conductivity, and MAX-CHANGE $= 2.0$, placing the restriction on $\log K$ requires that $(\log K)^{r+1}$, the estimate at the next parameter-estimation iteration, be between $(\log K)^r - 2.0(\log K)^r$ and $(\log K)^r + 2.0(\log K)^r$. If K at parameter-estimation iteration r is close to 1.0, say, $K = 1.1$, the restriction requires $(\log K)^{r+1}$ to be between -0.041 and 0.124, so that K^{r+1} is required to be within the narrow range 0.91 to 1.33. If K at parameter-estimation iteration r is far from 1.0, say, $K = 1 \times 10^{-4}$, the restriction requires that $(\log K)^{r+1}$ be between -12.00 and 4.00, so that K^{r+1} is allowed to vary within the very wide range of 1×10^{-12} to 1×10^4. More physically meaningful limitations are produced if the restriction is placed on the native parameter, which requires that K be between 0.0 and 3.3 in the first situation and between 0.0 and 3×10^{-4} in the second situation. In both

TABLE B.1 Quantities[a] Used to Test for Convergence and to Calculate Damping Parameter ρ_r for Parameter-Estimation Iterations

	Variable Used in Program and Brief Explanation		
	BDMXx	ADMXx	DMXx
Parameter category	A. Convergence test on the fractional change in the native parameter value[b]	B. Equation for ρ_r if the absolute value of quantity A is larger than MAX-CHANGE[c]	C. Fractional parameter change used to adjust ρ_r for oscillation control (Eq. (B.7))
Native	d_i^r/b_i^r (footnote d)	$\rho_r = \text{MAX-CHANGE}/(|d_i^r/b_i^r|)$ (footnote e)	$d_i^r/|b_i^r|$ (footnote d)
Transformed, $d_i^r > 0$	$\exp(d_i^r) - 1$	$\rho_r = \ln(\text{MAX-CHANGE} + 1)/d_i^r$ (footnote e)	$d_i^r/|b_i^r|$ (footnote d)
Transformed, $d_i^r < 0$	$\exp(d_i^r) - 1$	$\rho_r = \ln(-\text{MAX-CHANGE} + 1)/d_i^r$ (footnote d)	$d_i^r/|b_i^r|$ (footnote d)

[a] d_i^r, fractional change on the parameter value calculated by regression; b_i^r, value of the parameter; MAX-CHANGE, largest fractional change allowed for the parameter; called *max-allowed-change* in Chapter 5, Section 5.1.1.
[b] Largest absolute value needs to be less than a defined convergence criterion.
[c] Otherwise $\rho_r = 1.0$, except as needed for oscillation control. For each parameter-estimation iteration, the smallest of all ρ_r values is used and printed with the related parameter number in the output file.
[d] b_i^r is the native parameter value. If $b_i^r = 0.0$, these equal d_i^r.
[e] To enable parameter values to increase more quickly after being assigned very small native values, MAX-CHANGE is set to a larger number in some circumstances. See explanation in Chapter 5, Section 5.1.1.
[f] Only use if MAX-CHANGE < 1.0; otherwise, $\rho_r = 1.0$ except as determined for oscillation control.

situations, the lower limit of 0.0 is a result of estimating a log-transformed parameter and is always the lower limit for a log-transformed parameter when MAX-CHANGE ≥ 1.0.

To address this problem, a number of quantities can be calculated at each parameter-estimation iteration, as shown in Table B.1. The circumstances treated individually are (1) parameters that are not log-transformed, (2) parameters that are log-transformed and the regression is trying to increase their value ($d_i^r > 0$), and (3) parameters that are log-transformed and the regression is trying to decrease their value ($d_i^r < 0$). The objective that allows a single damping parameter to be chosen despite the individual circumstances is that the smallest of all values is needed, regardless of how it is calculated. The resulting value is used in Eq. (5.6b) to alter the magnitude of the change vector, leaving its direction undisturbed. MODFLOW-2000 always uses these equations; UCODE_2005 provides them as on option.

The equations in Table B.1 are derived as follows. For native parameters, the fractional change of the native parameter value of column A simply equals Eq. (5.7). For log-transformed parameters, the fractional change in the native

B.3 CALCULATING THE DAMPING PARAMETER

value equals $(\exp(b_i^{r+1}) - \exp(b_i^r))/\exp(b_i^r)$, or, equivalently, $(\exp(b_i^{r+1})/\exp(b_i^r)) - 1.0$. Substituting $\exp(d_i^r) = \exp(b_i^{r+1})/\exp(b_i^r)$, which is derived from Eq. (5.6b) with $\rho_r = 1.0$, yields

$$\text{fractional change in the native parameter value} = \exp(d_i^r) - 1.0 \tag{B.4}$$

In column B of Table B.1, the equation for native parameters is obvious, and the equations for log-transformed parameters are derived using Eq. (B.4). If $d_i^r > 0.0$, Eq. (B.4) produces a positive value and the MAX-CHANGE restriction requires that $\exp(\rho_r d_i^r) - 1.0 \leq$ MAX-CHANGE. If $d_i^r < 0.0$, Eq. (B.4) produces a negative value and the MAX-CHANGE restriction requires that $\exp(\rho_r d_i^r) - 1.0 \geq -$MAX-CHANGE. These requirements are satisfied if the following conditions are satisfied:

$$\rho_r = \min_{i=1,np} \begin{cases} 1.0 \\ \ln(\text{MAX-CHANGE}+1)/d_i^r, & d_i^r > 0.0 \\ \ln(-\text{MAX-CHANGE}+1)/d_i^r, & d_i^r < 0.0, \text{MAX-CHANGE} < 1.0 \end{cases} \tag{B.5}$$

The exception for MAX-CHANGE < 1.0 applies because, as mentioned previously, the exponential of a log-transformed parameter is always greater than 0.0 and can never decrease enough to require ρ_r to be less than 1.0 if MAX-CHANGE ≥ 1.0. Thus, if $d_i^r < 0$ for a log-transformed parameter and MAX-CHANGE ≥ 1.0, parameter i is excluded from consideration when calculating ρ_r.

A difficulty occurs in the above procedure if a parameter value is made much smaller than its starting value, and then the regression attempts to restore it. This situation can make the parameter estimation move slowly in the iterations following the iteration in which the parameter value is made small, because the damping parameter is being controlled by the small parameter value. This problem is addressed in MODFLOW-2000 and UCODE_2005 by allowing such small parameter values to increase by more than *max-allowed-change* (here we use MAX-CHANGE) would normally allow, and this is accomplished by assigning an increased value of MAX-CHANGE to parameter values that are small relative to their starting values. The increased MAX-CHANGE is calculated as

$$\text{MAX-CHANGE}^* = [|b/b_0| \times (\text{MAX-CHANGE} + 1.)^4]^{-1} \tag{B.6}$$

where b is the current native value of the parameter, and b_0 is the starting native value, as specified in the input file. If MAX-CHANGE* is smaller than MAX-CHANGE, the latter is used. The value of MAX-CHANGE* given different values of MAX-CHANGE and of $|b/b_0|$ is shown in Figure B.1. If it is expected that the smaller value is valid, restarting the regression with a smaller starting value will cause smaller MAX-CHANGE* values to be used in the lower range (Figure B.2). Modification is not made if MAX-CHANGE is specified to be less than 0.4 to allow the user to maintain control for small values of MAX-CHANGE.

Oscillation control is achieved using a slightly modified version of the method described by Cooley (1983b, p. 1274; 1993). Oscillation control is evaluated internal

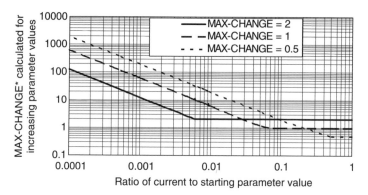

FIGURE B.1 The value of MAX-CHANGE* calculated given different values of MAX-CHANGE and the ratio of the current and starting parameter values ($|b/b_0|$).

to the regression, so it applies to log-transformed parameters where applicable. A preliminary damping parameter, ρ_r^*, is calculated as follows, where j_r identifies the parameter with the smallest ρ_r in iteration r.

$$\text{DMX}_r = d_i^r / |b_i^r|$$
$$\text{If } r = 0 \text{ or } j_r \neq j_{r-1}, \quad \rho_r^* = 1$$
$$\text{If } r > 0 \text{ and } j_r = j_{r-1},$$
$$s = \text{DMX}_r / (\rho_{r-1} \text{DMX}_{r-1}) \quad \text{(B.7)}$$
$$\text{If } s \geq -1, \quad \rho_r^* = (3+s)/(3+|s|)$$
$$\text{If } s < -1, \quad \rho_r^* = 1/(2|s|)$$

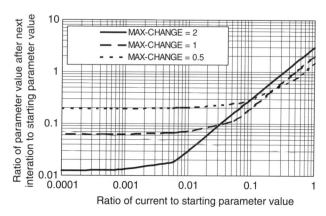

FIGURE B.2 The effect of MAX-CHANGE*, as indicated by the ratio of the current to the starting parameter value ($|b^r/b_0|$), and the ratio of the parameter value after the next parameter-estimation iteration to the starting parameter value ($|b^{r+1}/b_0|$). Using MAX-CHANGE*, it generally takes no more than three parameter-estimation iterations to restore a dramatically reduced parameter value.

B.4 SOLVING THE NORMAL EQUATIONS

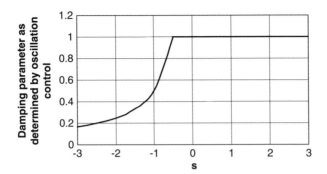

FIGURE B.3 The damping parameter produced through oscillation control based on s as calculated using Eq. (B.7).

The condition on j has been added to Cooley's method. The relationship between s defined in Eq. (B.7) and the damping parameter is shown in Figure B.3. The final damping parameter is the smaller of the value calculated using oscillation control or the value calculated to conform with the specified MAX-CHANGE.

MAX-CHANGE typically is larger than 1.0 and less than about 2.0. Use values less than 1.0 to reduce excessive parameter-value oscillations. Values less than 1.0 do not prohibit parameter values from changing sign because of the increase in MAX-CHANGE discussed above.

B.4 SOLVING THE NORMAL EQUATIONS

By using double precision as suggested by Stewart (1972, pp. 226–227), Eq. (5.6) has been solved accurately and efficiently in many applications using Cholesky LDL^T decomposition (Dennis and Schnabel, 1983, pp. 50–51). Exceptions were plagued by strong correlations between parameters or insensitive parameters, and were resolved by reparameterization. Dennis and Schnabel (1983, p. 221) and Seber and Wild (1989, p. 621) suggest that solving the alternative formulation $Xd = (y - y')$ using QR or singular-value decomposition (Dennis and Schnabel, 1983, pp. 49–51; Seber and Wild, 1989, pp. 680–681; Press et al., 1992, pp. 52–63) is more stable, but it is unclear whether or not they used the scaling and Marquardt parameter, which adds stability to Eq. (5.6). Press et al. (1992, pp. 515–520) suggest using singular-value decomposition for linear regression, but use Gauss–Jordon elimination to solve a variation of Eq. (5.6) that includes similar scaling and implementation of the Marquardt parameter for nonlinear regression. Considering the success experienced using Cholesky decomposition, it is used in UCODE_2005 and MODFLOW-2000.

B.5 REFERENCES

Cooley RL (1983). Incorporation of prior information on parameters into nonlinear regression groundwater flow models. 2. Applications. *Water Resources Research* **19**(3):662–676.

Cooley RL (1993). Regression modeling of ground-water flow, Supplement 1—Modifications to the computer code for nonlinear regression solution of steady-state ground-water flow problems. *U.S. Geological Survey Techniques of Water Resources Investigations*, Book 3, Chapter B4, Supplement 1.

Cooley RL, Hill MC (1992). A comparison of three Newton-like nonlinear least-squares methods for estimating parameters of ground-water flow models. In Russel TF, Ewing RE, Brebbia CA, Gray WG, Pinder GF, editors, *Proceeding of Computational Methods in Water Resources IX, Denver Colorado. Numerical Methods in Water Resources*. Elsevier, pp. 379–386.

Dennis JE, Gay DM, Welsch RE (1981). An adaptive nonlinear least-squares algorithm. *ACM Transactions on Mathematical Software* **7**(3):348–368.

Dennis JE, Schnabel RB (1983). *Numerical Methods for Unconstrained Optimization and Nonlinear Equations*. Englewood Cliffs, NJ: Prentice Hall.

Press WH, Flannery BP, Teukolsky SA, Vetterling WT (1992). *Numerical Recipes in Fortran*, 2nd ed. Cambridge, UK: Press Syndicate of the University of Cambridge.

Seber GAF, Wild CJ (1989). *Nonlinear Regression*. Hoboken, NJ: Wiley.

Stewart GW (1972). *Introduction to Matrix Computations*. New York: Academic Press.

APPENDIX C

TWO IMPORTANT PROPERTIES OF LINEAR REGRESSION AND THE EFFECTS OF NONLINEARITY

This appendix presents two basic properties of weighted linear regression, which are generally known as the Gauss–Markov theorem, in a manner that emphasizes the difficulties produced when the regression is nonlinear. More traditional derivations of the Gauss–Markov theorem can be found in Bard (1974) and Beck and Arnold (1977).

The two properties of concern are:

1. Parameters estimated by linear regression are unbiased.
2. The weight matrix needs to be defined in a particular way for the parameter estimates to have the smallest variance, and for the parameter variance–covariance matrix to be calculated using Eq. (7.1).

Needed definitions and identities are presented first, followed by the two proofs.

Effective Groundwater Model Calibration: With Analysis of Data, Sensitivities, Predictions, and Uncertainty. By Mary C. Hill and Claire R. Tiedeman
Published 2007 by John Wiley & Sons, Inc.

APPENDIX C: LINEAR REGRESSION AND THE EFFECTS OF NONLINEARITY

C.1 IDENTITIES NEEDED FOR THE PROOFS

C.1.1 True Linear Model

The true model is unknown and correctly represents the system of concern. A true linear model can be represented as

$$y = \beta_0 + \beta_1 X_1 + \beta_2 X_2 + \cdots + \beta_j X_j + \cdots + \beta_v X_v + \varepsilon, \quad E(\varepsilon) = 0 \quad \text{(C.1)}$$

where y = a measurement of the dependent variable (for groundwater models, hydraulic heads, flows, and so on);
β_j = true (unknown) parameter values;
X_j = independent variables (generally, location in three-dimensional space and time);
v = number of terms in the true model;
ε = true error, and needs to have a mean of zero, as shown, for regression to be valid.

C.1.2 True Nonlinear Model

The true nonlinear model cannot be represented as in Eq. (C.1) and requires the more general form $F(\boldsymbol{\beta}, \boldsymbol{\zeta}) + \varepsilon$, where F represents the form of the unknown nonlinear function, $\boldsymbol{\zeta}$ represents the independent variables, and the other symbols are as defined for Eq. (C.1). $\boldsymbol{\beta}$ is a vector of the true parameter values.

C.1.3 Linearized True Nonlinear Model

A linearized true nonlinear model is defined here for the purposes of this discussion. The model is linearized using a Taylor series expansion about the true parameter values and has the form of Eq. (C.1). The constant β_0 equals $y = F(\boldsymbol{\beta}, \boldsymbol{\zeta})$ evaluated at the independent variables associated with observation y. Each X_j is the derivative of the nonlinear model with respect to the jth parameter, evaluated at the true parameter values. Linearized models are discussed further next.

C.1.4 Approximate Linear Model

The approximate linear model is the model being developed to represent the system of concern and is the model to be calibrated. An approximate linear model can be represented as

$$y = b_0 + b_1 X_1 + b_2 X_2 + \cdots + b_j X_j + \cdots + b_n X_n + e = y' + e \quad \text{(C.2)}$$

where y = measurement of the dependent variable, as above;
b_j = estimated parameter values;

X_j = independent variables (generally, location in three-dimensional space and time);

n = number of terms in the approximate model;

e = residual;

y' = simulated equivalent of the measured dependent variable.

C.1.5 Approximate Nonlinear Model

As for the true nonlinear model, the approximate nonlinear model cannot be represented as in Eq. (C.1) and requires the more general form presented in Chapter 3, Section 3.4.2—that is, using vector notation, $y = f(\boldsymbol{b}, \boldsymbol{\xi}) + e$, where f represents the form of the approximate nonlinear function, $\boldsymbol{\xi}$ represents the independent variables, $\boldsymbol{b}$ is a vector of the estimated parameter values, and the other symbols are as defined for Eq. (C.2).

C.1.6 Linearized Approximate Nonlinear Model

The linearized approximate nonlinear model is produced using a Taylor series expansion about a defined set of parameter values, $\boldsymbol{b}'$, as discussed in Figure 5.1. The linearized approximate nonlinear model can be expressed in the form of Eq. (C.2). In this situation, however, the X_j are not independent variables; instead, they equal the derivatives of the approximate nonlinear model with respect to the parameter values, evaluated at $\boldsymbol{b}'$. There is a connection between using independent variables for the linear model and derivatives for the nonlinear model — for a linear model the derivatives equal the independent variables.

The derivatives were defined for Eq. (5.2) and have the following characteristics:

1. The derivatives generally include the independent variables and also include the effects of other aspects of the nonlinear model.
2. Because of model nonlinearity, the values of the derivatives depend on the parameter values in $\boldsymbol{b}'$, as demonstrated for Darcy's Law in Chapter 1, Section 1.4.1.
3. The derivatives generally are called sensitivities because they represent the sensitivity of the simulated value to a change in the parameter value.

By definition, linearized models reproduce the same simulated value at $\boldsymbol{b}'$ as the nonlinear model and often closely mimic the nonlinear model for nearby values of $\boldsymbol{b}$. As the linearized model is evaluated for values farther from $\boldsymbol{b}'$, simulated values vary from those of the approximate nonlinear model depending on its degree of nonlinearity. This deviation is apparent in Figures 5.1 and 5.3. In these figures, the linearized surfaces closely mimic the nonlinear surface near the parameter values about which the model is linearized, marked by an ×, and mimic it less well, and even poorly, for increasingly different sets of parameter values.

C.1.7 The importance of $\boldsymbol{X}$ and X

The different symbols, $\boldsymbol{X}$ and X, are used in Eqs. (C.1) and (C.2) because they may be different.

For linear problems they often are the same, but differences occur when the approximate model includes more or fewer terms than the true model (n does not equal v), or the terms are different. For nonlinear problems, variations in model structure can lead to different sensitivities and the sensitivities also generally vary depending on the set of parameter values about which the model is linearized. The differences between $\boldsymbol{X}_j$ and X_j become greater as the model and the optimized parameter values differ more from those of the true system.

Errors in measuring the independent variables could affect $\boldsymbol{X}_j$ and X_j (as considered by Fuller, 1987; Toy et al., 1993). In this book it is suggested that in some cases problems of inaccurate location can be integrated into the weighting. See Guideline 6 (Chapter 11) and the part of Section 15.2.1 (Chapter 15) on Guideline 6.

C.1.8 Considering Many Observations

Equations (C.1) and (C.2) use the variable y to indicate an observation. To identify many observations the index i is introduced. The ith observation used in the regression then can be expressed in terms of the true linear model as

$$y_i = \beta_0 + \beta_1 \boldsymbol{X}_{1i} + \beta_2 \boldsymbol{X}_{2i} + \cdots + \beta_j \boldsymbol{X}_{ji} + \cdots + \beta_v \boldsymbol{X}_{vi} + \varepsilon_i \qquad (C.3)$$

instead of Eq. (C.1), and in terms of the approximate linear or linearized model as

$$y_i = b_0 + b_1 X_{1i} + b_2 X_{2i} + \cdots + b_j X_{ij} \\ + \cdots + b_n X_{ni} + e_i = y'_i + e_i \qquad (C.4)$$

instead of Eq. (C.2).

All observations used in the regression together can be expressed in terms of the true linear model using matrix notation (vectors are bold lowercase or Greek letters, matrices are bold capital letters) as

$$\boldsymbol{y} = \boldsymbol{X}\boldsymbol{\beta} + \boldsymbol{\varepsilon} \qquad (C.5)$$

and in terms of the approximate linear or linearized model as

$$\boldsymbol{y} = \boldsymbol{X}\boldsymbol{b} + \boldsymbol{e} \qquad (C.6)$$

For the linearized approximate nonlinear model, each element of the $\boldsymbol{X}$ array is one of the derivatives, or sensitivities, discussed above. An expanded form of $\boldsymbol{X}$ was shown in Appendix B. $\boldsymbol{X}$ is called the sensitivity matrix or the Jacobian, as mentioned after Eq. (5.2b).

C.1.9 Normal Equations

Normal equations in parameter estimation are discussed in Chapter 5. For a linear model, the normal equations are

$$(X^T\omega X)b = X^T\omega y \text{ or } b = (X^T\omega X)^{-1}X^T\omega y \tag{C.7}$$

Despite some variation, the similarity between Eq. (C.7), which applies to an approximate linear model, and Eq. (5.4), which applies to an approximate nonlinear model, is apparent, with the major difference being that Eq. (C.7) produces the actual optimal parameter values after being evaluated just once, while Eq. (5.4) produces a vector that is used to update the parameter values, and optimal parameter values are obtained only after a number of parameter-estimation iterations. Because the iterative nature of the equations is not central to the issue addressed in the proofs later, Eq. (C.7) is used in the proofs.

C.1.10 Random Variables

The primary random variables in the above equations are the true errors ε. Then, noting that functions of random variables are random, y is random from Eq. (C.1), e is random from Eq. (C.2), and b is random from Eq. (C.7). Because for any step of the analysis the nonlinear model is linearized and X is evaluated for a defined set of parameters, X is not random.

C.1.11 Expected Value

The expected value can be taken of any term and is represented as $E(\bullet)$ or $E[\bullet]$, where the term appears within the parentheses or brackets. As noted in Eq. (C.1), ε has a mean of zero, so $E(\varepsilon) = 0$.

C.1.12 Variance–Covariance Matrix of a Vector

Property 2 requires the evaluation of the variance–covariance matrix of the vector of estimated parameters and the true errors. The variance–covariance matrix of any vector v is calculated as $E[(v - E(v))(v - E(v))^T]$.

C.2 PROOF OF PROPERTY 1: PARAMETERS ESTIMATED BY LINEAR REGRESSION ARE UNBIASED

Take the expected value of the optimized parameters, as calculated using Eq. (C.7):

$$E(b') = (X^T\omega X)^{-1}X^T\omega E(y) = (X^T\omega X)^{-1}X^T\omega X\beta \tag{C.8}$$

If $X = \mathcal{X}$,

$$(X^T\omega X)^{-1}X^T\omega \mathcal{X} = I \qquad (C.9)$$

where I is an identity matrix. Substituting Eq. (C.9) into Eq. (C.8) yields

$$E(b) = \beta \qquad (C.10)$$

Thus, if $X = \mathcal{X}$, the expected values of the estimates equal the true values, which means that the estimates are unbiased. In nonlinear models, even if the model is correct the noise in the observations is likely to produce parameter values that differ to some degree from the true parameter values, and the equality is unlikely to be true. Thus, unbiasedness is not guaranteed for nonlinear models, even if the model is correct (Also noted by, for example, Seber and Wild, 1989).

C.3 PROOF OF PROPERTY 2: THE WEIGHT MATRIX NEEDS TO BE DEFINED IN A PARTICULAR WAY FOR EQ. (7.1) TO APPLY AND FOR THE PARAMETER ESTIMATES TO HAVE THE SMALLEST VARIANCE

The variances of the parameter estimates equal the diagonal terms in the variance–covariance matrix of the parameters, which is calculated using the standard equation for the variance–covariance matrix of the terms of a vector discussed just before the proof of Property 1. Applying the equation to a vector of parameter values yields

$$V(b) = E[(b - E(b))(b - E(b))^T] \qquad (C.11)$$

replacing b with Eq. (C.7) and $E(b)$ with Eq. (C.10) yields

$$V(b) = E[((X^T\omega X)^{-1}X^T\omega y - \beta)((X^T\omega X)^{-1}X^T\omega y - \beta)^T] \qquad (C.12)$$

Expanding the product produces an equation with four terms:

$$V(b) = E[((X^T\omega X)^{-1}X^T\omega y)((X^T\omega X)^{-1}X^T\omega y)^T \\ - ((X^T\omega X)^{-1}X^T\omega y)\beta^T - \beta((X^T\omega X)^{-1}X^T\omega y)^T + \beta\beta^T] \qquad (C.13)$$

Rearrange the first term using (a) the matrix property $(AB)^T = B^TA^T$ and (b) $((X^T\omega X)^{-1}$ and ω are symmetric so the transpose equals $(X^T\omega X)^{-1}$ and ω, respectively. This yields

$$((X^T\omega X)^{-1}X^T\omega y)((X^T\omega y)^{-1}X^T\omega y)^T \\ = (X^T\omega X)^{-1}X^T\omega yy^T\omega X(X^T\omega X)^{-1} \qquad (C.14)$$

C.3 THE WEIGHT MATRIX

Take the expected value of each term and note that only y is stochastic to obtain:

$$V(b) = (X^T\omega X)^{-1} X^T \omega E[yy^T] \omega X (X^T \omega X)^{-1}$$
$$- (X^T\omega X)^{-1} X^T \omega E[y] \beta^T - \beta((X^T\omega X)^{-1} X^T \omega E[y])^T + \beta\beta^T \quad (C.15)$$

In the first term, apply $y = X\beta + \varepsilon$, so that

$$E[yy^T] = E[(X\beta + \varepsilon)(X\beta + \varepsilon)^T]$$
$$= E[(X\beta)(X\beta)^T + (X\beta)\varepsilon^T + \varepsilon X\beta^T + \varepsilon\varepsilon^T] \quad (C.16)$$

Taking the expected value of each term, and noting that only ε is stochastic and that the second and third terms of Eq. (C.16) equal zero because $E[\varepsilon] = 0$, produces

$$E[yy^T] = (X\beta)(X\beta^T)^T + E[\varepsilon\varepsilon^T] = X\beta\beta^T X^T + E[\varepsilon\varepsilon^T] \quad (C.17)$$

Note that $E[\varepsilon\varepsilon^T] = V(\varepsilon)$, the variance–covariance matrix of the true errors. This can be derived by applying the standard equation for calculating the variance–covariance matrix of a vector, so that $V(\varepsilon) = E[(\varepsilon - E(\varepsilon))(\varepsilon - E(\varepsilon))^T]$, and using $E(\varepsilon) = 0$.

Substituting these results into Eq. (C.15) yields

$$V(b) = (X^T\omega X)^{-1} X^T \omega X \beta\beta^T X^T \omega X (X^T\omega X)^{-1}$$
$$+ (X^T\omega X)^{-1} X^T \omega E[\varepsilon\varepsilon^T] \omega X (X^T\omega X)^{-1} \quad (C.18)$$
$$- ((X^T\omega X)^{-1} X^T \omega X \beta) \beta^T - \beta((X^T\omega X)^{-1} X^T \omega X \beta)^T + \beta\beta^T$$

If $X = X$, then $(X^T\omega X)^{-1} X^T \omega X = I$, which gives the following:

$$V(b) = \beta\beta^T + (X^T\omega X)^{-1} X^T \omega E[\varepsilon\varepsilon^T] \omega X (X^T\omega X)^{-1}$$
$$- \beta\beta^T - \beta\beta^T + \beta\beta^T \quad (C.19)$$

The $\beta\beta^T$ terms cancel, leaving

$$V(b) = (X^T\omega X)^{-1} X^T \omega E[\varepsilon\varepsilon^T] \omega X (X^T\omega X)^{-1} \quad (C.20)$$

If the weight matrix is defined such that

$$E[\varepsilon\varepsilon^T] = V(\varepsilon) = \sigma^2 \omega^{-1} \quad (C.21)$$

where σ^2 is the true common error variance, Eq. (C.20) reduces to

$$V(b) = \sigma^2(X^T\omega X)^{-1} \approx s^2(X^T\omega X)^{-1} \qquad (C.22)$$

where the last equal sign is approximate because s^2, the calculated error variance, is used to approximate σ^2, the unknown true common error variance. Equation (C.22) is the expression commonly used to calculate the variance–covariance matrix for the parameter values but really only applies if $X = X$, and the weights are defined based on Eq. (C.21).

If the equation for $V(b)$ cannot be simplified to Eq. (C.22), equations of the form (C.18) or (C.20) should be used to calculate the variance–covariance matrix of the parameter estimates, although it is unclear how to evaluate Eq. (C.18) because β is unknown. For linear problems, Eq. (C.19) always produces a larger variance for the parameters and simulated predictions than is produced by other possible equations (Bard, 1974; Beck and Arnold, 1977, pp. 232–234). Thus, the smallest variance parameter estimates are those for which Eq. (C.21) applies and, therefore, for which $X = X$ and the weighting is defined such that the ω is proportional to $V(\varepsilon)^{-1}$ (the variance–covariance matrix of the true unknown errors). Although approximate, linear theory provides the only available guidance for defining the weight matrix for nonlinear problems.

C.4 REFERENCES

Bard J (1974). *Nonlinear Parameter Estimation*. New York: Academic Press.

Beck JV, Arnold KJ (1977). *Parameter Estimation in Engineering and Science*. Hoboken, NJ: Wiley.

Toy TJ, Osterkamp WR, Renard KG (1993). Prediction by regression and intra-range data scatter in surface-process studies. *Environmental Geology* 22:121–128.

APPENDIX D

SELECTED STATISTICAL TABLES

TABLE D.1 Selected Web Sites for the Standard Normal Distribution (Mean = 0.0; Standard Deviation = 1,0) (accessed June 15, 2006)

http://www.anu.edu.au/nceph/surfstat/surfstat-home/tables/normal.php
http://www.statsoft.com/textbook/sttable.html#z
http://psych.colorado.edu/~mcclella/java/normal/tableNormal.html
http://www.fon.hum.uva.nl/Service/Statistics/NormalZ_distribution.html?Z=1.378

TABLE D.2 Selected Web Sites for the Student t-Distribution (accessed June 15, 2006)

http://www.statsoft.com/textbook/sttable.html#t
http://www.econtools.com/jevons/java/Graphics2D/tDist.html
http://www.itl.nist.gov/div898/handbook/eda/section3/eda3672.htm
http://www.stat.tamu.edu/~west/applets/tdemo.html

Effective Groundwater Model Calibration: *With Analysis of Data, Sensitivities, Predictions, and Uncertainty.* By Mary C. Hill and Claire R. Tiedeman
Published 2007 by John Wiley & Sons, Inc.

TABLE D.3 Critical Values of R_N^2 Below Which the Hypothesis that the Weighted Residuals Are Independent and Normally Distributed Is Rejected at the Stated Significance Level[a]

ND or ND + NPR	Significance Level 0.05	Significance Level 0.10	ND or ND + NPR	Significance Level 0.05	Significance Level 0.10
35	0.943	0.952			
50	0.953	0.963			
51	0.954	0.964	81	0.970	0.975
53	0.957	0.964	83	0.971	0.976
55	0.958	0.965	85	0.972	0.977
57	0.961	0.966	87	0.972	0.977
59	0.962	0.967	89	0.972	0.977
61	0.963	0.968	91	0.973	0.978
63	0.964	0.970	93	0.973	0.979
65	0.965	0.971	95	0.974	0.979
67	0.966	0.971	97	0.975	0.979
69	0.966	0.972	99	0.976	0.980
71	0.967	0.972	131	0.980	0.983
73	0.968	0.973	200	0.987	0.989
75	0.969	0.973			
77	0.969	0.974			
79	0.970	0.975			

[a] ND, the number of observations (N-OBSERVATIONS in the UCODE_2005 documentation); NPR, the number of prior information values (NPRIOR in the UCODE_2005 documentation).
Source: Shapiro and Francia (1972) and Brockwell and Davis (1987, p. 304), with permission.

TABLE D.4 Runs Test Probability Table[a]

	Cumulative Lower-Tail Probabilities for the Distribution of the Total Number of Runs in Samples of Size (n1, n2)						Cumulative Upper-Tail Probabilities for the Distribution of the Total Number of Runs in Samples of Size (n1, n2)											
	Number of Runs						Number of Runs											
n1, n2	2	3	4	5	6	7	9	10	11	12	13	14	15	16	17	18	19	20
3,7	0.017	0.083																
3,8	0.012	0.067																
3,9	0.009	0.055																
3,10	0.007	0.045																
4,6	0.010	0.048					0.024											
4,7	0.006	0.033					0.046											
4,8	0.004	0.024					0.071											
4,9	0.003	0.018	0.085				0.098											
4,10	0.002	0.014	0.068															
5,5	0.008	0.040					0.040	0.008	0.002									
5,6	0.004	0.024		0.095			0.089	0.024	0.008									
5,7	0.003	0.015	0.076					0.045	0.016									
5,8	0.002	0.010	0.054					0.071	0.028									
5,9	0.001	0.007	0.039					0.098	0.042									
5,10	0.001	0.005	0.029															
6,6	0.002	0.013	0.067					0.067	0.013	0.002								
6,7	0.001	0.008	0.043						0.034	0.008	0.001							
6,8	0.001	0.005	0.028	0.086					0.063	0.016	0.002							
6,9	0.000	0.003	0.019	0.063					0.098	0.028	0.006							

(*Continued*)

TABLE D.4 *Continued*

	Cumulative Lower-Tail Probabilities for the Distribution of the Total Number of Runs in Samples of Size (n1, n2)									Cumulative Upper-Tail Probabilities for the Distribution of the Total Number of Runs in Samples of Size (n1, n2)									
	Number of Runs									Number of Runs									
n1, n2	2	3	4	5	6	7	9	10	11	12	13	14	15	16	17	18	19	20
6,10	0.000	0.002	0.013	0.047						0.042	0.010							
7,7	0.001	0.004	0.025	0.078					0.078	0.025	0.004	0.001						
7,8	0.000	0.002	0.015	0.051						0.051	0.012	0.002						
7,9	0.000	0.001	0.010	0.035						0.084	0.025	0.006	0.001					
7,10	0.000	0.001	0.006	0.024	0.080						0.043	0.010	0.002					
8,8	0.000	0.001	0.009	0.032	0.100					0.100	0.032	0.009	0.001	0.000				
8,9	0.000	0.001	0.005	0.020	0.069						0.061	0.020	0.004	0.001	0.000			
8,10	0.000	0.000	0.003	0.013	0.048						0.097	0.036	0.010	0.002	0.000			
9,9	0.000	0.000	0.003	0.012	0.044							0.044	0.012	0.003	0.000	0.000		
9,10	0.000	0.000	0.002	0.008	0.029	0.077						0.077	0.026	0.008	0.001	0.000	0.000	
10,10	0.000	0.000	0.001	0.004	0.019	0.051							0.051	0.019	0.004	0.001	0.000	0.000

[a] For this book, n1 is the number of negative values in a set, and n2 is the number of positive values. For example, if a set contains 3 negative values (n1 = 3) and 10 positive values (n2 = 10), and 2 runs, there is only a probability of 0.007, or 0.7 percent, that the values could be arranged randomly. Probabilities are given for values of n1 and n2 that satisfy $3 \leq n1 \leq n2 \leq 10$, $10 \leq n1 + n2 \leq 20$, and tail probability ≤ 0.10. For $n1 \geq n2$, interchange n1 and n2. For values larger than 10, use a anormal distribution as described in Section 6.4.3.

Source: Draper and Smith (1998, p. 196), with permission.

APPENDIX D: SELECTED STATISTICAL TABLES

TABLE D.5 Selected Web Sites for the Chi-square Distribution (accessed June 15, 2006)

http://www.itl.nist.gov/div898/handbook/eda/section3/eda3674.htm
http//www.danielsoper.com/statcalc/calc12.aspx
http://faculty.vassar.edu/lowry/tabs.html#csq
http://www.vias.org/simulations/simusoft_distcalc.html

TABLE D.6 Percentage Points of the Bonferroni t Statistic[a] $t_B(n, 1.0 - \alpha/2k)$

$\alpha = 0.05$

k\n	2	3	4	5	6	7	8	9	10	15	20	25	30	35	40	45	50
5	3.17	3.54	3.81	4.04	4.22	4.38	4.53	4.66	4.78	5.25	5.60	5.89	6.15	6.36	6.56	6.70	6.86
7	2.84	3.13	3.34	3.50	3.64	3.76	3.86	3.95	4.03	4.36	4.56	4.78	4.95	5.09	5.21	5.31	5.40
10	2.64	2.87	3.04	3.17	3.28	3.37	3.45	3.52	3.58	3.83	4.01	4.15	4.27	4.37	4.45	4.53	4.59
12	2.56	2.78	2.94	3.06	3.15	3.24	3.31	3.37	3.43	3.65	3.80	3.93	4.04	4.13	4.20	4.26	4.32
15	2.49	2.69	2.84	2.95	3.04	3.11	3.18	3.24	3.29	3.48	3.62	3.74	3.82	3.90	3.97	4.02	4.07
20	2.42	2.61	2.75	2.85	2.93	3.00	3.06	3.11	3.16	3.33	3.46	3.55	3.63	3.70	3.76	3.80	3.85
24	2.39	2.58	2.70	2.80	2.88	2.94	3.00	3.05	3.09	3.26	3.38	3.47	3.54	3.61	3.66	3.70	3.74
30	2.36	2.54	2.66	2.75	2.83	2.89	2.94	2.99	3.03	3.19	3.30	3.39	3.46	3.52	3.57	3.61	3.65
40	2.33	2.50	2.62	2.71	2.78	2.84	2.89	2.93	2.97	3.12	3.23	3.31	3.38	3.43	3.48	3.51	3.55
60	2.30	2.47	2.58	2.66	2.73	2.79	2.84	2.88	2.92	3.06	3.16	3.24	3.30	3.34	3.39	3.42	3.46
120	2.27	2.43	2.54	2.62	2.68	2.74	2.79	2.83	2.86	2.99	3.09	3.16	3.22	3.27	3.31	3.34	3.37
∞	2.24	2.39	2.50	2.58	2.64	2.69	2.74	2.77	2.81	2.94	3.02	3.09	3.15	3.19	3.23	3.26	3.29

TABLE D.6 Continued

$\alpha = 0.01$

k \ n	2	3	4	5	6	7	8	9	10	15	20	25	30	35	40	45	50
5	4.78	5.25	5.60	5.89	6.15	6.36	6.56	6.70	6.86	7.51	8.00	8.37	8.68	8.95	9.19	9.41	9.68
7	4.03	4.36	4.59	4.78	4.95	5.09	5.21	5.31	5.40	5.79	6.08	6.30	6.49	6.67	6.83	6.93	7.06
10	3.58	3.83	4.01	4.15	4.27	4.37	4.45	4.53	4.59	4.86	5.06	5.20	5.33	5.44	5.52	5.60	5.70
12	3.43	3.65	3.80	3.93	4.04	4.13	4.20	4.26	4.32	4.56	4.73	4.86	4.95	5.04	5.12	5.20	5.27
15	3.29	3.48	3.62	3.74	3.82	3.90	3.97	4.02	4.07	4.29	4.42	4.53	4.61	4.71	4.78	4.84	4.90
20	3.16	3.33	3.46	3.55	3.63	3.70	3.76	3.80	3.85	4.03	4.15	4.25	4.33	4.39	4.46	4.52	4.56
24	3.09	3.26	3.38	3.47	3.54	3.61	3.66	3.70	3.74	3.91	4.04	4.1[b]	4.2[b]	4.3[b]	4.3[b]	4.3[b]	4.4[b]
30	3.03	3.19	3.30	3.39	3.46	3.52	3.57	3.61	3.65	3.80	3.90	3.98	4.13	4.26	4.1[b]	4.2[b]	4.2[b]
40	2.97	3.12	3.23	3.31	3.38	3.43	3.48	3.51	3.55	3.70	3.79	3.88	3.93	3.97	4.01	4.1[b]	4.1[b]
60	2.92	3.06	3.16	3.24	3.30	3.34	3.39	3.42	3.46	3.59	3.69	3.76	3.81	3.84	3.89	3.93	3.97
120	2.86	2.99	3.09	3.16	3.22	3.27	3.31	3.34	3.37	3.50	3.58	3.64	3.69	3.73	3.77	3.80	3.83
∞	2.81	2.94	3.02	3.09	3.15	3.19	3.23	3.26	3.29	3.40	3.48	3.54	3.59	3.63	3.66	3.69	3.72

[a] n is the degrees of freedom, k is the number of simultaneous intervals, α is the significance level.
[b] Obtained by graphical interpolation.

Source: Miller (1981, p. 238), with permission.

TABLE D.7 Selected Web Sites for the F Statistic (accessed June 15, 2006)

http://www.itl.nist.gov/div898/handbook/eda/section3/eda3673.htm
http//davidmlane.com/hyperstat/F_table.html
http://fsweb.berry.edu/academic/education/vbissonnette/tables/f.pdf
http://calculators.stat.ucla.edu/cedf/f/fcalc.php

D.1 REFERENCES

Brockwell PJ, Davis RA (1987). *Time Series, Theory and Methods*. New York: Springer-Verlag.

Draper NR, Smith H (1998). *Applied Regression Analysis*, 3rd ed. Hoboken, NJ: Wiley.

Miller RG Jr (1981). *Simultaneous Statistical Inference*, 2nd ed. New York: Springer-Verlag.

Shapiro SS, Francia RS (1972). An approximate analysis of variance test for normality. *Journal of the American Statistical Association* **63**:1343–1372.

REFERENCES

Akaike H (1973). Information theory as an extension of the maximum likelihood principle. In Petrov BN, editor, *Proceedings of the Second International Symposium on Information Theory*. Akademiai Kiado, Budapest, pp. 267–281.

Akaike H (1974). A new look at statistical model identification. *Institute of Electrical and Electronics Engineers Transactions on Automatic Control* **AC-19**(6):716–723.

Akaike H (1978). Time series analysis and control through parametric models. In Findley DF, editor, *Applied Time Series Analysis*. New York: Academic Press, pp. 1–25.

Alley WM, Emery PA (1986). Groundwater model of the Blue River basin, Nebraska—twenty years later. *Journal of Hydrology* **85**:225–249.

Anderman ER, Hill MC (1998). Improving ground-water flow model calibration with the Advective-Transport Observation (ADV2) Package to MODFLOW2000. *U.S Geological Survey Fact Sheet FS-059–98*; reprinted 2001. Available at http://water.usgs.gov/nrp/gwsoftware/modflow2000/modflow2000.html.

Anderman ER, Hill MC (1999). A new multi stage ground-water transport inverse method: presentation, evaluation, and implications. *Water Resources Research* **35**(4):1053–1063.

Anderman ER, Hill MC (2001). MODFLOW-2000, the U.S. Geological Survey modular ground-water model—documentation of the ADVective-transport observation (ADV2) Package, version 2. *U.S Geological Survey Open-File Report 01–54*.

Anderman ER, Hill MC, Poeter EP (1996). Two-dimensional advective transport in ground-water flow parameter estimation. *Ground Water* **34**(6):1001–1009.

Andersen PF, Lu S (2003). A post audit of a model-designed ground water extraction system. *Ground Water* **41**(2):212–218.

Effective Groundwater Model Calibration: With Analysis of Data, Sensitivities, Predictions, and Uncertainty. By Mary C. Hill and Claire R. Tiedeman
Published 2007 by John Wiley & Sons, Inc.

Anderson MP (2005). Heat as a ground water tracer. *Ground Water* **43**(6):951–968.

Anderson MP, Woessner WW (1992). *Applied Groundwater Modeling*. San Diego, CA: Academic Press.

Aster RC, Borchers B, Thurber CH (2005). *Parameter Estimation and Inverse Problems*, Amsterdam: Elsevier Academic Press.

Backus GE (1988). Bayesian inference in geomagnetism. *Geophysical Journal* **92**:125–142.

Ballio F, Guadagnini A (2004). Convergence assessment of numerical Monte Carlo simulations in groundwater hydrology. *Water Resources Research* **40**, W04603. DOI: 10.1029/2003WR002876.

Banta ER, Poeter EP, Doherty J, Hill MC (2006). JUPITER: Joint Universal Parameter Identification and Evaluation of Reliability—an application programming interface (API) for model analysis. *U.S. Geological Survey Techniques and Methods TM6E-1*. http://water.usgs.gov/nrp/gwsoftware/jupiter/jupiter_api.html

Bard J (1974). *Nonlinear Parameter Estimation*. New York: Academic Press.

Barlebo HC, Hill MC, Rosbjerg D (1996). Identification of groundwater parameters at Columbus, Mississippi, using a three-dimensional inverse flow and transport model. In: van der Heidje P, Kovar K, editors, *Calibration and Reliability in Groundwater Modeling. Proceedings of the Symposium at the ModelCARE Conference*, Sept. 1996, Golden, CO. *International Association of Hydrologic Sciences Publ.* **237**:189–198.

Barlebo HC, Hill MC, Rosbjerg D, Jensen KH (1998). On concentration data and dimensionality in groundwater transport models. *Nordic Hydrology* **29**:149–178.

Barlebo HC, Hill MC, Rosbjerg D (2004). Investigating the Macrodispersion Experiment (MADE) site in Columbus, Mississippi, using a three-dimensional inverse flow and transport model. *Water Resources Research* **40**, W04211. DOI: 10.1029/2002WR001935.

Barrash W, Clemo T (2002). Hierarchical geostatistics and multi-facies systems: Boise Hydrogeophysical Research Site, Boise, Idaho. *Water Resources Research* **38**(10): 1196. DOI: 10.1029/2002WR001436.

Barth GR, Hill MC (2005a). Numerical methods for improving sensitivity analysis and parameter estimation of virus transport simulated using sorptive-reactive processes. *Journal of Contaminant Hydrology* **76**(3–4):251–277.

Barth GR, Hill MC (2005b). Parameter and observation importance in modeling virus transport in saturated porous media—investigations in a homogenous system. *Journal of Contaminant Hydrology* **80**(3–4):107–129.

Barth GR, Hill MC, Illangasekare TH, Rajaram H (2001). Predictive modeling of flow and transport in a two-dimensional intermediate-scale, heterogeneous porous media. *Water Resources Research* **37**(10):2503–2512.

Bates DM, Watts DG (1980). Relative curvature measures of nonlinearity. *Journal of the Royal Statistical Society, Series B* **43**:1–25.

Beale EML (1960). Confidence regions in nonlinear estimation. *Journal of the Royal Statistical Society, Series B* **2**(1):41–76.

Beck JV, Arnold KJ (1977). *Parameter Estimation in Engineering and Science*. Hoboken, NJ: Wiley.

Beckie R, Harvey CF (2002). What does a slug test measure: an investigation of instrument response and the effects of heterogeneity. *Water Resources Research* **38**(12), 1290. DOI: 10.1029/2001WR001072.

Bedford T, Cook RM (2001). *Probabilistic Risk Analysis, Foundations and Methods*. Cambridge, UK: Cambridge University Press.

REFERENCES

Belcher W, editor (2004). Evaluation of the Death Valley regional ground-water flow system (DVRFS), Nevada and California. *U.S. Geological Survey Scientific Investigations Report 2004–5205*.

Belsley DA, Kuh E, Welsch RE (1980). *Regression Diagnostics, Identifying Influential Data and Source of Collinearity*. Hoboken, NJ: Wiley.

Benjamin JR, Cornell CA (1970). *Probability, Statistics, and Decision for Civil Engineers*. New York: McGraw Hill.

Bentley LR (1997). Influence of the regularization matrix on parameter estimates. *Advances in Water Resources* **20**(4):231–247.

Beven K, Binley A (1992). The future of distributed models, model calibration and uncertainty prediction. *Hydrological Processes* **6**(3):279–298.

Binley A, Beven K (2003). Vadose zone flow model uncertainty as conditioned on geophysical data. *Ground Water* **41**(2):119–127.

Bolstad WM (2004). *Introduction to Bayesian Statistics*. Hoboken, NJ: Wiley.

Box GEP, Jenkins GM, Reinsel G (1994). *Time Series Analysis, Forecasting and Control*, 3rd ed. San Francisco: Holden-Day.

Bras RL, Tucker GE, Teles V. (2003) Six myths about mathematical modeling in geomorphology. In Wilcock PR, Iverson RM, editors, *Predictions in Geomorphology*. Washington, DC: American Geophysical Union, Geophysical Monograph 135, pp. 63–79.

Bravo HR, Jiang F, Hunt RJ (2002). Using groundwater temperature data to constrain parameter estimation in a groundwater flow model of a wetland system. *Water Resources Research* **38**(8), 1153. DOI: 10.1029/2000WR000172.

Bredehoeft JD (2003). From models to performance assessment—the conceptualization problem. *Ground Water* **41**(5):571–577.

Bredehoeft JD (2005). Modeling: the conceptual model problem—surprise. *Hydrogeology Journal* **13**(1):37–46.

Bredehoeft JD, Konikow LF (1993). Ground-water models: validate or invalidate. *Ground Water* **31**(2):178–179.

Brockwell PJ, Davis RA (1987). Time Series, *Theory and Methods*. New York: Springer-Verlag.

Brooks R, Lerner DN, Tobias A (1994). Determining the range of predictions of a groundwater model which arises from alternative calibrations. *Water Resources Research* **30**(11):2993–3000.

Burnham KP, Anderson DR (2002). *Model Selection and Multimodel Inference: A Practical Information-Theoretic Approach*. New York: Springer-Verlag.

Burnham KP, Anderson DR (2004). Multimodel inference, understanding AIC and BIC in model selection. *Sociological Methods and Research* **33**(2):261–304.

Burow KR, Constantz J, Fujii R (2005). Heat as a tracer to estimate dissolved organic carbon flux from a restored wetland. *Ground Water* **43**(4):545–556.

Carle SF, LaBolle EM, Weissman GS, van Brocklin D, Fogg GE (1998). Conditional simulation of hydrofacies architecture: a transition probability/Markov chain approach. In Fraser GS, Davis JM, editors, *Hydrogeologic Models of Sedimentary Aquifers, Concepts in Hydrogeology No. 1*. SEPM (Society for Sedimentary Geology) Special Publication, pp. 147–170.

Carrera J, Neuman SP (1986). Estimation of aquifer parameters under transient and steady-state conditions. *Water Resources Research* **22**(2):199–242.

Carrera J, Alcolea A, Medina A, Hildago J, Slooten LJ (2005). Inverse problem in hydrogeology. *Hydrogeology Journal* **13**(1):206–222.

Carter RW, Anderson IE (1963). Accuracy of current meter measurements. *American Society of Civil Engineers Journal* **89**(HV4):105–115.

Chen J, Rubin Y (2003). An effective Bayesian model for lithofacies estimation using geophysical data. *Water Resources Research* **39**(5):1118. DOI: 10.1029/2002WR001666.

Chernick MR (1999). *Bootstrap Methods, A Practitioner's Guide*. Wiley Series in Probability and Statistics. Hoboken, NJ: Wiley.

Christensen S (1997). On the strategy of estimating regional-scale transmissivity fields. *Ground Water* **35**(1):131–139.

Christensen S, Cooley RL (1999). Simultaneous confidence intervals for a steady-state leaky aquifer groundwater flow model. *Advances in Water Resources Special Section on Model Calibration and Reliability Evaluation for Groundwater Systems* **22**(8):807–817.

Christensen S, Cooley RL (2005). User guide to the UNC process and three utility programs for computation of nonlinear confidence and prediction intervals using MODFLOW-2000 *U.S. Geological Survey Techniques and Methods Report 2004–1349*.

Christensen S, Rasmussen KR, Moeller K (1998). Prediction of regional ground-water flow to streams. *Ground Water* **36**(2):351–360.

Clifton PM, Neuman SP (1982). Effects of kriging and inverse modeling on conditional simulation of the Avra Valley aquifer in southern Arizona. *Water Resources Research* **18**(4):1215–1234.

Constable SC, Parker RL, Constable CG (1987). Occam's inversion, a practical algorithm for generating smooth models from electromagnetic sounding data. *Geophysics* **52**:289–300.

Cook RD (1977a). Detection of influential observations in linear regression. *Technometrics* **19**:15–18.

Cook RD (1977b). Letter to the editor. *Technometrics* **19**:348.

Cook RD, Weisberg S (1982). *Residuals and Influence in Regression*. New York: Chapman and Hall.

Cook RD, Weisberg S (1999). *Applied Regression Including Computing and Graphics*. Hoboken, NJ: Wiley.

Cooley RL (1977). A method of estimating parameters and assessing reliability for models of steady-state groundwater flow, 1. Theory and numerical properties. *Water Resources Research* **13**(2):318–324.

Cooley RL (1979). A method of estimating parameters and assessing reliability for models of steady state groundwater flow. 2. Application of statistical analysis. *Water Resources Research* **15**(3):603–617.

Cooley RL (1983a). Incorporation of prior information on parameters into nonlinear regression groundwater flow models. 2. Applications. *Water Resources Research* **19**(3):662–676.

Cooley RL (1983b). Some new procedures for numerical solution of variably saturated flow problems. *Water Resources Research* **19**(5):1271–1285.

Cooley RL (1985). A comparison of several methods of solving nonlinear regression groundwater flow problems. *Water Resources Research* **21**(10):1525–1538.

Cooley RL (1993). Regression modeling of ground-water flow, Supplement 1—Modifications to the computer code for nonlinear regression solution of steady-state ground-water flow problems. *U.S Geological Survey Techniques of Water Resources Investigations*, Book 3, Chapter B4, Supplement 1.

Cooley RL (1997). Confidence intervals for ground-water models using linearization, likelihood, and bootstrap methods. *Ground Water* **35**(5):869–880.

Cooley RL (2004). A theory for modeling ground-water flow in heterogeneous media. *U.S. Geological Survey Professional Paper 1679*.

Cooley RL, Hill MC (1992). A comparison of three Newton-like nonlinear least-squares methods for estimating parameters of ground-water flow models. In Russell TF, Ewing RE, Brebbia CA, Gray WG, Pinder GF, editors, *Computational Methods in Water Resources*, 9th ed., Volume 1: Numerical Methods in Water Resources. New York: Elsevier, pp. 379–386.

Cooley RL, Naff RL (1990). Regression modeling of ground-water flow. *U.S. Geological Survey Techniques in Water-Resources Investigations*, Book 3, Chapter B4.

Cooley RL, Sinclair PJ (1976). Uniqueness of a model of steady-state ground-water flow. *Journal of Hydrology* **31**:245–269.

Cooley RL, Konikow LF, Naff RL (1986). Nonlinear regression groundwater flow modeling of a deep regional aquifer system. *Water Resources Research* **22**(13):1759–1778.

Copty N, Rubin Y, Mavko G (1993). Geophysical–hydrological identification of field permeabilities through Bayesian updating. *Water Resources Research* **29**(8):2813–2825.

Craven P, Whaba G (1979). Smoothing noisy data with Spline functions: estimating the correct degree of smoothing by the method of generalized cross-validation. *Numerische Mathematik* **31**:377–403.

D'Agnese FA, Faunt CC, Turner AK (1996). Using remote sensing and GIS techniques to estimate discharge and recharge fluxes for the Death Valley regional ground-water flow system, Nevada and California, USA. In Kovar K, Nachtnebel HP, editors, *Proceedings of HydroGIS '96: Application of Geographic Information Systems in Hydrology and Water Resources Management*, April 1996, Vienna, Austria. International Association of Hydrologic Sciences Publ. **235**:503–511.

D'Agnese FA, Faunt CC, Turner AK, Hill MC (1997). Hydrogeologic evaluation and numerical simulation of the Death Valley regional ground-water flow system, Nevada and California. *U.S. Geological Survey Water-Resources Investigation Report 96–4300*.

D'Agnese FA, Faunt CC, Hill MC, Turner AK (1999). Death Valley regional ground-water flow model calibration using optimal parameter estimation methods and geoscientific information systems. Invited paper, *Advances in Water Resources Special Section on Model Calibration and Reliability Evaluation for Ground-Water Systems* **22**(8):777–790.

D'Agnese FA, O'Brien GM, Faunt CC, Belcher WR, San Juan C (2002). A three-dimensional numerical model of predevelopment conditions in the Death Valley regional ground-water flow system, Nevada and California. *U.S. Geological Survey Water-Resources Investigation Report 02–4102*.

Dam D, Christensen S (2003). Including geophysical data in ground water model calibration. *Ground Water* **41**(2):178–189.

Davis JC (2002). *Statistics and Data Analysis in Geology*, 3rd ed., Hoboken, NJ: Wiley.

Day-Lewis F D, Lane JW Jr, Gorelick SM (2004). Combined interpretation of radar, hydraulic and tracer data from a fractured-rock aquifer. *Hydrogeology Journal* **14**(1–2):1–14.

Deb, K (2001). *Multi-objective Optimization Using Evolutionary Algorithms*. Hoboken, NJ: Wiley.

De Marsily G, Delay F, Goncalves J, Renard PH, Teles V, Violette S (2005). Dealing with spatial heterogeneity. *Hydrogeology Journal* **13**(1):161–183.

Dennis JE, Schnabel RB (1996). *Numerical Methods for Unconstrained Optimization and Nonlinear Equations*. Philadelphia: Society for Industrial and Applied Mathematics.

Doherty J (1994). *PEST*. Corinda, Australia: Watermark Computing.

Doherty J (2003). Ground water model calibration using pilot points and regularization. *Ground Water* **41**(2):170–177.

Doherty J (2005). *PEST Version 9.01*. Corinda, Australia: Watermark Computing. Available at http://www.sspa.com/PEST/index.html.

Donaldson JR, Schnabel RB (1987). Computational experience with confidence regions and confidence intervals for nonlinear least squares. *Technometrics* **29**(1):67–87.

Draper NR, Smith H (1981). *Applied Regression Analysis*, 2nd ed. Hoboken, NJ: Wiley.

Draper NR, Smith H (1998). *Applied Regression Analysis*, 3rd ed. Hoboken, NJ: Wiley.

Drécourt JP, Madsen H (2002). Uncertainty estimation in groundwater modelling using Kalman filtering. In Kovar K, Hrkal, Z, editors, *Calibration and Reliability in Groundwater Modeling. Proceedings of the ModelCARE Conference*, Sept. 2002, Prague, the Czech Republic, pp. 306–309. Available at http://projects.dhi.dk/daihm/Files/ModelCare2002_UncertaintyReduction.pdf.

Dynamic Graphics (2006). Earth Vision web site. Available at http://www.dgi.com/earthvision/. Accessed June 7, 2006.

Earth Decision Sciences (2006). GOCAD web site. Available at http://www.earthdecision.com/. Accessed June 7, 2006.

Eberts SM, George LL (2000). Regional ground-water flow and geochemistry in the midwestern basins and arches aquifer system in parts of Indiana, Ohio, Michigan, and Illinois. *U.S. Geological Survey Professional Paper 1323-C*.

Efron B (1982). *The Jackknife, the Bootstrap, and Other Resampling Plans*. Philadelphia: Society for Industrial and Applied Mathematics.

Efron B, Tibshirani R (1993). *An Introduction to the Bootstrap*. Boca Raton, FL: CRC Press.

Ehrgott M (2000). *Multicriteria Optimization*. New York: Springer-Verlag.

Environmental Modeling Systems, Inc. (2004). GMS web site. Available at http://www.ems-i.com/. Accessed June 7, 2006.

Eppstein MJ, Dougherty DE (1996). Simultaneous estimation of transmissivity values and zonation. *Water Resources Research* **32**(11):3321–3336.

Essaid HI, Cozzarelli IM, Eganhouse RP, Herkelrath WN, Bekins BA, Delin GN (2003). Inverse modeling of BTEX dissolution and biodegradation at the Bemidji, MN crude-oil spill site. *Journal of Contaminant Hydrology* **67**(1–4):269–299.

Evers S, Lerner DN (1998). How uncertain is our estimate of a wellhead protection zone? *Ground Water* **36**(1):49–57.

Faunt CC, D'Agnese FA, Turner AK (1993). Characterizing the three-dimensional hydrogeologic framework model for the Death Valley region, southern Nevada and California, USA. In Kovar K, Nachtnebel HP, editors, *Proceedings of HydroGIS '93: Application of Geographic Information Systems in Hydrology and Water Resources*, Vienna, Austria. *International Association of Hydrologic Sciences Publ.* **211**:227–234.

Faunt CC, Blainey JB, Hill MC, D'Agnese FA, O'Brien GA (2004). Transient numerical model of ground-water flow. In Belcher W, editor, *Evaluation of the Death Valley*

Regional Ground-water Flow System (DVRFS), Nevada and California. U.S Geological Survey Scientific Investigations Report 2004–5205.

Fazal MA, Imaizumi M, Ishida S, Kawachi T, Tsuchihara T (2005). Estimating groundwater recharge using the SMAR conceptual model calibrated by genetic algorithm. *Journal of Hydrology* **303**(1–4):56–78.

Feehley CE, Zheng C, Molz FJ (2000). A dual-domain mass transfer approach for modeling solute transport in heterogeneous porous media, application to the MADE site. *Water Resources Research* **36**(9):2501–2515.

Feyen L, Beven KJ, De Smedt F, Freer J (2001). Stochastic capture zone delineation within the generalized likelihood uncertainty estimation methodology, conditioning on head observations. *Water Resources Research* **37**(3):625–638.

Feyen L, Gómez-Hernández JJ, Ribeiro PJ, Beven KJ, De Smedt FA (2003). Bayesian approach to stochastic capture zone delineation incorporating tracer arrival times, conductivity measurements, and hydraulic head observation. *Water Resources Research* **39**(5), 1126. DOI: 10.1029/2002WR001544.

Foglia L, Mehl SW, Hill MC, Burlando P (in press). Evaluation of alternative groundwater models using information criteria, linear statistics, cross-validation, and predictions. *Ground Water*.

Forsythe GE, Strauss EG (1955). On best conditioned matrices. *Proceedings of the American Mathematical Society* **10**(3):340–345.

Franke OL, Reilly TE, Bennett GD (1987). Definition of boundary and initial conditions in the analysis of saturated ground-water flow systems—an introduction. *U.S. Geological Survey Techniques of Water-Resources Investigations*, Book 3, Chapter B5.

Frind EO, Muhammad DS, Molson JW (2002). Delineation of three-dimensional well capture zones for complex multi-aquifer systems. *Ground Water* **40**(6):586–598.

Fuller WA (1987). *Measurement Error Models*. New York: John Wiley and Sons.

Gaganis P, Karapanagioti HK, Burganos VN (2002). Modeling multicomponent NAPL transport in the unsaturated zone with the constituent averaging technique. *Advances in Water Resources* **25**(7):723–732.

Gailey RM, Gorelick SM, Crowe AS (1991). Coupled process parameter estimation and prediction uncertainty using hydraulic head and concentration data. *Advances in Water Resources* **14**(5):301–314.

Gannett MW, Lite KE (2004). Simulation of regional ground-water flow in the Upper Deschutes Basin, Oregon. *U.S. Geological Survey Water-Resources Investigations Report 03–4195*. Available at http://water.usgs.gov/pubs/wri/wri034195/.

Gelhar LW (1993). *Stochastic Subsurface Hydrology*. Englewood Cliffs, NJ: Prentice Hall.

Ghandi RK, Hopkins GD, Goltz MN, Gorelick SM, McCarty PL (2002a). Full-scale demonstration of in situ cometabolic biodegradation of trichloroethylene in groundwater, 1. Dynamics of a recirculating well system. *Water Resources Research* **38**(4), 1039. DOI: 10.1029/2001WR000379.

Ghandi RK, Hopkins GD, Goltz MN, Gorelick SM, McCarty PL (2002b). Full-scale demonstration of in situ cometabolic biodegradation of trichloroethylene in groundwater. 2. Comprehensive analysis of field data using reactive transport modeling. *Water Resources Research* **38**(4), 1040. DOI: 10.1029/2001WR000380.

Glasgow HS, Fortney MD, Lee J, Graettinger AJ, Reeves HW (2003). MODFLOW-2000 head uncertainty, a first-order second-moment method. *Ground Water* **41**(3):342–350.

Gomez-Hernandez JJ (2006). Complexity. *Ground Water.* DOI: 10.1111/j.1745-6584.2006.00222.x/.

Gwo J-P, Toran LE, Morris MD, Wilson GV (1996). Subsurface stormflow modeling with sensitivity analysis using a Latin-hypercube sampling technique. *Ground Water* **34**(5): 811–818.

Halford KJ, Hanson RT (2002). User guide for the drawdown-limited, multinode well (MNW) package for the U.S. Geological Survey's modular three-dimensional finite-difference groundwater flow model, versions MODFLOW-96 and MODFLOW-2000. *U.S. Geological Survey Open-File Report 02–293.*

Hanson RT (1996). Postaudit of head and transmissivity estimates and ground-water flow models of Avra Valley, Arizona. *U.S. Geological Survey Water-Resources Investigation Report 96–4045.*

Harbaugh AW (2002). A data input program (MFI2K) for the U.S. Geological Survey modular ground-water model (MODFLOW-2000). *U.S. Geological Survey Open-File Report 02–41.*

Harbaugh AW, Banta ER, Hill MC, McDonald MG (2000). MODFLOW-2000, The U.S. Geological Survey modular ground-water model—User guide to modularization concepts and the ground-water flow process. *U.S. Geological Survey Open-File Report 00–92.*

Harry DL (2003). Numerical modeling strategies revisited. *EOS Transactions, American Geophysical Union* **84**:100.

Harvey JW, Wagner BJ, Bencala KE (1996). Evaluating the reliability of the stream tracer approach to characterize stream–subsurface water exchange. *Water Resources Research* **32**(8):2441–2451.

Hazen A (1914). Storage to be provided in the impounding reservoirs for municipal water supply. *Transactions of the American Society of Civil Engineers* **77**:1547–1550.

Heidari M, Ranjithan SR (1998). A hybrid optimization approach to the estimation of distributed parameters in two-dimensional confined aquifers. *Journal of the American Water Resources Association* **34**(4):909–920.

Helsel DR, Hirsch RM (2002). Statistical methods in water resources. *U.S. Geological Survey Techniques in Water Resources*, Book 4, Chapter A3. Available at http://pubs.water.usgs.gov/twri4a3.

Hill MC (1989). An analysis of accuracy of approximate, simultaneous, nonlinear confidence intervals on hydraulic heads in analytical and numerical test cases. *Water Resources Research* **25**(2):177–190.

Hill MC (1990). Relative efficiency of four parameter-estimation methods in steady-state and transient ground-water flow models. In Gambolati G, Rinaldo A, Brebbia CA, Gray WG, Pinder GF, editors, *Proceedings of the 8th Computational Methods in Subsurface Hydrology, International Conference on Computational Methods in Water Resources*, June 1990 Venice, Italy, pp. 103–108.

Hill MC (1992). A computer program (MODFLOWP) for estimating parameters of a transient, three-dimensional, ground-water flow model using nonlinear regression. *U.S. Geological Survey Open-File Report 91–484.*

Hill MC (1994). Five computer programs for testing weighted residuals and calculating linear confidence and prediction intervals on results from the ground-water parameter estimation computer program MODFLOWP. *U.S. Geological Survey Open-File Report 93–481.*

Hill MC (1998). Methods and guidelines for effective model calibration. *U.S. Geological Survey Water-Resources Investigations Report 98–4005*. Available at http://pubs.water.usgs.gov/wri984005/.

Hill MC (2006). The practical use of simplicity in developing ground-water models. *Ground Water*. DOI: 10.1111/j.1745-6584.2006.00227.x/.

Hill MC, Østerby O (2003). Determining extreme parameter correlation in ground water models. *Ground Water* **41**(4):420–430.

Hill MC, Cooley RL, Pollock DW (1998). A controlled experiment in ground-water flow model calibration using nonlinear regression. *Ground Water* **36**(3):520–535.

Hill MC, Banta ER, Harbaugh AW, Anderman ER (2000). MODFLOW-2000, The U.S. Geological Survey modular ground-water model—User guide to the observation, sensitivity, and parameter-estimation processes. *U.S. Geological Survey Open-File Report 00–184*.

Hill MC, Ely DM, Tiedeman CR, D'Agnese FA, Faunt CC, O'Brien GA (2001). Preliminary evaluation of the importance of existing hydraulic-head observation locations to advective-transport predictions, Death Valley regional flow system, California and Nevada. *U.S. Geological Survey Water-Resources Investigations Report 00–4282*. Available at http://water.usgs.gov/pubs/wri/wri004282/.

Hill MC, Middlemis H, Hulme P, Poeter EP, Riegger J, Neuman SP, Williams H, Anderson M (2004). Brief overview of selected groundwater modelling guidelines, in Kovar K, Hrkal Z., eds, Proceedings of the conference Finite-Element Models, MODFLOW, and More-Solving Groundwater Problems. Carlsbad, Czech Republic, September 13–16, 2004, pp. 105–120.

Hoffmann J, Galloway DL, Zebker HA (2003). Inverse modeling of interbed storage parameters using land subsidence observations, Antelope Valley, California. *Water Resources Research* **39**(2), 1031. DOI: 10.1029/2001WRR001252.

Holtschlag DJ, Luukkonen CL, Nicholas JR (1996). Simulation of ground-water flow in the Saginaw aquifer, Clinton, Eaton, and Ingham Counties, Michigan. *U.S. Geological Survey Water-Supply Paper 2480*.

Høybye JA (1998). Model error propagation and data collection design: an application in water quality modelling. *Water, Air, and Soil Pollution* **103**(1–4):101–119.

Hsieh PA, Winston, RB (2002). User's guide to ModelViewer, a program for three-dimensional visualization of ground-water model results. *U.S. Geological Survey Open-File Report 02–106*. Available at http://water.usgs.gov/nrp/gwsoftware/modelviewer/ModelViewer.html.

Huber PJ (1981). *Robust Statistics*. Hoboken, NJ: Wiley.

Hunt RJ, Lin Y, Krohelski JT, Juckem PF (2000). Simulation of the shallow hydrologic system in the vicinity of the middle Genesee Lake, Wisconsin, using analytic elements and parameter estimation. *U.S. Geological Survey Water-Resources Investigations Report 00–4136*.

Hunt RJ, Feinstein DT, Pint CD, Anderson MP (2006). The importance of diverse data types to calibrate a watershed model of the Trout Lake Basin, northern Wisconsin, USA. *Journal of Hydrology* **321**(1–4):286–296.

Hunt RJ, Steuer JJ (2000). Simulation of the recharge area for Frederick Springs, Dane County, Wisconsin. *U.S. Geological Survey Water-Resources Investigations Report 00–4172*.

Hvilshøj S, Jensen KH (2000). Single-well dipole tests: parameter estimation and field testing. *Ground Water* **38**(1):53–62.

Hyndman DW, Gorelick SM (1996). Estimating lithologic and transport properties in three dimensions using seismic and tracer data, The Kesterton aquifer. *Water Resources Research* **32**(9):2659–2670.

Hyndman DW, Harris JM, Gorelick SM (1994). Coupled seismic and tracer test inversion for aquifer property characterization. *Water Resources Research* **30**(7):1965–1977.

Hyndman DW, Harris JM, Gorelick SM (2000). Inferring the relation between seismic slowness and hydraulic conductivity in heterogeneous aquifers. *Water Resources Research* **36**(8):2121–2132.

Jacobson EA (1985). A statistical parameter estimation method using singular value decomposition with application to Avra Valley aquifer in southern Arizona [unpublished PhD dissertation]. Phoenix: University of Arizona, Department of Hydrology and Water Resources.

Jacques D, Simunek J, Timmerman A, Feyen J (2002). Calibration of Richards' and convection–dispersion equations to field-scale water flow and solute transport under rainfall conditions. *Journal of Hydrology* **259**(1–4):15–31.

Julian HE, Boggs JM, Zheng C, Feehley CE (2001). Numerical simulation of a natural gradient tracer experiment for the Natural Attenuation Study: flow and physical transport. *Ground Water* **39**(4):534–545.

Kashyap RL (1982). Optimal choice of AR and MA parts in autoregressive moving average models. *IEEE Transactions on Pattern Analysis and Machine Intelligence (PAMI)* **4**(2): 99–104.

Kass EK, Raftery AE (1995). Bayes factors. *Journal of the American Statistical Association* **90**(430):773–795.

Kean JW, Smith JD (2005). Generation and verification of theoretical rating curves in the Whitewater River basin, Kansas. *Journal of Geophysical Research* **110**, F04012. DOI: 10.1029/2004JF000250.

Keating EH, Vesselinov VV, Kwicklis E, Zhiming L (2003). Coupling basin- and site-scale inverse models of the Espanola aquifer. *Ground Water* **41**(2):200–211.

Keidser A, Rosbjerg D (1991). A comparison of four inverse approaches to groundwater flow and transport parameter identification. *Water Resources Research* **27**(9): 2219–2232.

Kelson VA, Hunt RJ, Haitjema HM (2002). Improving a regional model using reduced complexity and parameter estimation. *Ground Water* **40**(2):132–143.

Kitanidis PK (1995). Quasi-linear geostatistical theory for inversing. *Water Resources Research* **31**(10):2411–2419.

Kitanidis PK (1997). *Introduction to Geostatistics: Applications in Hydrogeology.* Cambridge, UK: Cambridge University Press.

Knopman DS, Voss CI (1988). Further comments on sensitivities, parameter estimation and sampling design. *Water Resources Research* **24**(2):225–238.

Knopman DS, Voss CI (1989). Multiobjective sampling design for parameter estimation and model discrimination in studies of solute transport. *Water Resources Research* **25**(10):2245–2258.

Koltermann, CE, Gorelick, SM (1996). Heterogeneity in sedimentary deposits: a review of structure-imitating, process-imitating, and descriptive approaches. *Water Resources Research* **32**(9), 2617–2658.

Konikow LF (1986). Predictive accuracy of a ground-water model: lessons from a post-audit. *Ground Water* **24**(2):173–184.

Konikow LF, Bredehoeft JD (1992). Ground-water models cannot be validated. *Advances in Water Resources* **15**(1):75–83.

Konikow LF, Person M (1985). Assessment of long-term salinity changes in an irrigated stream-aquifer system. *Water Resources Research* **21**(11):1611–1624.

Kuiper LK (1994). Nonlinear-regression flow model of the Gulf Coast aquifer systems in the south-central United States. *U.S. Geological Survey Water-Resources Investigations Report 93–4020*.

Landmark (2006). Stratworks 3D web page. Accessed June 7, 2006 at http://www.lgc.com/.

LeBlanc DR, Celia MA (1991). Density-induced downward movement of solutes during a natural-gradient tracer test, Cape Cod, Massachusetts, U.S. Geological Survey Toxic Substances Hydrology Program, Monterey, California. *U.S. Geological Survey Water Resources Investigation Report 91–4034*, pp. 10–14.

Lebbe LC (1999). *Hydraulic Parameter Identification*. Berlin: Springer-Verlag.

Levy J, Ludy EE (2000). Uncertainty quantification for delineation of wellhead protection areas using the Gauss–Hermite quadrature approach. *Ground Water* **38**(1):63–75.

Levy J, Clayton MK, Chesters G (1998). Using an approximation of the three-point Gauss–Hermite quadrature formula for model prediction and quantification of uncertainty. *Hydrogeology Journal* **6**(4):457–468.

Linssen HN (1975). Nonlinearity measures—a case study. *Statistica Neerlandica* **29**:93–99.

Loaiciga HA, Marino MA (1986). Estimation and inference in the inverse problem. In *Proceeding of Water Forum '86, World Water Issues in Evolution*, Aug. 4–6, 1986, Long Beach, California. American Society of Civil Engineers, pp. 973–980.

Looney SW, Gulledge TR (1985a). Use of the correlation coefficient with normal probability plots. *The American Statistician* **39**:75–79.

Looney SW, Gulledge TR (1985b). Probability plotting positions and goodness of fit for the normal distribution. *The Statistician* **34**:297–303.

Mahar PS, Datta B (2001). Optimal identification of ground-water pollution sources and parameter estimation. *Journal of Water Resources Planning and Management* **127**(1):20–29.

Manning CE, Ingebritsen SE (1999). Permeability of the continental crust: the implications of geothermal data and metamorphic systems. *Reviews of Geophysics* **37**(1):127–150.

Margulis SA, McLaughlin D, Entekhabi D, Dunne S (2002). Land data assimilation and estimation of soil moisture using measurements from the Southern Great Plains 1997 field experiment. *Water Resources Research* **38**(12): 1299 DOI. 10.1029/2001WR001114.

Marquardt DW (1963). An algorithm for least-squares estimation of nonlinear parameters. *Journal for the Society of Industrial and Applied Mathematics* **11**(2):431–441.

McAda DP (1999). Simulation of a long-term aquifer test conducted near the Rio Grande, Albuquerque, New Mexico. *U.S. Geological Survey Water-Resources Investigations Report 99–4260*.

McDonald MG, Harbaugh AW (1988). A modular three-dimensional finite-difference ground-water flow model. *U.S. Geological Survey Techniques of Water Resources Investigations*, Book 6, Chapter A1. Available at http://water.usgs.gov/pubs/twri/twri6a1.

McKenna SA, Poeter EP (1995). Field example of data fusion in site characterization. *Water Resources Research* **31**(12):3229–3240. Available at http://www.mines.edu/~epoeter/pubs/1995/wrr-field.

McKinney DC, Loucks DP (1992). Network design for predicting groundwater contamination. *Water Resources Research* **28**(1):133–147.

McLaughlin D (2002). An integrated approach to hydrologic data assimilation: interpolation, smoothing, and filtering. *Advances in Water Resources* **25**(8–12):1275–1286.

McLaughlin D, Wood EF (1988). A distributed parameter approach for evaluating the accuracy of groundwater model predictions: 2. Application to groundwater flow. *Water Resources Research* **24**(7):1048–1060.

McLaughlin D, Townley LR (1996). A reassessment of the groundwater inverse problem. *Water Resources Research* **32**(5):1131–1161.

Medina A, Carrera J (1996). Coupled estimation of flow and transport parameters. *Water Resources Research* **32**(10):3063–3076.

Mehl SW, Hill MC (2001). A comparison of solute-transport solution techniques and their effect on sensitivity analysis and inverse modeling results. *Ground Water* **39**(2): 300–307.

Mehl SW, Hill MC (2002). Evaluation of a local grid refinement method for steady-state block-centered finite-difference groundwater models. In Hassanizadeh SM, Schotting RJ, Gray WG, Pinder GF, editors, *Proceedings of the XIVth International Conference on Computer Methods in Water Resources Conference. Developments in Water Science*, Vol. 47. Delft, The Netherlands: Elsevier, pp. 367–374.

Mehl SW, Hill MC (2003). Locally refined block-centered finite-difference groundwater models. Evaluation of parameter sensitivity and the consequences for inverse modelling and predictions. In Kovar K, Hrkal Z, editors, *Calibration and Reliability in Groundwater Modeling: A Few Steps Closer to Reality. International Association of Hydrologic Sciences Publ.* **277**:227–232.

Mehl SW, Hill MC (2005). MODFLOW-2005, the U.S. Geological Survey modular groundwater model—Documentation of shared node local grid refinement (LGR) and the boundary flow and head (BFH) package. *U.S. Geological Survey Techniques and Methods Report 6-A12*. Available at http://water.usgs.gov/nrp/gwsoftware/modflow2005_lgr/mflgr.html.

Melching CS, Yen BC, Wenzel HG Jr. (1990). A reliability estimation in modeling watershed runoff with uncertainties. *Water Resources Research* **26**(10):2275–2286.

Menke W (1989). *Geophysical Data Analysis: Discrete Inverse Theory*, 2nd ed. San Diego, California: Academic Press.

Merry AG, Martin PJ, Meyer P, Harvey DJM (2003). Assessing the calibration and predictive sensitivity of model parameters. In Kovar K, Hrkal Z, editors, *Proceedings of the ModelCARE2002, Prague, Czech Republic, Calibration and Reliability in Groundwater Modeling: A Few Steps Closer to Reality. International Association of Hydrologic Sciences Publ.* **277**:233–238.

Meyer PD, Ye M, Neuman SP, Cantrell KJ (2004) Combined estimation of hydrogeologic conceptual model and parameter uncertainty. U.S. Nuclear Regulatory Commission NUREG/CR-6843, Pacific Northwest Laboratory PNNL-14534.

Miller RG Jr (1981). *Simultaneous Statistical Inference*, 2nd ed. New York: Springer-Verlag.

Minsker B, editor (2003). *Long-Term Ground-Water Monitoring, The State of the Art*. American Society of Civil Engineers, stock number 40678.

Molson JW, Frind EO (2002). *WATFLOW/WTC User Guide and Documentation, Version 3*. Waterloo, Ontario, Canada: Department of Earth Sciences, University of Waterloo.

Moore C, Doherty J (2005). Role of the calibration process in reducing model predictive error. *Water Resources Research* **41**, DOI: 10.1029/2004WR003501.

Moore C, Doherty J (2006). The cost of uniqueness in groundwater model calibration. *Advances in Water Resources* **29**(4):605–623.

Morse BS, Pohll G, Huntington J, Rodriguez Castillo R (2003). Stochastic capture zone analysis of an arsenic-contaminated well using the generalized likelihood uncertainty estimator (GLUE) methodology. *Water Resources Research* **39**(6):1151. DOI: 10.1029/2002WR001470.

Murray AB (2002). Seeking explanation affects numerical modeling strategies. *EOS Transactions; American Geophysical Union* **83**:418–419.

Murray AB (2003). Contrasting the goals, strategies, and predictions associated with simplified numerical models and detailed simulation. In Iverson RM, Wilcock P, editors, *Predictions in Geomorphology. American Geophysical Union Geophysical Monograph* **135**:151–165.

National Research Council (2002). *Report of a Workshop of Predictability and Limits-to-Prediction in Hydrologic Systems*. Washington, DC: National Academic Press.

Neuman SP (1973). Calibration of distributed parameter groundwater flow models viewed as a multiple-objective decision process under uncertainty. *Water Resources Research* **9**(4):1006–1021.

Neuman SP (1982). Statistical characterization of aquifer heterogeneities—an overview. In Narasimhan TN, editor, *Recent Trends in Hydrology. Geological Society of America Special Paper* **189**:81–102.

Neuman SP (2003). Accounting for conceptual model uncertainty via maximum likelihood model averaging. In Kovar K, Hrkal Z, editors, *Proceedings of the 4th International Conference on Calibration and Reliability in Groundwater Modelling (ModelCARE 2002)*. Charles University, Prague, Czech Republic. IAHS Publication **277**:303–313.

Neuman SP, Jacobson EA (1984). Analysis of nonintrinsic spatial variability by residual kriging with application to regional groundwater levels. *Mathematical Geology* **16**(5): 499–521.

Nishikawa T, Yeh WW-G (1989). Optimal pumping test design for the parameter identification of groundwater systems. *Water Resources Research* **25**(7):1737–1747.

Niswonger RG, Prudic DE, Pohll G, Constantz J (2005). Incorporating seepage losses into the unsteady streamflow equations for simulating intermittent flow along mountain front streams. *Water Resources Research* **41**, W06006, DOI: 10.1029/2004WR003677.

Nunoo C, Mrawira D (2004). Shuffled complex evolution algorithms in infrastructure works programming. *Journal of Computing in Civil Engineering* **18**(3):257–266.

Olsthoorn TN (1995). Effective parameter optimization for ground-water model calibration. *Ground Water* **33**(1):42–48.

Oreskes N. (2000). Why believe a computer? Models, measures, and meaning in the natural system. In Jill Schneiderman, editor, *The Earth Around Us*. New York: W.H. Freedman, pp. 70–82.

Ott L (1993). *An Introduction to Statistical Methods and Data Analysis*, 4th ed. Belmont, CA: Duxbury Press.

Paola C, Mullin J, Ellis C, Mohrig DC, Swenson JB, Parker G, Hickson T, Heller PL, Pratson L, Syvitski J, Sheets B, Strong N (2001). Experimental stratigraphy. *GSA Today*, **11**:4–9.

Parker JC, Islam M (2000). Inverse modeling to estimate LNAPL plume release timing. *Journal of Contaminant Hydrology* **45**(3–4):303–327.

Parker RL (1994). *Geophysical Inverse Theory*. Princeton, NJ: Princeton University Press.

Parkhurst DL (1995). User's guide to PHREEQC, a computer program for speciation, reaction-path, advective-transport, and inverse geochemical calculations. *U.S. Geological Survey Water-Resources Investigation Report 95–4227*.

Pavelko MT (2004). Estimates of hydraulic properties from a one-dimensional numerical model of vertical aquifer-system deformation, Lorenzi Site, Las Vegas, Nevada. *U.S. Geological Survey Water-Resources Investigations Report 03–4083*.

Poeter EP, Anderson DR (2004). Multi-model ranking and inference in groundwater modeling. In Kovar K, Hrkal Z, editors, *Proceedings of the Conference on Finite-Element Models, MODFLOW, and More-Solving Groundwater Problems*, Sept. 13–16, 2004, Carlsbad, Czech Republic, pp. 85–89.

Poeter EP, Anderson D (2005). Multi-model ranking and inference in groundwater modeling. *Ground Water* **43**(4):597–605.

Poeter, EP, Gaylord, DR (1990). Influence of aquifer heterogeneity on contamination transport at the Hanford Site. *Ground Water* **28**(6):900–909.

Poeter EP, Hill MC (1996). Unrealistic parameter estimates in inverse modeling: a problem or a benefit for model calibration? *Proceedings of the ModelCARE Conference*, Sept. 1996, Golden, CO. *International Association of Hydrological Sciences Publication* **237**: 277–285.

Poeter EP, Hill MC (1997). Inverse models: a necessary next step in groundwater modeling. *Ground Water* **35**(2):250–260.

Poeter EP, McKenna SA (1995). Reducing uncertainty associated with groundwater flow and transport predictions. *Ground Water* **33**(6):899–904. Available at http://www.mines.edu/~epoeter/pubs/1995/gw-reducing.

Poeter EP, Hill MC, Banta ER, Mehl SW (2005). UCODE_2005 and three post-processors—computer codes for universal sensitivity analysis, inverse modeling, and uncertainty evaluation. *U.S. Geological Survey Techniques and Methods Report TM 6-A11*.

Poeter EP, Hill MC (in press). MMA—computer code for MultiModel Analysis. *U.S. Geological Survey Techniques and Methods TM6E-3*.

Pollock DW (1994). User's guide for MODPATH/MODPATH-PLOT, Version 3: a particle tracking post-processing package for MODFLOW, the U.S. Geological Survey finite-difference ground-water flow model. *U.S. Geological Survey Open-File Report 94–464*.

Popper KR (1982). *The Open Universe*. London: Hutchinson.

Pint CD, Hunt RJ, Anderson MP (2003). Flow path delineation and ground water age, Allequash Basin, Wisconsin. *Ground Water* **41**(7):895–902.

Prommer H, Barry DA, Zheng C (2003). MODFLOW/MT3DMS-based multicomponent transport modeling. *Ground Water* **41**(2):247–257.

RamaRao BS, LaVenue MA, Marsily G, Marietta MG (1995). Pilot point methodology for automated calibration of an ensemble of conditionally simulated transmissivity fields. 1; Theory and computational experiments. *Water Resources Research* **31**(3):475–493.

Raper JF (1989). The 3-dimensional geoscientific mapping and modeling system: A conceptual design. In Raper JF, editor, *Three-dimensional Applications in Geographic Information Systems*. London: Taylor and Francis, pp. 11–19.

Rawlings JO (1988). *Applied Regression Analysis*. Pacific Grove, CA: Wadsworth and Brooks.

Reed P, Minsker BS, Goldberg DE (2003). Simplifying multiobjective optimization, an automated design methodology for the nondominated sorted genetic algorithm—II. *Water Resources Research* **39**(7):1196. DOI: 10.1029/2002WR001483.

REFERENCES

Reeves HW, Lee J, Dowding CH, Graettinger AJ (2000). Reliability-based evaluation of groundwater remediation strategies. In Stauffer F, Kinzelbach W, Kovar K, Hoehn E, editors, *Calibration and Reliability in Groundwater Modelling—Coping with Uncertainty, Proceedings of the ModelCARE'99 Conference*, Sept. 1999, Zürich. *International Association of Hydrological Sciences Publ.* **265**:304–309.

Refsgaard JC, Henrikson HJ (2004). Modelling guidelines—terminology and guiding principles. *Advances in Water Resource* **27**:71–82.

Reichard EG, Meadows JK (1992). Evaluation of a ground-water flow and transport model of the upper Coachella Valley, California. *U.S. Geological Survey Water-Resources Investigation Report 91–4142*.

Reid LB (1996). *A Functional Inverse Approach for Three-Dimensional Characterization of Subsurface Contamination* [dissertation]. Cambridge (MA): Massachusetts Institute of Technology, Department of Civil Engineering. Available from: University Microfilms, Ann Arbor, MI.

Reilly TE, Franke OL, Bennett GD (1987). The principle of superposition and its application in ground-water hydraulics. *U.S Geological Survey Techniques of Water Resources Investigations*, Book 3, Chapter B6.

Rosenberry DO (1990). Effect of sensor error on interpretation of long-term water-level data. *Ground Water* **28**(6):927–936.

Ross M, Parent M, Lefebvre R (2005). 3D geologic framework models for regional hydrogeology and land-use management: a case study from a Quaternary basin of southwestern Quebec, Canada. *Hydrogeology Journal* **13**(5–6):690–707.

Ross WH (1987). The geometry of case deletion and assessment of influence in nonlinear regression. *The Canadian Journal of Statistics* **15**(2):91–103.

Rubin Y, Mavko G, Harris J (1992). Mapping permeability in heterogeneous aquifers using hydrologic and seismic data. *Water Resources Research* **28**(7):1809–1816.

Rubin Y, Bellin A, Lawrence AE (2003). On the use of block-effective macrodispersion for numerical simulations of transport in heterogeneous formations. *Water Resources Research* **39**(9):1242. DOI:10.1029/2002WR001727.

Saltelli A, Chan K, Scott EM (2000). *Sensitivity Analysis*. Hoboken, NJ: Wiley.

Saltelli A, Tarantola S, Campolongo F, Ratto M (2004). *Sensitivity Analysis in Practice*. Hoboken, NJ: Wiley.

Saiers JE, Genereux DP, Bolster CH (2004). Influence of calibration methodology on ground water flow predictions. *Ground Water* **42**(1):32–44.

San Juan CA, Belcher WR, Laczniak RJ, Putnam HM (2004). Hydrologic components for model development. In Belcher W, editor, *Evaluation of the Death Valley Regional Ground-water Flow System (DVRFS), Nevada and California. U.S. Geological Survey Scientific Investigations Report 2004–5205*.

Sanford WE, Plummer LN, McAda DP, Bexfield LM, Anderholm SK (2004a). Use of environmental tracers to estimate parameters for a predevelopment ground-water flow model of the Middle Rio Grande Basin, New Mexico. *U.S. Geological Survey Water-Resources Investigations Report 03–4286*.

Sanford WE, Plummer LN, McAda DP, Bexfield LM, Anderholm SK (2004b). Hydrochemical tracers in the Middle Rio Grande Basin, USA: 2. Calibration of a ground-water flow model. *Hydrogeology Journal* **12**(4):389–407.

Schäfer D, Dahmke A, Kolditz, O, Teutsch G (2003). "Virtual aquifers": a concept for the evaluation of exploration, remediation and monitoring strategies. In Kovar K, Hrkal Z, editors, *Calibration and Reliability in Groundwater Modeling. Proceedings of the ModelCARE Conference*, Sept. 2002 Prague, Czech Republic. *International Association of Hydrologic Sciences Publication* **277**:52–59.

Scheibe TD, Chien Yi-Ju (2003). An evaluation of conditioning data for solute transport prediction. *Ground Water* **41**(2):128–141.

Scott DT, Goosef MN, Bencala KE, Runkel RL (2003). Automated calibration of a stream solute transport model. Implications for interpretation of biogeochemical parameters. *Journal of the North American Benthological Society* **22**(4):492–510.

Seber GAF, Wild CJ (1989). *Nonlinear Regression*. Hoboken, NJ: Wiley.

Shapiro SS, Francia RS (1972). An approximate analysis of variance test for normality. *Journal of the American Statistical Association* **63**:1343–1372.

Shoemaker WB (2004). Important observations and parameters for a salt water intrusion model. *Ground Water* **42**(6):829–840.

Skinner DC (1999). *Introduction to Decision Analysis*, 2nd ed. Gainesville, FL: Probabilistic Publishing.

Smith T, Hoversten M, Gasperkiva E, Morrison F (1999). Sharp boundary inversion of 2-D magnetotelluric data. *Geophysical Prospecting* **47**:469–486; text only. Available at http://appliedgeophysics.berkeley.edu:7057/geoengineering/appliedgeophysics/research/sharpinverse.html.

Solomatine DP, Dibike YB, Kukuric N (1999). Automatic calibration of groundwater models using global optimization techniques. *Hydrological Sciences Journal* **44**(6): 879–894.

Sonnenborg TO, Engesgaard P, Rosbjerg D (1996). Contaminant transport at a waste residue deposit: 1. Inverse flow and nonreactive transport modeling. *Water Resources Research* **32**(4):925–938.

Starn JJ, Stone JR, Mullaney JR (2000). Delineation and analysis of uncertainty of contributing areas to wells at the Southbury Training School, Southbury, Connecticut. *U.S. Geological Survey Water-Resources Investigations Report 00–4158*.

Statnikov RB, Matusov JB (1995). *Multicriteria Optimization and Engineering*. New York: Chapman and Hall.

Stauffer F, Hendricks Franssen HJ, Kinzelbach W (2004). Semi-analytical uncertainty estimation of well catchments: conditioning by head and transmissivity data. *Water Resources Research* **40**, W08305, DOI: 10.1029/2004WR003320.

Stewart M, Langevin C (1999). Post audit of a numerical prediction of wellfield drawdown in a semiconfined aquifer system. *Ground Water* **37**(2):245–252.

Sugiura N (1978). Further analysis of the data by Akaike's information criterion and the finite corrections. *Communications in Statistics, Theory and Methods* **A7**:13–26.

Sulieman H, McLellan PJ, Bacon DW (2001). A profile-based approach to parametric sensitivity analysis of nonlinear regression models. *Technometrics* **43**(4):425–433.

Sun N-Z (1994). *Inverse Problems in Ground-water Modeling*. Boston: Kluwer Academic Publishers.

Sun N-Z, Yeh WW-G (1985). Identification of parameter structure in groundwater inverse problem. *Water Resources Research* **21**(6):869–883.

Sun N-Z, Yeh WW-G (1990a). Coupled inverse problems in groundwater modeling. 1: Sensitivity analysis and parameter identification. *Water Resources Research* **26**(10): 2507–2525.

Sun N-Z, Yeh WW-G (1990b). Coupled inverse problems in groundwater modeling. 2: Identifiability and experimental design. *Water Resources Research* **26**(10):2527–2540.

Sun N-Z, Yeh WW-G (1992). A stochastic inverse solution for transient groundwater flow: parameter identification and reliability analysis. *Water Resources Research* **28**(12): 3269–3280.

Sykes JF, Wilson JL, Andrews RW (1985). Sensitivity analysis for steady state groundwater flow using adjoint operators. *Water Resources Research* **21**(3):359–371.

Tarantola A (2005). *Inverse Problem Theory*. Philadelphia: Society for Industrial and Applied Mathematics.

Theil H (1963). On the use of incomplete prior information in regression analysis. *American Statistical Association Journal* **58**(302):401–414.

Tiedeman CR, Gorelick SM (1993). Analysis of uncertainty in optimal groundwater contaminant capture design. *Water Resources Research* **29**(7):2139–2153.

Tiedeman CR, Goode DJ, Hsieh PA (1997). Numerical simulation of ground-water flow through glacial deposits and crystalline rock in the Mirror Lake area, Grafton County, New Hampshire. *U.S. Geological Survey Professional Paper 1572*.

Tiedeman CR, Goode DJ, Hsieh PA (1998a). Characterizing a ground-water basin in a New England mountain-and-valley terrain. *Ground Water* **36**(4):611–620.

Tiedeman CR, Kernodle JM, McAda DP (1998b). Application of nonlinear-regression methods to a ground-water flow model of the Albuquerque Basin, New Mexico. *U.S. Geological Survey Water-Resources Investigations Report 98-4172*.

Tiedeman CR, Hill MC, D'Agnese FA, Faunt CC (2003). Methods for using groundwater model predictions to guide hydrogeologic data collection, with application to the Death Valley regional ground-water flow system. *Water Resources Research* **39**(1), 1010. DOI: 10.1029/2001WR001255.

Tiedeman CR, Ely DM, Hill MC, O'Brien GM (2004). A method for evaluating the importance of system state observations to model predictions, with application to the Death Valley regional groundwater flow system. *Water Resources Research* **40**, W12411. DOI: 10.1029/2004WR003313.

Tikhonov AN, Arsenin VY (1977). *Solution of Ill-Posed Problems*. New York: Winston and Sons.

Tonkin MJ, Doherty J (2005). A hybrid regularized inversion methodology for highly parameterized environmental models. *Water Resources Research*, **41**, Wl0412. DOI: 10.1029/2005WR003995.

Tonkin MJ, Tiedeman CR, Ely DM, Hill MC (in press). Documentation of OPR-PPR, a computer program for assessing data importance to model predictions using linear statistics. *U.S. Geological Survey Techniques and Methods TM6-E2*.

Townley LR, Wilson JL (1985). Computationally efficient algorithms for parameter estimation and uncertainty propagation in numerical models of groundwater flow. *Water Resources Research* **21**(12):1851–1860.

Tsai FT-C, Sun N-Z, Yeh WW-G (2003a). A combinatorial optimization scheme for parameter structure identification in ground water modeling. *Ground Water* **41**(2): 156–169.

Tsai FT-C, Sun N-Z, Yeh WW-G (2003b). Global–local optimization for parameter structure identification in three-dimensional groundwater modeling. *Water Resources Research* **39**(2):1043. DOI: 10.1029/2001WR001135.

Tung CP, Chou CA (2002). Application of tabu search to groundwater parameter zonation. *Journal of the American Water Resources Association* **38**(4):1115–1126.

U.S. Geological Survey (1980). Accuracy specifications for topographic mapping. In *Technical Instructions of the National Mapping Division*. Reston, VA: USGS, Chapter 1B4, pp. 1–13.

Valstar JR, Minnema B (2003). Using a Bayesian estimation algorithm for estimating parameter uncertainty and optimization of monitoring networks. In Kovar K, Hrkal Z, editors, *Proceedings of the ModelCARE2002*, Prague, the Czech Republic, *Calibration and Reliability in Groundwater Modeling: A Few Steps Closer to Reality*. International Association of Hydrologic Sciences Publication **277**:380–385.

Valstar JR, McLaughlin D, Stroet CBM, van Geer FC (2004). A representer-based inverse method for groundwater flow and transport applications. *Water Resources Research* **40**(5), W05116. DOI: 10.1029/2002WR002922.

Van Leeuwen M, Butler AP, Stroet CBM, Tompkins JA (2000). Stochastic determination of well capture zones conditioned on regular grids of transmissivity measurements. *Water Resources Research* **36**(4):949–957.

van Loon EE, Troch PA (2002). Tikhonov regularization as a tool for assimilating soil moisture data in distributed hydrological models. *Hydrological Processes* **16**(2):531–556.

Vasco DW, Data-Gupta A, Long JCS (1997). Resolution and uncertainty in hydrologic characterization. *Water Resources Research* **33**(3):379–397.

Vecchia AV, Cooley RL (1987). Simultaneous confidence and prediction intervals for nonlinear regression models with application to a groundwater flow model. *Water Resources Research* **22**(2):95–108.

Vose D (2000). *Risk Analysis, A Quantitative Guide*, 2nd ed. West Essex, UK: Wiley.

Vrugt JA, Gupta HV, Bastidas LA, Bouten W, Sorooshian S (2003). Effective and efficient algorithm for multiobjective optimization of hydrologic models. *Water Resources Research* **39**(8):1214. DOI: 10.1029/2002WR001746.

Wagner BJ, Gorelick SM (1987). Optimal groundwater quality management under parameter uncertainty. *Water Resources Research* **23**(7):1162–1174.

Wagner BJ (1992). Simultaneous parameter estimation and contaminant source characterization for coupled groundwater flow and contaminant transport modeling. *Journal of Hydrology* **135**(1–4):275–303.

Wagner BJ (1995). Sampling design methods for groundwater modeling under uncertainty. *Water Resources Research* **31**(10):2581–2591.

Wagner BJ (1999). Evaluating data worth for ground-water management under uncertainty. *Journal of Water Resources Planning and Management* **125**(5):281–288.

Weiss R, Smith L (1998). Efficient and responsible use of prior information in inverse methods. *Ground Water* **36**(2):151–163.

Weissman GS, Carle SF, Fogg GE (1999). Three dimensional hydrofacies modeling based on soil surveys and transition probability geostatistics. *Water Resources Research* **35**(6):1761–1770.

Wilcock PR, Iverson RM (2003). *Predictions in Geomorphology*. Washington, D.C.: American Geophysical Union, Geophysical Monograph 135.

Winston RB (2000). Graphical User Interface for MODFLOW, Version 4. *U.S. Geological Survey Open-File Report 00–315*. Available at http://water.usgs.gov/nrp/gwsoftware/GW_Chart/GW_Chart.html.

Woodbury AD, Smith L (1988). Simultaneous inversion of hydrogeologic and thermal data 2. Incorporation of thermal data. *Water Resources Research* **24**(3):356–372.

Woodbury AD, Ulrych TJ (2000). A full-Bayesian approach to the groundwater inverse problem for steady state flow. *Water Resources Research* **36**(8):2081–2093.

Xiang Y, Sykes JF, Thomson NR (1993). A composite L_1 parameter estimator for model fitting in groundwater flow and solute transport simulation. *Water Resources Research* **29**(6):1661–1673.

Yager RM (1993). Estimation of hydraulic conductivity in a riverbed and aquifer system near the Susquehanna River, Broome County, New York. *U.S. Geological Survey Water-Supply Paper 2387*.

Yager RM (1996). Simulated three-dimensional ground-water flow in the Lockport Group, a fractured dolomite aquifer near Niagra Falls, New York. *U.S. Geological Survey Water-Supply Paper 2487*. Available at http://pubs.er.usgs.gov/pubs/wsp/wsp2487.

Yager RM (1998). Detecting influential observations in nonlinear regression modeling of ground-water flow. *Water Resources Research* **34**(7):1623–1633.

Yager RM (2002). Simulated transport and biodegradation of chlorinated ethenes in a fractured dolomite aquifer near Niagara Falls, New York. *U.S. Geological Survey Water-Resources Investigations Report 00-4275*.

Yager RM (2004). Effects of model sensitivity and nonlinearity on nonlinear regression of ground water flow. *Ground Water* **42**(3):390–400.

Yager RM, Miller TS, Kappel WM (2001). Simulated effects of 1994 salt-mine collapse on ground-water flow and land subsidence in a glacial aquifer system, Livingston County, New York. *U.S. Geological Survey Professional Paper 1611*.

Yeh T-C, Jin M, Hanna S (1996). An iterative stochastic inverse method, conditional effective transmissivity and hydraulic head fields. *Water Resources Research* **32**(1):85–92.

Yeh T-CJ, Gutjahr AL, Jim M (1995). An iterative cokriging-like technique for ground-water flow modeling. *Ground Water* **33**(1):33–41.

Yeh WW-G (1986). Review of parameter identification procedures in ground-water hydrology—the inverse problem. *Water Resources Research* **22**(2):95–108.

Yeh WW-G, Yoon YS (1981). Aquifer parameter identification with optimum dimension in parameterization. *Water Resources Research* **17**(3):664–672.

Yeh WW-G, Sun N-Z (1990). Variational sensitivity analysis, data requirements, and parameter identification in a leaky aquifer system. *Water Resources Research* **26**(9):1927–1938.

Zhang H, Schwartz FW, Wood WW, Garabedian SP, LeBlanc DR (1998). Simulation of variable-density flow and transport of reactive and nonreactive solutes during a tracer test at Cape Cod, Massachusetts. *Water Resources Research* **34**(1):67–82.

Zhang Y, Gable CW, Person M (2006). Equivalent hydraulic conductivity of an experimental stratigraphy: Implications for basin-scale flow simulations, *Water Resources Research* **42**, W05404. DOI: 10.1029/2005WR004720.

Zhang Y, Pinder GF (2003). Latin hypercube lattice sample selection strategy for correlated random hydraulic conductivity fields. *Water Resources Research* **39**(8):1226. DOI: 10.1029/2002WR001822.

Zheng C (1994). Analysis of particle tracking errors associated with spatial discretization. *Ground Water* **32**(5):821–828.

Zheng C (2005). *MT3DMS v5 Supplemental User's Guide*, Technical Report to the U.S. Army Engineer Research and Development Center. Department of Geological Sciences, University of Alabama.

Zheng C, Bennett GD (2002). *Applied Contaminant Transport Modeling*, 2nd ed. Hoboken, NJ: Wiley.

Zheng C, Gorelick SM (2003). Analysis of solute transport in flow fields influenced by preferential flowpaths at the decimeter scale. *Ground Water* **41**(2):142–155.

Zheng C, Wang PP (1996). Parameter structure identification using tabu search and simulated annealing. *Advances in Water Resources* **19**(4):215–224.

Zheng C, Wang PP (1999). MT3DMS, *A modular Three-Dimensional Multi-Species Transport Model for Simulation of Advection, Dispersion and Chemical Reactions of Contaminants in Groundwater Systems; Documentation and User's Guide*. U.S. Army Engineer Research and Development Center Contract Report SERDP-99-1, Vicksburg, MS. Available at http://hydro.geo.ua.edu/mt3d/.

Zheng C, Wang PP (2003). *MGO: A Modular Groundwater Optimizer Incorporating MODFLOW/MT3DMS; Documentation and User's Guide*. University of Alabama and Groundwater Systems Research Ltd. Available at http://www.frtr.gov/estcp/source_codes.htm#optimizationmgo.

INDEX

Page references followed by t indicate material in tables.

Accuracy, defined, 14
Accurate models, attributes of, 310
Accurate simulated results, requirements for, 30–32
Additive errors, 31, 32
ADIFOR program, 19
Adjoint-states, 47, 77
Advective transport
 analysis of, 193–195
 calculation of, 195, 366
 simulating with particle tracking methods, 220
Advective-Transport Observation Package (ADV) of MODFLOW-2000, 24, 194, 195, 257, 366. *See also* MODFLOW-2000 program
 calculation of particle paths by, 195, 366
 particle projection by, 209–211, 220, 257
Advective-transport predictions, 162, 163, 164, 302, 365–366
 opr statistics for, 368–370
 ppr statistics for, 366–368

Advective-transport predictions, in Exercises, 195–196, 211, 254–255
 linear confidence intervals on, 207–208, 257–259
 nonlinear confidence intervals on, 209–212, 257–259
 opr statistics for, 202–204, 205–207
 ppr statistics for, 199–202
Age observations, 287
AIC_c statistic, 98–99
 in calculating alternative model weighting, 189
 in Exercises, 115
 use of, 265t, 310, 311
AIC statistic, 98–99
 in Exercises, 115
Alternative models, 140, 189, 241t. *See also* Guideline 8
 application of, 313–314
 considering, 308–314
 developing, 309–310
 simulating predictions with, 312

Effective Groundwater Model Calibration: With Analysis of Data, Sensitivities, Predictions, and Uncertainty. By Mary C. Hill and Claire R. Tiedeman
Published 2007 by John Wiley & Sons, Inc.

428 INDEX

Ambiguous measurement errors, 300
ANIV (vertical anisotropy) parameters, 279, 280
Approximate critical values for linear simultaneous intervals, 177
Approximate likelihood-function approach, 187
Approximate linear model, 392–393
Approximate nonlinear model, 393
Aquifer test, long-term transient, 229
Aquifer tests, use of results from, 33, 94, 274, 277
Arid environments, groundwater systems in, 217, 287
Atmospheric systems, initial conditions for, 213
Average weighted residual, 101, 115, 117

Backward differences, 47
Base 10 logarithms, 79
Bayesian, 10, 34, 174, 305
BEALE-2000 program, 143
  output file for, 156, 253
BEALE2-2K program, 190
Beale's measure, 142. *See also* Modified Beale's measure
Bedrock hydraulic conductivity, 313–314
Best fit, parameter values that produce, 6, 7, 77, 137, 315, 343–344. *See also* Optimal parameters
Bias, 13–14
  in observations or prior information, 30–31, 215, 286, 304. *See also* Model bias
BIC statistic, 99
  in Exercises, 115
  use of, 265t, 310, 311
Biodegradation models, calibrating, 226–227
Bonferroni simultaneous intervals, 177, 208. *See also* Simultaneous confidence intervals
  critical values for, 176t, 177
Bonferroni t statistic, percentage points of, 404–405t
Bootstrap methods, 140, 170,
Boundary conditions, parameterization of, 57, 216, 280
Breakthrough curves (BTC), 221, 222, 224

Calculated error variance, 95. *See also* Standard error of regression; True error variance
  in Exercises, 113, 114
  expected value of, 96
  interpreting, 96–98, 303–304
Calibrated models. *See also* Model calibration
  determining whether predicted values are contradicted by, 337–338
  testing using omitted data and postaudits, 339
  use in identifying system properties important to predictions, 160
  use in identifying observations important to predictions, 170
  versus predictive models, 8
Cape Cod groundwater flow model, 280–281
Capture zones, Monte Carlo evaluations of, 342
Carbon-14 measurements, 287
Central differences, 47
Central-limit theorem, 380
Chi-square (χ^2) distribution, 96–97, 114
  selected Web Sites for, 403t
χ^2 test statistic, 97
Clustered observations, 45, 285
Coefficient of variation, fitted, 95–96
Coefficient of variation, in calculating weights, 32, 34, 294, 297, 303
  examples of, 40, 90, 361
  issues for graphs of weighted residuals, 101–104, 310, 362
  for concentration observations, 224–225
  for streamflow gain or loss, 297–298
  using to interpret confidence interval on calculated error variance, 97
Coefficient of variation, of parameter estimates, 127, 129t
  in Exercises, 152t
Combined intrinsic model nonlinearity measures, 190
  for confidence intervals, 191–192
  for correction factors, 192–193
Common error variance, 29, 398

INDEX   **429**

Composite scaled sensitivities (*css*), 48,
    50–51
  effects of nonlinearity on, 281–283
  in model development, 278–280, 328
  for evaluating information observations
      provide about parameters, 60–62,
      89, 240, 251, 263t, 265t
  for identifying observations important to
      predictions, 170–171, 204–205,
      266t, 335
  for identifying observations to
      improve simulated processes,
      331–333
  for identifying parameters important to
      predictions, 162–163, 196–199,
      256–257, 266t
  for identifying the need for improved
      parameter estimates, 165–166
  plotting, 60
  reevaluating, 124–125, 145–146
  _sc output file, 197
Comprehensive regression problem, 278
Computer execution time. *See* Execution
    time
Concentration observations, 42, 223, 226,
    370–373. *See also* Transport
    observations
  alternatives to using point
      concentrations, 225
  weighting, 32, 224–225
Concentration predictions, 338, 342–344
Concentrations, defining for contaminant
    sources, 218–219
Concentrations, simulating, 221. *See also*
    Transport Conceptual models,
    alternative, 309, 313–314, 352t
  development of, 261t, 272, 354–357
  importance of, 264
  and predictions, 313–314, 334, 341
Confidence intervals
  assumptions underlying, 138, 180, 340
  on calculated error variance, 96–98, 114
  calculating, 138, 139, 176, 178, 180, 341
  combined intrinsic model nonlinearity
      measure for, 191–192
  comparing to reasonable parameter
      values, 140–141, 153, 251–252,
      324–325, 373
  definition, 138, 175

  for determining weights, 200, 295–296,
      305, 358–360
  on differences, 182–184
  individual, 138–139, 175, 176–178.
      See also Individual confidence
      intervals
  linear, 138, 176–177, 341. *See also*
      Linear confidence intervals
  on the native equivalent of log-
      transformed parameters, 130
  nonlinear, 139, 177–181, 341. *See also*
      Nonlinear confidence intervals
  on parameters, 138–139, 151–153, 251,
      313, 324–325, 373
  on predictions, 176–181, 208–212,
      257–259, 341
  relation of Monte Carlo analysis to,
      187–188
  relation to hypothesis testing, 328
  simultaneous, 175, 176–178, 180, 341.
      See also Simultaneous confidence
      intervals
  using the Theis example to understand,
      181–182
  using to replace traditional sensitivity
      analysis, 184–185
Confidence regions. *See* Parameter
    confidence regions
Constant-Head Boundary (CHD)
    parameters, 57
Constant-over-time observation
    errors, 216
Constrained minimization method, 219
Contaminant transport. *See also* Transport
    predictions involving, 193, 365. *See also*
      Advective-transport predictions
    simulating, 218
Contoured sensitivity maps, 55
Contour maps
  of objective function, 35–37, 82,
      75, 181
  of one-percent scaled sensitivities, 55,
      64, 237–239, 241–242. *See also*
      One-percent scaled sensitivity
      contour maps
  of *opr* statistics, 208
Convergence of nonlinear regression
  encouraging, 306–308
  quantities used to test for, 386t

Convergence criteria, 76–77
Convergence problems. *See also*
  Guideline 7
  diagnosing, 88–89, 306–308
  improving by log-transforming
    parameters, 79
  reasons for, 51, 215, 221,
    306–308, 322
Cook's *D* influence statistic, 134–136
  critical values for, 136
  in Exercises, 146, 147t
  use of, 228t, 263t, 265t, 327, 364–365
Correction factors that account for
  unrepresented heterogeneity,
  192–193
Correlated-over-time observation
  errors, 216
Correlation coefficient matrix, 53
Correlation coefficient, parameter. *See*
  Parameter correlation coefficients
Correlation coefficient *R* between weighted
  observations and weighted simulated
  values, 105–106
  in Exercises, 115–117
Correlation coefficient R_N^2, 109, 110–111
  critical values of, 111, 400t
  in Exercises, 119–120, 249–250
Correlation coefficients
  considering all, 155
  for sample data, 127–128, 129t
Correlations
  between observation errors, 28, 35, 216,
    284, 298, 376–381
  between weighted residuals, expected,
    109, 111–112, 119, 123
Coupled estimation methodology,
  223–224
Covariances, 127. *See also* Parameter
  variance–covariance matrix;
  Variance–covariance matrix
  of parameters, 126–127
  of sample data, 127–128, 129t
  for weight matrices, 35, 298
Critical values
  Bonferroni simultaneous, 176–177
  of Cook's *D*, 136, 146, 364
  of *css*, 51
  for calculating confidence
    intervals, 176, 178

  for calculating weights, 295
  of *DFBETAS*, 136, 146
  of intrinsic model nonlinearity, 145
  of modified Beale's measure, 144
  of R_N^2, 111, 120, 250, 400t
  of runs statistic, 107–108, 118, 249
  Scheffe simultaneous, 176–177, 178
  of total model nonlinearity, 144–145
css. *See* Composite scaled sensitivities
CTB statistic, 50
Cumulative probabilities, calculating,
  109–110

Damping, in the Gauss–Newton method,
  72–73, 83
Damping parameter, 75, 79, 88
  calculating, 73–74, 385–389
Darcy's Law, 12, 36, 82, 149, 393
Data. *See also* Geophysical data; Hard
  data; Observation data; Potential
  new data (Guidelines 11 and 12);
  Soft data
  effect of transient processes on, 215
  improving use of, 1, 260, 264
  including many kinds of in regressions,
    284–288
  omitting from model calibration, 338
  scarcity of, 6, 9, 78, 225
Data assessment strategies, 1
Data assimilation, 273, 284, 353–357
Data collection
  to improve simulated processes,
    330–334
  methods to guide, 159–174
  relation to modeling objectives, 329
  using predictions to guide, 159–174,
    200, 207, 334–336, 365–370
Data collection strategies, 159–160, 173,
  267, 330
Data clustering. *See* Clustered observations
Data error, analyzing, 34
Data fusion, 273
Data interpretation, diagnosing error
  in, 140
Data management, in model development,
  274–277, 353–357
Data needs assessment, utility of, 6–7
Data noise, 51
Data variability, managing, 272

INDEX   **431**

Death Valley regional groundwater
  flow system (DVRFS) model, 226,
     347–370
  composite scaled sensitivities in,
     278–280
  evaluating system information in,
     353–357
  flowchart of data used to develop,
     354, 355
  hydrogeologic framework conceptual
     and digital models for, 354–356
  including many kinds of observations in,
     357–358
  model fit to observations in, 362–364
  parameter estimates for, 324–325
  identifying new data to improve
     predictions in, 365–370
  reasonable parameter ranges for,
     324–325
  weights that reflect errors in, 358–362
Defined parameters
  deciding which to estimate, 45–46
  designing, 45, 278
  including all, 131, 134, 164, 180–181,
     289
Dependent variable, 6, 376
  errors, uncorrelated, 377–378
Designed inconsistencies, 286
Deterministic methods
  evaluating prediction uncertainty using,
     337–339
  for alternative model development,
     186, 309
Dewatering, effect on transmissivity, 272
DFBETAS influence statistic, 50, 134, 136
  use of, 146–147, 228t, 263t, 265t, 306t
  in identifying model error, 326–327
Diagnostic statistics, 7
Diagonal matrix, 380
Diagonal weight matrix, 26–28, 31,
     34, 294
  assumptions required for, 376–381
  correlation coefficient R and, 106
  dimensionless scaled sensitivities
     and, 48
  using differencing to achieve, 216
  weighted residuals and, 35
Differences, calculating weights on,
     297–298

Differences, forward, backward, or
     central, 47
Differences, for predictions
  calculation of, 182
  confidence or prediction intervals on,
     183
  spatial, 184
  standard deviation of, 182–183
Differencing, temporal, 216
Digital elevation models (DEMs), 355
  simulated equivalents and, 322
Dimensionless scaled sensitivities (*dss*),
     48–50, 61t, 148t
  for CHD parameters, 57
  for diagnosing cause of poor model fit,
     320–321
  for evaluating information observations
     provide about parameters, 60–62,
     263t, 265t, 327
  for identifying observations important to
     predictions, 170–171, 204–205,
     266t, 355
  for identifying observations to improve
     simulated processes, 331–333
  for identifying parameters important to
     predictions, 162–163
  use of, 49
Dimensionless scaled sensitivities values,
     large, 174
Direct inverse modeling, 11–12
Discharge. *See* Groundwater discharge
Dispersivity values, in transport modeling,
     221
Dissolution rate parameter, 226
Dominant errors, 381
Dottie plots, 140
Double-dogleg trust region approach, 68
Drawdown observations
  defined as differences, 182–183, 216
  fitted standard deviation for, 246
  in demonstrating modified
     Gauss–Newton method, 70, 74–75
  in Exercises, 231–233
  potential, 332–333
  weighted residuals for, 246–249
dss. *See* Dimensionless scaled sensitivities
dss-css-pss-pcc, using together, 335
DVRFS model. *See* Death Valley regional
     groundwater flow system model

Earth systems, well-posed regressions in, 277–278
Earthvision software, 274, 355
Effective hydraulic-conductivity values, 325. *See also* Hydraulic conductivity
Effective porosity parameters, in Exercises, 195, 199
 prediction scaled sensitivites for, 198
 parameter-prediction statistic for, 200–201
Effluent transport simulation, 23, 24, 193–194
Eigenvalues, 132
Eigenvectors, 10, 11, 132
Error analysis, defining weights on the basis of, 31, 304, 376–381
Error components, for observations
 accumulating, 296–297
 assumptions about, 377–381
 quantifying, 295–296
Errors. *See also* Measurement error; Model error; Observation errors, True errors
 accurate simulated results and, 30
 in data interpretation and model construction, diagnosing, 140–141
 fitting, 270
 independent and normally distributed, 296–297
 with a mean of zero, 377
 normally distributed, 380–381
 uncorrelated, 377–378
 variance–covariance matrices of, 380
 weighting and, 31–32, 34–35, 216, 224, 293–305, 376–381
 weights that reflect, 358–362
Error, study of elevation error, 295–296
Estimated parameter values. *See* Parameter estimates
Estimates
 accurate, definition, 14
 precise, definition, 13
 reliable, definition, 14
Estimates, precision of, 126, 127, 138
 classification of, 165–166
 evaluating using nonlinear confidence intervals, 154–155
 evaluating using standard deviations, linear confidence intervals, and coefficients of variation, 151–153

ETM (maximum evapotranspiration) parameters, 279, 280
Evapotranspiration (ET), 356
Exact critical values, linear simultaneous intervals, 177
Execution time issues, 345–346
Execution times
 improving, 346
 for global-search methods, 77
 for forward models, 271–272
 for Monte Carlo methods, 186–187
 for nonlinear confidence intervals, 139, 180–181
 reducing, 45–46
 for regression simulations, 346
 for transport models, 220–221
Exercises, 2–3, 24–25
 advective transport prediction, 195–196, 254–255
 composite scaled sensitivities (css), 60–62, 145–146, 204–205, 240, 243, 250–251, 256
 confidence intervals on parameters, 151–153, 154–155, 251–252
 confidence intervals on predictions, 207–212, 257–259
 Cook's D, 146–147
 DFBETAS, 146–147
 dimensionless scaled sensitivities (dss), 60–61, 204–205
 fitted error statistics, 114, 246
 graphical analyses of model fit, 115–123, 246–250
 groundwater management problem used for, 21–24
 influence statistics, 146–148
 leverage statistics, 66, 146–147
 linearity measures, 155–157, 253
 model fit, 40, 113–123, 235, 244–250
 modified Beale's measure, 155–156, 252
 modified Gauss–Newton method, 80–87
 nonlinear regression, 80–92
 nonlinearity measures, 156–157
 normal probability graphs, 119–123
 objective-function surfaces, 92
 observation definition, 38–40, 231
 observation–prediction (opr) statistic, 202–204, 205–207

INDEX **433**

parameter correlation coefficients
(pcc), 60–62, 148–149 196–199,
204–207, 243, 244, 252–253,
256–257
parameter definition, 36, 38, 230–231
parameter estimates, 91, 245–246
parameter estimation, 87–92, 243–244
parameter-prediction (ppr) statistic,
199–202
prediction scaled sensitivities (pss),
196–199, 256
prior information, 90–91
reasonable ranges, 153, 251–252
runs statistic, 117–118
sensitivity analysis, 60–66, 235–243
simulate heads, 25, 229
uniqueness of parameter estimates,
148–151
weighted residuals, 115–119, 246–250
weighting, 39–40, 231–235
Existing model recalibration, strategies for,
227–228
Expected value, 395
of calculated error variance, 96,
113–114
of error, 294, 300
of optimized parameters, 395
of standard error of the regression, 96,
113–114

Failed regressions, information available
from, 307–308
Field applications, 277, 278–281,
287–288, 293, 313–314, 317–318,
319–320, 324–325, 347–351,
353–379
Field systems, developing alternative
models for, 309–310
Field work, using observations to guide, 45,
369. *See also* Data collection
First order, second moment (FOSM)
methods, 170, 174
Fit-consistent statistics, 97
Fit-independent statistics, 46–56, 263t,
265–266t, 278
advantages and limitations of, 56–60
insights about *opr* statistic from,
173–174
integration of by *opr* statistic, 171

as measures of leverage, 46
for sensitivity analysis, 46–56
use in determining parameters
supported by observations, 45
use in encouraging well-posed
regression, 278
Fitted coefficient of variation, 95–96
Fitted error statistics, 95–96, 114,
use of, 265t
Fitted standard deviation, 95–96,
114, 246
Flow and transport parameters, coupled
estimation of, 223–224
Flow data. *See* Flow observations
Flow model. *See* Groundwater model(s)
Flow observations
calculating weights on, 297–298
contribution toward simulating
advective-transport predictions, 202
in Exercises, 38–39, 231
importance in regression, 284
issues for weighting of, 301–303
in objective function, 27
in reducing parameter correlations,
81–82
residuals for, in Death Valley regional
flow system model, 362–364
Flow parameters, errors in, 56
Flow residuals. *See* Flow observations
Flow system. *See* Groundwater flow system
Flow system dynamics, using sensitivity
maps to understand, 63–65, 235
Flow-system observations
in calibrating transport models, 217
use with transport observations,
223–224
Flow system properties, in Exercises, 38t,
231t
Forward differences, 47
Forward model execution times. *See*
Execution times
Forward model nonlinearities, managing,
271–272
Forward model solution, difficulties
attaining, 143–144
Fractional parameter value changes,
73–74, 386–387
Fractured-rock aquifer, 275–276, 277,
313–314, 353

Frequency domain analysis, 215
F statistic, selected Web sites for, 406t
Full weight matrices, 28, 34–35, 294
  correlation coefficient R and, 106
  dimensionless scaled sensitivities and, 49
  weighted residuals and, 35

Gauss–Markov theorem, 391
Gauss–Newton nonlinear regression method, 67–68, 70, 71–72. *See also* Modified Gauss–Newton method
  difficulties with, 71–72, 75
Gauss–Newton nonlinear regression normal equations, 69, 70–71
  deriving, 87
GCV measure, 95
General Head Boundary (GHB) Package of MODFLOW-2000 parameters, 279, 280
Geochemistry observations, 287
Geographic information system products, 354
Geographic information systems (GIS), 274
Geophysical data
  as observations or to support prior information, 291–292t
  use of, 272, 293
Geoscientific information system (GSIS) products, 354, 356
Global optimization methods, 77
Global-search methods, 77–78
GMS software, 274, 355
GOCAD software, 274, 355
Gradient methods, 67, 68, 77–78. *See also* Modified Gauss–Newton method
Graphical analyses
  of model fit and weighted residuals, 99–113, 316–320, 362–364
  in Exercises, 115–119, 246–250
  and utility of inverse modeling, 7
Graphs. *See* Graphical analyses
Grindsted landfill
  groundwater system at, 370–373
  flow and transport model of, 226, 370–373
  two- and three-dimensional models for, 372

Groundwater data, variability in, 272
Groundwater discharge, 320, 356. *See also* Flow observations
Groundwater flow model. *See* Groundwater model(s)
Groundwater flow problems, understanding, 15
Ground-Water Flow Process. *See* MODFLOW-2000 program
Groundwater flow simulations, 24
  confined layers in, 271
  in MODFLOW-2000, 19
  transient stress and, 214
Groundwater flow system, 21
  changes in, 214
  characteristics of, 23–24
  differences in pumping scenarios in, 182–183
Groundwater hydrology, model complexities in, 268–269
Groundwater inverse modeling, 11–12
Groundwater inverse problems, level of parameterization for, 10
Groundwater management
  problem, 21–24
  purpose and strategy of, 23
Groundwater model(s). *See also* Groundwater modeling; Models
  calibration using different types of observations for, 287–288
  development, 25 OK
  hydrogeology in, 273–277, 354–357
  observation clustering in, 285
  parameterization in, 5, 10–11, 290–291
  prior information in, 289–290
  scale issues in, 219–220
  scaling by parameter value and, 57
  systematic misfit in, 94
  well-posed regression for, 280–281
Groundwater modeling. *See also* Groundwater model(s)
  differencing in, 216
  Monte Carlo methods in, 342
  pioneers of using regression methods in, 9
Groundwater monitoring network design, 174
Groundwater recharge estimates, 356

INDEX   **435**

Groundwater system conceptual model,
    356–357. *See also* Conceptual
    models
Groundwater system numerical models,
    357. *See also* Numerical models
Groundwater systems
    data assimilation in, 273
    fundamental aspects of, 2
    heat as a tracer in, 288
    hydrogeologic data in, 160, 273–274,
        354–357
    interpolated hydraulic heads in,
        284–285
    plumes in, 42, 280–281, 372–373
    selected field investigations of,
        348–351t
    synthetic models of, 352t
Groundwater temperature data, 287–288
Guideline 1 (Principle of parsimony), 44,
    95, 97, 218, 264, 265t, 268–272, 273,
    306, 345
Guideline 2 (Broad range of system data),
    219, 228t, 264, 265t, 272–277, 278,
    284, 288, 291, 306, 309, 321, 330,
    353–357, 371–373
Guideline 3 (Well-posed regression),
    44, 46, 56, 60, 137, 228t, 265t,
    277–283, 300, 306, 307, 321, 327,
    328, 333
Guideline 4 (Many kinds of observations),
    44, 45, 214, 228t, 264, 265t, 273,
    284–288, 371–373
Guideline 5 (Careful use of prior infor-
    mation), 33, 34, 50, 80, 137, 265t, 274,
    278, 288–293, 325, 357–358
Guideline 6 (Weights that reflect errors),
    28, 30, 31, 34, 35, 39, 49, 79, 101, 131,
    144, 165, 176, 200, 214, 216, 228t,
    284, 290, 291–305, 316, 323, 331,
    358–362, 377, 378, 380, 381, 394
Guideline 7 (Convergence), 306–308
Guideline 8 (Alternative models), 43, 77,
    186, 264, 303, 328, 330, 342
Guideline 9 (Model fit), 100, 108, 111,
    228t, 265–266t, 309–311, 316–323,
    326–327, 329, 330, 362–364, 369
Guideline 10 (Parameter values), 124, 137,
    228t, 266t, 290, 309–311, 323–328,
    329, 330, 364, 373

Guideline 11 (New data to improve model),
    34, 46, 60, 228t, 265t, 267, 306,
    330–334
Guideline 12 (New data to improve pre-
    dictions), 48, 228t, 264, 266t, 267,
    306, 320, 332, 334–336, 365–370
Guideline 13 (Prediction uncertainty using
    deterministic methods), 266t, 320,
    337–339
Guideline 14 (Prediction uncertainty using
    statistical methods), 43, 186, 266t,
    267, 309, 311, 312, 333, 334,
    339–344
Guidelines for effective modeling, 17t,
    261–262t, 265–266t, 308
    implementation of, 264–267
    purpose, 263–264
    relation to previous work, 264
    using and testing, 345–373

Hard data, 260
Head-change observations, transient,
    360–362
Head observations. *See* Hydraulic-head
    observations
Heads. *See* Hydraulic heads
Heat transport models, 215, 288
Hydraulic conductivity
    Darcy's Law and, 12, 36
    and geophysical data, 291
    in groundwater system for Exercises, 24,
        38, 81,
    measurements of, 45, 94, 221, 272
    prior information and, 33, 305
    representing, 10, 42, 264, 272, 274, 310,
        313, 325
    ranges of, 325
Hydraulic-conductivity parameter(s), 81
    confidence intervals on, 153, 252, 324
    defining, 10–11, 42, 44
    in Exercises, 38, 81
    log transforming, 87
Hydraulic gradient
    hydraulic-conductivity representation
        and, 42
    hydraulic-head error and, 359
    interpolated observations and, 285
Hydraulic-head data. *See* Hydraulic-head
    observations

Hydraulic-head observations. *See also* Hydraulic heads; Hydraulic-head weighted residuals
  calculating weights on, 295–296, 358
  clustered, 285
  confidence intervals on, 180
  errors in, 216, 296–297, 358–362, 377–381
  evaluating using *opr* statistic, 368–369
  interpolated, 284–285
  in Exercises, 38–39, 231
  model construction and, 42
  multilayer, 321–322
  in objective function, 27
  potential, 204–207, 331–332
  temporal differencing of, 216
Hydraulic-head prediction(s) calculating differences for, 182
Hydraulic-head residuals. *See* Hydraulic head weighted residuals
Hydraulic heads
  Darcy's Law and, 12–13
  linear relationships of, 235–240
  maps of one-percent scaled sensitivities for, 55–56, 64, 235, 237–242, 333
  simulation of, 271–272
  sources of error related to, 296
Hydraulic-head weighted residuals, spatial randomness of, 117, 318, 363
Hydraulic properties, assigning to framework digital model, 356
Hydrocarbon dissolution/biodegradation model, applying regression to, 226–227
Hydrogeologic boundary position/location errors, 359
Hydrogeologic data, use of, 273–275. *See also* Hydraulic conductivity
Hydrogeologic framework conceptual model, 354–355
Hydrogeologic framework digital model, 355–356
Hydrographs, 183
Hydrologic data, assimilation of, 273
Hydrologic models, 169, 339
Hypothesis testing, 328

Important observations. *See also* Observation–prediction (*opr*) statistic
  to parameter estimates, 132–136, 327–328, 364–365
  to predictions, 170–173, 334, 368–370
Important parameters, to predictions, 160–170, 196–202, 366–368. *See also* Parameter–prediction (*ppr*) statistic; Prediction scaled sensitivities (*pss*)
Important processes
  capturing, 270
  omission of, 321
Improved information, and *ppr* statistic, 168–169
Improved parameter estimation, classification of the need for, 165–166
Inconsistencies
  between observations and model construction, 286, 306
  in initial conditions, 213–214
Independent variables, evaluating weighted residuals against, 106–108, 117, 248, 318, 319, 332–333, 363–364
Indirect inverse modeling, 11–12
Individual confidence intervals, 341. *See also* Confidence intervals
  in Exercises, 151–155, 208–211, 251–252
  on parameters, 138–139
  on predictions, 175, 176–178
Individual fit-consistent statistics, 97
Individual prediction intervals, 175
Inferential methods, to compute nonlinear needs work confidence intervals, 139
Inferential statistics, 7
  quantifying parameter uncertainty using, 137–139, 151–153, 154–155, 251–252
  quantifying prediction uncertainty using, 174–185, 207–212, 257–259, 341
Influence statistics, 46, 133–136, 172. *See also* Cook's *D* influence statistic; *DFBETAS* influence statistic
  diagnosing the cause of poor model fit and, 320
  diagnosing the cause of unreasonable parameter estimates and, 326–327

INDEX   **437**

evaluating the importance of
  observations using, 146–147,
  327, 364–365
nonlinearity of, 135
Initial conditions, for transient models,
  213–214
Insensitive parameters, 33, 45–46, 289,
  290, 306–307t
Insensitivity, 6
Instability, 6
Interpolated observations, 284–285
Interpolation methods, for parameterization,
  10, 11
Intrinsic model nonlinearity, 144, 145,
  156–157
  combined. *See* Combined intrinsic
    model nonlinearity
Inverse modeling, 6
  direct and indirect, 11–12
  model calibration with, 3–8
  problems revealed by, 7
  use of parameter limits in, 80
  utility of, 6–7
Inverse models
  execution times for, 346
  log transformations in, 79
  nonlinearity of, 58
Investigate Objective Function mode. *See*
  UCODE_2005 program

Jackknifing method, 170, 172
Jacobian matrix, 69

Kashyap's measure, 95, 310
K (hydraulic conductivity) parameters.
  See Hydraulic-conductivity
  parameter(s)
Kriged hydraulic-head measurements,
  284
Kriging, 284, 309, 342

L_1 norm objective function, 29
L_1 norm of sensitivities, 50
Lagrangian methods, 221
Landfill. *See* Exercises, groundwater
  management problem used for;
  Grindsted landfill
Large weights, using, 301–303
Latin hypercube sampling, 186

Least-squares objective function. *See*
  Weighted least-squares objective
  function; Weighted least-squares
  objective-function surfaces
Levenberg–Marquardt method, 68
Leverage, 46
  calculating, 134
Leverage statistics, 54, 133–134
  advantages and limitations of, 59–60
  for evaluating potential new data, 332
  for identifying observations important to
    parameters, 263t, 327
  in Exercises, 66, 146–147
  use of, 265t, 306t
Likelihood confidence regions, 178
Limits, on estimated parameter values,
  80, 140
Linear confidence intervals, 138, 176–177,
  341. *See also* Confidence intervals
  assumptions underlying, 180, 340
  comparing to nonlinear intervals,
    179–180
  comparing to reasonable parameter
    values, 140–142
  on differences, 183
  in Exercises, 151–153, 208–210,
    251–252, 257–259
  for log-transformed parameters, 130
  overlapping, 328
  on parameters, 138
  on predictions, 176–177
  relation of Monte Carlo analysis to,
    187–188
  testing linearity for, 143–145
Linear individual confidence intervals. *See*
  Individual confidence intervals;
  Linear confidence intervals
Linear inferential statistical methods,
  137–138
Linearity, testing, 142–145, 155–157,
  189–193, 253
Linearized approximate nonlinear model,
  393
Linearized objective function, 69–70
Linearized objective-function surface, 71,
  74–76, 181–182
Linearized parameter confidence region,
  189
Linearized true nonlinear model, 392

Linear models, 12–13
Linear prediction intervals, 176–177
Linear prior information equations, 32
Linear regression, 2
  observation influence in, 133–134
  observation leverage in, 133–134
  important properties of, 391–398
  use in illustrating noise-sensitivity interaction, 51
  versus nonlinear regression, 70
Linear simultaneous confidence intervals. *See* Linear confidence intervals; Simultaneous confidence intervals
Linear uncertainty analysis, 9. *See also* Linear confidence intervals; Linear prediction intervals; First order, second moment (FOSM) methods
LINEAR_UNCERTAINTY program, 177, 184
Linssen's measure, 142
Local sensitivity methods, 9, 47
Location uncertainty, 359
Log-transformed parameters, 78–80, 305
  calculating damping parameter for, 385–388
  dss for, 49
  in Exercises, 87
  prior information and, 305
  statistics for, 130

Madison aquifer groundwater model, 300
Maggia Valley models, 310, 311
Maps, using independent variables and the runs statistic, 106–108, 319
Markov chain Monte Carlo method, 186
Marquardt method, 68
Marquardt parameter, 72
  calculating, 73
  difference in PEST, 19
Mathematical models, 1
Matrices. *See also* Errors, Full weight matrices; Parameter correlation coefficient matrices; Parameter variance–covariance matrix; Variance–covariance matrix; Weight matrices
Jacobian, 69
  for nonlinear regression, 383–384
  sensitivity, 69, 71, 132, 384

Max-allowed-change, 73, 74, 76, 385
Max-calculated-change, 74, 76, 307, 308
  convergence and, 88
MAX-CHANGE (MaxChange) variable, 73, 83, 385–389
  effect of, 388
Maximum-likelihood objective function, 29, 95, 98, 99, 374
  derivation of, 375–376
  relation to least-squares objective function, 376
Measured values, observations as sums of or differences between, 297–298
Measurement error, 300–301, 359–360
Method of characteristics (MOC), 223
Methods, using and testing, 1–2, 345–373
MFI2K program, 3
MGO program, 78
Microsoft Excel, 109
Mirror Lake groundwater flow system, alternative models for, 313–314
Model accuracy, encouraging convergence via, 306–308
Model bias, 13–14, 315. *See also* Model error
Model calibration, 289
  execution time and, 345
  flowchart for, 4
  guidelines for, 17t
  inconsistencies and, 214
  with inverse modeling, 3–8
  issues fundamental to, 4–8
  steps in, 4
  techniques, 2
  of transient models, 213–217
  of transport models, 217–227
Model construction
  addressing problems with, 321
  modifying, 322
  observations inconsistent with, 286–287
  regression match and, 320–321
  use of observations in, 42–43
Model development
  highly parameterized, 291
  measurements in, 1–2
  parsimony in, 270–271
  stages in, 308

INDEX 439

Model development guidelines, 268–314.
  See also Guidelines; Modeling
  guidelines
Model error, 296, 300–301
  as a cause of model fit problems, 315
  detecting, 16, 80, 303–304, 323–327
  and model bias, 315
  Death Valley model and, 358
  model fit and, 97, 285, 303–304
  graphs showing evidence of, 100,
    103, 105
  graphs showing no evidence of, 102
  Grindsted transport model and, 373
  indications of, 97, 107, 141, 262
  omission of processes and, 321
  parameter estimates in detecting,
    323–326
  postaudits and, 339
  weighting and, 294, 296, 298, 300–301,
    377–379
Model fit. *See also* Overfitting;
  Guideline 9; Systematic model misfit
  analysis of, 316
  determining, 316–320
  evaluating, 93–123, 315, 316–323
  evaluating measures of, 244–246
  evaluating using starting parameter
    values, 40
  examining for observations important to
    model purpose, 320
  graphical analyses of, 99–113, 246–250
  magnitude of residuals and weighted
    residuals in, 93–94
  measures of, 94–99
  poor model fit, diagnosing cause,
    320–323, 327
  statistics consistent with, 97
  statistics related to, 246
Model-fit convergence criterion, 76–77
Model-fit statistics, 95–99, 269
Model fit to observations, investigating,
  362–364
Model inconsistencies, identifying the
  cause of, 286–287
Modeling guidelines, 260–267. *See also*
  Guidelines; Model development
  guidelines
  categories of, 2
  comprehensive nature of, 9

  effective implementation of, 264–267
  placement of predictions in, 267
  purpose of, 263–264
  questions, statistics, graphs, figures, and
    tables related to, 265–266t
Modeling process, links and associated
  methods in, 263
Modeling protocol, 263
Modeling steps, flexible application of, 267
Model input files, defining observations in,
  38–39
Model inputs, 5
  determination of, 8
  additional for transport model
    calibration, 225–226
Model layers, confined, 271–272
MODEL_LINEARITY_ADV program,
  190
MODEL_LINEARITY program, 143
Model nonlinearity. *See also* Linearity,
  testing; Nonlinearity
  methods for detecting, 144–145
  effect on analysis, 44
  effect on regression, 391–398
  managing, 271–272
  methods that account for, 170
  testing, 189–193
Model nonuniqueness, testing, 151
Model parameters, implementing improved
  information on, 169
Model predictions. *See also* Predictions
  evaluating, 158–170
  importance of model parameters to, 169
  use of, 158–159
Model predictive capabilities, testing,
  337–339
Model purpose, 261t
  observations important to, 320
Model recalibration, issues related to, 227,
  228t
Models. *See also* Alternative models;
  Death Valley regional groundwater
  flow system (DVRFS) model;
  Existing model recalibration;
  Modeling guidelines; Transient model
  calibration; Transport model
  calibration
  ability to simulate predictions, 256
  accurate, 30

Models (*Continued*)
  biased, 30–31
  complex, 269, 291
  with different levels of complexity, 270
  discriminating among, 98–99, 310–311
  levels of confidence in, 174–193, 338–339
  nonlinear, 12–13
  predictive versus calibrated, 8
  quantitative links provided by, 7–8
  refutability of, 270
  resolving problems with, 111
  simple, 268–270
  transparency of, 7, 270, 278
Model sensitivity analysis, 160. *See also* Sensitivity analysis
Model simplification,
  in Guideline 1, 268–277
  example of, 277
Model testing
  methods, 391–398
  Guidelines 9 and 10, 315–328, 362–365
Model uncertainty, reduction of, 227
ModelViewer program, 3
MODFLOW-2000 program, 3, 18–21, 25, 112. *See also* Advective-Transport Observation Package (ADV); General Head Boundary (GHB) Package, Multi-Node Well (MNW) Package
  capabilities of, 19–21
  correlation coefficients calculated by, 62t, 63t, 148t, 149t, 199t, 206t, 244t, 253t, 257t
  familiarity with, 15
  full weight matrices and, 28
  for graphs of observed versus simulated values, 105
  Ground-Water Flow (GWF) Process, 3, 18–21, 25
  hydraulic conductivities in, 81
  linear prior information equations supported by, 32
  listing of weighted residuals by, 93
  LIST output file showing model fit, 236
  LIST output file showing particle path, 196, 255
  Marquardt parameter in, 73

Observation (OBS) Process, 3, 18, 19, 21, 25, 358
Parameter-Estimation (PES) Process, 3, 18, 19, 25
  parameter correlation coefficient matrix calculated by, 148
  plot-symbol variables in, 101
  prediction scaled sensitivities calculated by, 197
  regression simulations in, 346
  runs statistic information printed by, 108
  runs statistic test results via, 107–108
Sensitivity (SEN) Process, 3, 18, 19, 25
  scaled sensitivities in, 161, 162
  scaling of sensitivities difficulty, 57
  sensitivities calculation by sensitivity-equation method, 47, 280
  sensitivities calculated by, 61t
Uncertainty (UNC) Process, 112, 139, 154, 190
  runs statistic in, 108, 118, 249
MODFLOWP program, 280, 342
Modified Beale's measure, 142–145, 189, 190
  calculation of, 143
  comparing with critical values, 144
  using to test model linearity, 155–156, 253
Modified Gauss–Newton iterations, statistics from, 89, 90, 150, 245
Modified Gauss–Newton method, 68–77. *See also* Nonlinear regression
  application to a two-parameter problem, 80–87
  calculation details of, 383–390
  convergence of, 77
  convergence criteria for, 76–77
  example of, 74–76
  normal equations for, 68–74
  performance of, 82–83
Modified method of characteristics (MMOC), 223
Modified weighted simulated values for graphical analysis, 104
Monitoring network design, 174
Monod kinetics, 226
Monte Carlo analyses, 140
  advantages of, 188
  elements of, 185–187

individual intervals calculated using, 175
least-squares objective function in, 188
parameter sampling, 186, 188–189
quantifying prediction uncertainty using, 185, 341–344
relation to linear and nonlinear confidence intervals, 187–188
results, analyzing and displaying, 187
results to save from, 187
runs, number of, 186–187
using the Theis example to understand, 188–189
Multi-Model Analysis (MMA) program, 189
Multi-Node Well (MNW) Package of MODFLOW-2000, 322
Multiobjective optimization, 29, 30, 78
alternative weighting schemes in, 299
Multiple parameters, improved information on, 169, 330–334
Multiple parameter value sets, calculating statistics for, 333–334
Multiplication arrays, 20, 277
Multiplicative errors, 32

Native parameter values, 79
calculating statistics for, 130
in Exercises, 87
Natural-gradient tracer transport, model calibration of, 225
Natural logarithms, 79
Negative runs test statistic, 108
Network design. *See* Monitoring network design
Nevada Test Site, potential contaminants from, 347
New data. *See also* Data collection; Guideline 11; Guideline 12
to improve predictions, 334–336, 365–370
to improve simulated processes, 330–334
to test model predictive capabilities, 339
New observations, importance to predictions, 170–173, 334–336

95-percent confidence intervals, 176–181, 251, 252. *See also* Confidence intervals
computed, 257
linear, 138, 176–177, 208
nonlinear, 139, 177–181, 209, 210
Nonaqueous phase liquids (NAPLs), 218
Nonintrinsic nonlinearity, 135
Nonlinear confidence intervals, 139, 177–181. *See also* Confidence intervals
calculating on advective transport components, 209–212
characteristics of, 179–180
in Exercises, 154–155, 209–212
on parameters, 139
on predictions, 177–181
relation of linear intervals to, 180, 341
relation of Monte Carlo analysis to, 187–188
Nonlinear function, linearizing, 69
Nonlinear individual confidence intervals, 138–139. *See also* Confidence intervals
Nonlinear intervals, 341. *See also* Nonlinear confidence intervals; Nonlinear prediction intervals
calculating, 178, 180
investigating characteristics of, 179
Nonlinearity, 12. *See also* Model nonlinearity
of influence measures, 135
effects on parameter correlation coefficients, 58, 281–283
effects on scaled sensitivities, 57, 281–283
methods that ignore, 9–10
Nonlinear models, 12–13
Nonlinear objective-function surface, 37, 58, 70, 74, 75, 82, 139, 179, 181–182
Nonlinear prediction intervals, 177–181
Nonlinear regression, 67. *See also* Modified Gauss–Newton method
difference with linear regression, 70
encouraging convergence of, 306–308
estimating parameters by, 43
execution times for, 245–246
in Exercises, 82–86, 87–92, 243–244

Nonlinear regression (*Continued*)
  in geophysics, 9
  predictions and, 160, 344
  vectors and matrices for, 383–384
  weighting issues in, 298–305
Nonlinear regression theory, 2
Nonlinear simultaneous intervals, 175. *See also* Nonlinear confidence intervals; Simultaneous confidence intervals
Nonoptimal parameter values, variance–covariance matrix with, 126, 131
Nonrandom-appearing residuals, testing, 111–112
Nonsimulated processes, effect on observations, 322–323
Nonsimulated transient stress errors, 360
Nonuniqueness problem, 6
Nonunique parameter estimates, 137
  detecting, 58, 149–151
Normal equations, 68–74, 395
  deriving, 87
  iterative form of, 70
  quasi-Newton updating of the, 384–385
  solving, 389
Normality, 30. *See also* Normal probability distribution
Normally distributed errors, 380–381
Normally distributed weighted residuals, acceptable deviations from, 119–123
Normal order statistics, correlation between ordered weighted residuals and, 108–111, 250
Normal probability axis, 109
Normal probability distribution, 118
  assumed for linear intervals, 30, 138, 177
  assumed for the maximum likelihood objective function, 374
  not assumed to estimate parameter values, 30
Normal probability graphs, 108–111. *See also* Normal order statistics
  common problems with, 110
  in Exercises, 119–123, 249–250
Normal quantiles (normal score), 109
Numerical dispersion, 221–223

Numerical issues, in transport model calibration, 220–223
Numerical model(s)
  developing, 354
  for Exercises, 21, 23
  groundwater system, 357
  parameters in, 269

Objective function(s), 6, 26. *See also* Maximum-likelihood objective function; Weighted least-squares objective function
  alternative, 28–30
  comparing observed and simulated values using, 26–40
  L_1 norm, 29
  least-squares, 29, 30
  linearized, 69–70
  maximum-likelihood, 29
Objective function issues, 374–382
Objective-function surfaces, 35–36, 37, 58, 181
  linearized, 71, 75, 181
  constructing, 35
  data sets for constructing, 81
  in Exercises, 81–82
  irregular, 78
  relation to parameter correlation coefficients, 58, 81–82
  usefulness of, 36
  using to explore regression performance, 92
Objective function trade-offs, in multiobjective optimization, 78
Objective-function values
  in Exercises, 91, 113
  as a measure of overall model fit, 95
Observation data. *See* Observations
Observation errors, 300. *See also* Weighting(s), Weights
  classification of, 216
  correlations between, 298
  random-over-time, 216
  uncorrelated, 378
Observation–parameter combinations
  dimensionless scaled sensitivities of, 48–50
  DFBETAS values of, 147
  leverage statistics of, 54

INDEX   **443**

Observation(s)–parameter(s)–
 prediction(s) sequence/triad, 7–8, 46,
 160, 170, 263t
fit-independent statistics to evaluate, 42
Observation–prediction (*opr*) statistic,
 131, 170, 171–173, 335, 336. *See also*
 OPR-PPR program
 calculating, 173
 in Exercises, 205–207
 fit-independent, 173–174
 insights about, 173–174
 large values of, 173
 prediction standard deviation and, 172
 strengths and weaknesses of, 173
 using, 202, 205–207, 368–370
Observation possibilities, 284
Observations (observed values), 4, 260. *See*
 also Guideline 4; Potential new
 observations
 advantageous use of, 270
 alternatives to using point concentration
  measurements as, 225
 assigning large weights to, 301
 clustered, 45, 285
 comparing the relative importance of, 48
 concentration, 42
 considering many, 394
 consistent with simulated processes, 215
 in direct inverse modeling, 12
 that dominate parameter estimation, 364
 effect of non-simulated processes on,
  322–323
 errors in, 35
 errors in the weighting of, 323
 evaluating information provided
  by, 43–44, 306–308
 in Exercises, 38–40, 231–235
 hydraulic-head, 42
 important to estimated parameter values,
  132–136
 including in regressions, 45, 284–288
 including many kinds of in regression,
  357
 inconsistent with model construction,
  286–287
 interpolated, 284–285
 leverage statistics for, 54
 potential new data to support, 335–336
 regularization on, 10

simulated equivalents of, 28. *See also*
 Simulated equivalents, of
 observations
sources of error related to, 296–297
time-consistent, 215
total error for, 378–380
transient, 214–216
transport, 217
unbiased, 30–31
use in adjusting parameter
 values, 43–44
using to calibrate groundwater flow and
 transport models, 287–288
using predictions to guide
 collection of, 170–174
versus prior information, 288–293
weighting, 26–27, 215, 293–305. *See*
 also Weighting(s), Weights
Observation (OBS) Process. *See*
 MODFLOW-2000 program
Observation uncertainty, represented as
 proportional to concentration,
 224–225
Observation well, quantification of error in
 the elevation of, 295–296
Observed temporal effects, 214–215
Observed values. *See also* Observation(s)
 comparing with simulated
  values, 26–40
Off-diagonal covariance terms, 298
Omitted data, using, 338–339
One-percent prediction scaled sensitivities,
 163. *See also* Prediction scaled
 sensitivities (*pss*)
One-percent scaled sensitivities (*1ss*), 48,
 54–56, 161
 for instructional purposes, 56
 using groundwater flow system physics
  to understand, 235
One-percent scaled sensitivity contour
 maps, 55, 63–64
 in Exercises, 235–240
OPR-PPR program, 169, 173, 195. *See also*
 Observation–prediction (*opr*)
 statistic; Parameter–prediction (*ppr*)
 statistic
 use in calculating *opr* statistic, 202
 use in calculating parameter–prediction
  statistic, 200

Optimality of parameter estimates, 137
Optimal parameters, 26. *See also*
    Optimized parameter estimates
    reasonable, implications of, 141, 311
    test for reasonableness, 140–142, 153, 251–252
    unreasonable, as indication of model error, 323–326
    unreasonable, diagnosing causes of, 326–327
Optimization, 67
    advantage of, 337
    methods, 67–68, 77–78
    multiobjective, 78
Optimized parameter values. *See also*
    Optimal parameters
    evaluating (Guideline 10), 323–328
    variance–covariance matrix with, 125–126
Ordinary regression, 27
Oscillation control, 387–388
Overall model fit
    in Exercises, 113–115, 244–246
    measures of, 94–99
    selected statistics related to, 114
Overfitting, 269, 303–304
Overshoot, 72

Parallel processing, 187, 346
Parameter change vector, accuracy of, 72
Parameter complexity, 11, 290–291
Parameter confidence region, 139, 178, 179
    and Beale's measure, 142
    exact, 178
    likelihood, 178
Parameter correlation coefficient matrices, 53, 149t, 199t, 253t, 257t
    calculated using potential observations, 206t
    calculated using sensitivity-equation and perturbation sensitivities, 21, 59, 62t, 63t
Parameter correlation coefficients (pcc), 21, 48, 51–54, 128–130, 160, 164, 204t
    advantages and limitations of, 58–59
    assessing parameter value uniqueness using, 60, 137, 148–149
    in collecting observation data, 170–171

combining parameters, effect on performance, 59
composite scaled sensitivities (css) used with, 51, 165–166, 243, 265–257, 278–281
prediction scaled sensitivities (pss) used with, 60–63, 165–166, 256–257
effects of nonlinearity on, 281–283
in Exercises, 62–63, 243–244, 252–253
evaluating potential new data and, 332
maximum percent reduction in, caused by new observation, 209
in model development, 280–281
potential observations and, 205
without and with predictions, 162–165
relation of objective-function surfaces to, 58, 81–82
Parameter correlations
    effects on ppr statistic, 368
    scaled sensitivities do not reflect, 56
Parameter definition
    modifying, 321
    use of observations in, 42–43
Parameter design. *See also* Parameterization
    role of hydrogeologic framework in, 272–277
    role of observations in, 45
Parameter estimates, 91. *See also*
    Parameter estimates, unreasonable;
    Parameter estimation
    checking against reasonable values, 140–142, 153
    evaluating, 124–137, 250–253
    evaluating optimality of, 137
    evaluating uniqueness of, 137, 148–149, 165–166
    evaluating uncertainty of, 137–140
    in Exercises, 91, 246
    identifying observations important to, 132–136, 327–328
    in model recalibration, 227
    nonunique, 137
Parameter estimates, unrealistic. *See*
    Parameter estimates, unreasonable
Parameter estimates, unreasonable, 140–142
    data collection, motivation for, 330
    in Exercises, 153
    Grindsted landfill model and, 373

INDEX   **445**

and limits on parameter values, 80
model error, caused by, 215, 219
model error, as indicator of, 262,
  286–287, 311, 322, 323–327,
  315–316
Monte Carlo, criterion in, 343
prior information, as reason for, 261,
  289–290
predictions, as criterion for
  unlikely, 338
recharge representation and, 327
Parameter estimation. *See also* Nonlinear
  regression
  execution times for, 243
  formal, 68–80, 308
  iterations, 70, 71
  observations that dominate, 46–55, 364
  for transport models, 224–225
  review of methods for, 3–12
Parameter-Estimation (PES) Process. *See
  also* MODFLOW-2000 program
Parameter importance, rankings of, 200
Parameter importance to predictions,
  methods for identifying, 160–170
Parameterization(s), 4–6
  many-parameter, 11
  simple, 11
Parameter–prediction (*ppr*) statistic, 160,
  162, 166–170, 172, 366–368. *See
  also* OPR-PPR program
  drawback of, 167
  equation for, 167
  in Exercises, 199–202
  first-order second-moment equation for
    prediction uncertainty and, 170
  for multiple parameters, 169
Parameter ranges, reasonable. *See also*
  Parameter estimates, unreasonable
Parameters, 3
  changing the number of, 328
  defining, 44
  defining a range of reasonable values for,
    87
  log-transformed, 78–80, 144
  prior information on, 10, 90, 91
  problematic, 289
  using independent information on,
    325–326
Parameter standard deviation, 57

Parameter statistics, in evaluating optimal
  parameter estimates, 145–155
Parameter uncertainty. *See also* Confidence
  intervals; Parameter confidence region
  alternate methods for evaluating, 132
  minimizing, 46
  quantifying, 137–140, 323
Parameter value(s). *See also* Guideline 10;
  Parameter estimates
  convergence and, 76
  coordinated changes in, 185
  estimated, 6, 53, 67–92, 124–137
  generated, problems with, 143–144
  generation for Beale's measure, 143
  obtaining data related to, 330
  scaling by, 56–57
  sensitivity calculation for, 46
  simultaneously estimated, 219
  using observations to adjust, 43–44
  weighting system information on,
    304–305
Parameter value range, choosing, 188
Parameter variance–covariance matrix,
  124. *See also* Variance–covariance
  matrix
  dependence on defined
    parameters, 278
  eigenvalues and eigenvectors of, 132
  equations for, 125, 164, 167, 172, 191,
    398
  in Monte Carlo methods, 186
  potential new observations and, 126, 332
  prediction uncertainty and, 126, 132,
    159, 164, 168, 176, 303
  singular-value decomposition
    of, 132
  statistics from, 125–132
Parameter variance–covariance matrix,
  versions of, 125–126
  with all defined parameters, 126, 131
  with alternate observation sets, 126,
    131–132
  with nonoptimal parameter values, 126,
    131
  with optimized parameter values,
    125–126, 130–131
  with predictions, 126, 132
  when to use, 130–132
Parameter variances, 126

Parsimonious models, 11, 264. *See also* Guideline 1
  execution times and, 345–346
  Guideline 1, 17, 261, 268–272
  highly parameterized models and, 290–291
  parameter correlation coefficients and, 59, 280
  processes simulated and, 270
  prior information in, 290–293
  system information and, 272
Particle tracking. *See* Advective transport
pcc. *See* Parameter correlation coefficients (*pcc*)
Perturbation methods, calculating sensitivities by
  accuracy, 19, 59, 60, 62t, 63t
  accuracy with Lagangian transport solutions, 221
  execution time, 243, 346
  flexibility, 19
  used with sensitivity-equation sensitivities, 21
PEST program
  capabilities of, 18–21
  exercises and, 3, 18, 25
  confidence intervals, linear, 138
  confidence intervals, nonlinear, 139, 178
  Marquardt parameter in, 19, 73
  objective-function plots, data for, 36
  observations and, 32, 221, 284, 287
  parameter definition options, 81, 142
  parameterization using SVD, 10, 11
  perturbation sensitivities and, 47, 59, 346
  predictions and, 159, 338
  regression parameter controls variable and, 74, 76
  regression simulations in, 346
  regularization capability of, 10, 290, 305
  sensitivity maps and, 55
  superparameter method using SVD, 10, 11, 219
Porosity. *See* Effective porosity
Postaudits
  model Guidelines and, 262, 263
  model testing and, 320
  as a deterministic uncertainty method, 337–339

Potential new data, 17. *See also* Observation–prediction (*opr*) statistic; Parameter–prediction (*ppr*) statistic; Potential new data (Guidelines 11 and 12)
  in Exercises, 23, 204–207
  evaluation of, 44
  importance to parameters, 46
  parameter-variance–covariance matrix and, 132
  scaled sensitivities, parameter correlation coefficients, and, 171
Potential new data (Guidelines 11 and 12), 262, 329–336. *See also* Guideline 11; Guideline 12
  model accuracy needed for analysis, 329
  fit-independent statistics and, 329, 331–334
  for improving predictions, 334–336
  for improving simulated processes, features, and properties, 330–334
  to support observations, 335–336
Potential new observations
  assessing the likely importance of, 204–205
  evaluating, 333
  using *opr* to evaluate importance to predictions, 205–207
PowerPoint files, 2, 3
PPCC statistic, 110. *See also* Correlation coefficient R_N^2
Precision, defined, 13
Predicted values, contradiction by the calibrated model, 337–338
Prediction accuracy. *See also* Confidence intervals; Prediction intervals, Prediction uncertainty
  dependence on observations, 339
  evaluation using deterministic methods, 337–339
  model error and, 339
  observation weighting and, 301
Prediction improvement, identifying new data for, 334–336. *See also* Potential new data
Prediction intervals, 175–176, 341
  calculating, 176
  individual, 175
  linear, 176

INDEX

nonlinear, 177
predictions and, 175–176
simultaneous, 175
Predictions
accuracy of, 14, 340
determining weights for, 164–165
identifying new data to improve, 365–370
importance of observations to, 170
importance of parameters to, 166–167
precise, 13
reliable, 14
simulating, 158–159
testing model nonlinearity with respect to, 189–193
use of recalibrated model to update, 254–259
using to guide data collection, 159–170
Prediction scaled sensitivities (pss), 159, 160–162
calculation of, 161–162
in collecting observation data, 170–171
comparing with composite scaled sensitivities, 256
in conjunction with composite scaled sensitivities, 162
in Exercises, 198
parameter importance and, 196–198, 366–368
used with parameter correlation coefficients, 165–166
Prediction simulation, with alternative models, 312
Prediction standard deviation, 167
calculated with improved information, 168–169
calculating, 159
opr statistic and, 172
use in monitoring network design, 174
Prediction uncertainty, 174–189. See also Guideline 13; Guideline 14
calculation of, 168
effect of removing or adding one observation on, 172–173
evaluation using deterministic methods, 337–339
evaluation using Monte Carlo analysis, 185–189, 340, 341–344

evaluation using statistical inference, 174–185, 341, 339–344
Guidelines 13 and 14, 337–344
model averaged using MMA, 189
first-order second-moment equation for, 170
quantifying using alternative models, 189
Prediction uncertainty measurement, using inferential statistics, 207–208
Predictive models, versus calibrated models, 8
Predictor-corrector (P-C) method, 223
Prior estimates, 33. See also Prior information
scaled for graphical analysis, 104
Prior information. See also Guideline 5
in activating all parameters to evaluate model predictions, 131
assigning, 90
assigning large weights to, 301
biased, 30–31
careful use of, 288–293, 357–358
equations, 33
extensive use of, 289–290
importance of, 146
issues related to, 32–34
parameters and, 293
smoothing and, 33
use of, 10, 33–34
versus observations, 288–293, 357–358
versus regularization, 10
weighting on, 168, 200, 304
Prior weights, specifying, 168
Probability graphs, normal, 108–111. See also Normal probability graphs
Problem constraint (Guideline 2), use of system information in, 272–277, 353–357
Process selection, in calibrating transport models, 217–218
Projected particles, examples of, 209–212
Pumpage
in Exercises, 21, 193
uncertainties in amount and location of, 361
Pumping rate parameters, 230–231

Quasi-Newton updating, of normal
    equations, 384–385

R. See Correlation coefficient R
R_N^2. See Correlation coefficient R_N^2
Random sampling, by Monte Carlo
    analysis, 140
Reading assignments, suggested, 14–15
Reasonable parameter values. See also
    Parameter ranges, reasonable;
    Parameter estimates, unreasonable
  checking parameter estimates against,
    140–142
  important characteristics of, 141–142
Recalibration, of existing models,
    227–228
Recharge parameters
  confidence intervals for, 251
  correlation with hydraulic-conductivity
    parameters, 81, 82
  estimated using age and geochemistry
    observations, 287
  in Exercises, 24, 38, 81, 240
  for a transient problem, 217
Regional-scale data, integrating, 354
Regional-scale models, postaudits
    of, 339
Regional steady-state measurements, 357
Redundancy, correlations indicating, 35
Regression. See also Linear regression;
    Nonlinear regression; Regression
    methods
  clustered observations in, 285
  goal of, 36
  including many kinds of data in,
    284–288
  Monte Carlo method integration with,
    342
  observations to include in, 45
  omitting prior information from, 289
  parameter values in, 128
  role of observations in, 132–136
  using to determine whether predicted
    values are contradicted by
    calibrated model, 337–339
Regression convergence, problems with,
    221, 280, 306–308
Regression fitting process, correlations
    produced by, 111

Regression methods, 67
  pioneers of using in groundwater
    modeling, 9
Regression performance. See also
    Regression convergence
  evaluation of, 308
  using objective-function surfaces to
    explore, 92
  weighting, dependence on, 303t
Regression, well-posed and comprehensive,
    277–281
Regression simulations, execution times
    for, 346
Regression statistics, relation to sample
    statistics, 127–130
Regularization, 289, 290. See also PEST
    program
  highly parameterized models and, 290
  limits on parameters and, 80
  sensitivity analysis and, 145
  model transparency and, 278
  prior information and, 10, 91, 153
  stabilizing highly parameterized
    models, 10
  weighting and, 304–305
Reliability, defined, 14
Representer method, 10, 219, 328
RESAN-2000 program, 112
Residual analysis. See also Weighted
    residuals
  background for methods presented,
    9
  in evaluating model fit, 93–94
RESIDUAL_ANALYSIS program, 112
Residuals. See also Residual analysis;
    Weighted residuals
  calculating, 35
Results, presentation of, 2
Robust regression methods, 304
Runs statistic, 107
  in exercises, 118, 249
  maps of, 319
  ordering of observations
    and, 107–108
  probability table, 401–402t

Sample statistics
  equations for calculating, 129t
  relation to regression statistics, 127–130

INDEX

Scaled sensitivities, 48. *See also* Composite scaled sensitivities (*css*); Dimensionless scaled sensitivities (*dss*); Fit-independent statistics; Leverage statistics; One-percent scaled sensitivities (*1ss*); Parameter correlation coefficients (*pcc*)
  advantages and limitations of, 56–58
  for evaluating potential new data, 331–332
  relationship to parameter values, 57
  statistics related to, 133
Scale issues, in transport model calibration, 219–220, 289, 304
Scaling. *See also* Scaled sensitivities
  alternative, 57
  in the Gauss–Newton method, 72
  by parameter value, 56–57
Scheffé critical values, 177. *See also* Simultaneous confidence intervals
Sensitivities, 47
  calculating in MODFLOW-2000, 19
  calculating in PEST and UCODE_2005, 19–20
  calculation and use of, 47
  differences in, 184
  execution time and, 346
  local, 47
  parallel calculation of, 346
Sensitivity analysis, 1, 41–44. *See also* Scaled sensitivities; Fit-independent statistics
  in Exercises, 60–66, 235–243
  in previous work, 9
  scaling of, 48
  traditional, purpose of, 184
  traditional, using confidence intervals to replace, 184–185
  transient model, for the, 235–243
  utility of, 6–7, 57–58
Sensitivity (SEN) Process. *See* MODFLOW-2000 program
Sensitivity-equation method, 19
Sensitivity-equation sensitivities, 19, 47, 280
  accuracy of, 19, 47, 59
  accuracy, conditions which diminish, 59
  accuracy, consequences, 59
  ADIFOR and, 19
  custom coding and, 19
  grid sensitivities produced by, 54–55
Sensitivity maps, contoured. *See* One-percent scaled sensitivity contour maps
Sensitivity matrix, 69, 71, 384
Sensitivity precision, effect on parameter correlation coefficients, 59
Sensitivity statistics, using, 46
Sequential estimation strategy for transport, 223–224
Sequential indicator simulation method, 219
Shuffle Complex Evolution (SCE) method, 77, 78
Simple models, 268–272. *See also* Parsimonious models
  utility of, 270
Simplifying assumptions, 325
Simulated equivalents, of observations, 12, 28, 321–322. *See also* Simulated values
  incorrect calculation of, 300, 321–322, 329
  linearization of, 70, 143, 181
  omitted observations and, 306
  sensitivities of, 47
  sensitivity maps and, 55
  uniqueness and, 137
Simulated processes
  observations consistent with, 215
  potential new data for improving, 330–334
  time discretization of, 215
Simulated results, requirements for accurate, 30–32
Simulated stresses, variation in time, 217
Simulated systems, collection of data that characterize, 160, 272–277
Simulated values, 33. *See also* Simulated equivalents, of observations
  checking, 39
  comparing with observed values, 214
  importance of checking, 36–38
  linearized estimates of, 69–70, 143
  matching observed values to, 26–40
  one-percent scaled sensitivities and, 55
  sensitivities and, 47
  weighted, 310

Simulated values (*Continued*)
    weighted or unweighted observations versus, 105–106
    weighted residuals versus, 247
Simultaneous confidence intervals, 175–177. *See also* Confidence intervals; Simultaneous prediction intervals
    in Exercises, 208–212
    linear and nonlinear, 180, 211, 257–258
    on predictions, 175–181
Simultaneous estimation
    of flow and transport parameters, 223–224
    of source location and history, 219
Simultaneous intervals, 341. *See also* Simultaneous confidence intervals
Simultaneous prediction intervals, 175
Singular value decomposition (SVD) method
    as alternative to parameter correlation coefficients, 132
    superparameter method of PEST, 10, 219
Soft data, 260
Software, 3. *See also* Earthvision software; GMS software; GOCAD software; LINEAR_UNCERTAINTY program; MODEL_LINEARITY entries; MODFLOW_2000 program; Multi-Model Analysis (MMA) program; PEST program; RESAN-2000 program; RESIDUAL_ANALYSIS program; StratWorks 3D software; UCODE_2005 program; Visualization software
Solute transport. *See* Transport entries
Source characteristics and geometry *See also* Transport entries
    simultaneous estimation of, 219
    in calibrating transport models, 218–219
Specific storage parameter(s), in Exercises, 231
Specific storage values, prior information from field data, 33
Spring-flow observations, 363–364
Standard deviation(s), 127
    converting judgments about errors to, 34
    decreases associated with *ppr* statistics, 200
    of errors in concentrations, 224
    for differences, 182, 297
    evaluating estimate precision using, 151–153
    examples of, for DVRFS observations, 361
    increases associated with *opr* statistics, 203–204
    as indicator of importance of observations to parameters and predictions, 263
    iterative procedure for decreasing, 169
    fitted, 95, 97, 246. *See also* Fitted standard deviation
    for log-transformed parameter estimates, 79, 130
    for parameter estimates, 126–127
    for predictions, 159
    for predictions, and monitoring network design, 174
    not additive, 297
    reflective of unrealistic error, 304
    use by MODFLOW-2000 and UCODE_2005, 295t
    used to judge changes in parameter estimates, 151
    weighting and, 34, 198, 204, 299, 358, 360, 361
Standard error of the regression, 95, 304. *See also* Calculated error variance
    expected value of, 113
    grid line definition for graphs, 103
    interpreting, 96, 97
Standard normal distribution, selected web sites for, 399t
Standard normal statistics, 109
Starting parameter values, in Exercises, 91, 246
    evaluating steady-state model fit using, 40
    evaluating transient model fit using, 235
    varying, 83
Statistical consistency, testing weighted residuals for, 100
Statistical inference, quantifying prediction uncertainty using, 339–344

INDEX **451**

Statistical tables, 399–406
Steady-state model, 23, 24. *See also*
    Exercises
    calculating sensitivities for, 60
    cell-by-cell and boundary fluxes of, 65
    evaluate potential landfill advective
        transport using, 193–195
    fit, evaluating, 115–123
    linearity, testing, 155–156
    observations, 38t, 358
    parameter definition for, 36–38
    parameter values, estimated, 87–92,
        152t
    regression, 118
    simulated volumetric flows in, 24
    simulation, 229
    with starting parameter values, 60–66
Steepest descent direction, 71–73, 77
Stochastic methods, 11, 263, 264,
    309, 347
    field examples, 350
StratWorks 3D software, 274
Streamflow gain(s)
    nonlinear confidence intervals on
        predictions of, 341
    observed and simulated, 317
Streamflow gain/loss observations, errors
    in, 35
Streamflow gauging stations,
    measurements between, 297–298
Streamflow observations, biased, 31
Student t-distribution, selected web sites
    for, 399t
Student t-statistic, 138
Subset definition, for residuals, 94
Superparameter method, 219
Superparameters, 10
Superposition principle, 235, 240
Surface-water studies, 335
Synthetic test cases, 347, 352t
System data, using. *See also* Guideline 2
System dynamics
    potential new data to improve,
        334–335
    sensitivities as a reflection of, 65
System information, 260
    evaluating, 353–357
    methods to guide collection, 335
    use in defining model structure, 273

use in problem constraint, 272–277
using a broad range of, 371–373
weighting, 304–305
Systematic error, 322–323
Systematic model misfit, 94
    importance of testing for, 101

Temperature observations, 215, 287–288
Temporal differencing, 216
Temporal effects, observed and simulated,
    214–215
Tests of methods and guidelines,
    347–373
Theim equation, 69–70
Theis equation, 74–75
Theis example
    for understanding linear and nonlinear
        confidence intervals, 181–182
    for understanding Monte Carlo methods,
        188–189
    for understanding the modified
        Gauss–Newton method, 74–75
Thermodynamic parameters, 226
Three-dimensional hydrogeologic data,
    analyzing, 274, 354–356
Three-dimensional simulation, versus
    two-dimensional simulation,
    372–373
Time-consistent observations, 215
Time-dependent deterministic error
    components, 378
Time-step size issues, in transport models,
    221
TolPar variable, 76. *See also*
    UCODE_2005 Program
Total model nonlinearity, 144–145
    in Exercises, 156–157
Total variation diminishing (TVD) method,
    223
Tractable models, obtaining, 9–11,
    226–227, 290
Traditional sensitivity analyses,
    weaknesses of, 184–185
Transient groundwater flow problems,
    errors in, 376–381
Transient head-change observations, 216,
    360–362
Transient model calibration, strategies for,
    213–217

Transient model, 25. *See also* Exercises
  calculating observation weights for, 231–235
  defining observations for, 231–234
  fit, evaluating, 246–250
  flow observations for, 234t
  hydraulic-head observations for, 232–233t
  parameters in, 230–231
  predictions for, 254–259
  regression for, 243–244
  sensitivity analysis for, 235–243
  starting, estimated, and true parameter values for, 246t
  weighted residuals for, 247–250
Transient model initial conditions, 213–214
Transient model inputs, additional, 216–217
Transient observations, 214–216
  weighting, 216
Transient processes, effect on observations, 215, 360
Transient simulations, confined layers in, 271–272
Transitional probability method, 219
Transmissivity, effect of dewatering on, 272
Transmissivity parameters, 36, 69–70. *See also* Hydraulic conductivity parameter(s)
Transport. *See also* Advective transport; Contaminant transport
  options for simulating, 218
  predicting, 217
  simulating advection and dispersion, 220–223
Transport and flow-system observations, simultaneous use of, 223–224
Transport model(s). *See also* Heat transport models
  advection and dispersion in, 220–223
  defining source characteristics for, 218
  execution time issues for, 220
  Lagrangian methods and, 221
  selecting processes to include, 217–218
  simulated well concentrations, histograms of, 343
  simulations, system state at the beginning of, 213–214
  solution methods and numerical dispersion, 221–223
  unresolved and misrepresented features in, 219
Transport model calibration
  applications of, 226–227, 288, 370–373
  estimating source characteristics for, 219
  insensitivity and correlation problems in, 226
  numerical dispersion and, 221–233
  numerical issues in, 220–223
  point concentrations in, 225
  scale issues in, 219–220
  strategies for, 217–227
  transport observations in, 223–225
Transport observations, 223–225. *See also* Concentration observations
Transport-step size, 221
Trends, evaluation of using runs statistic, 108, 317
Trial-and-error methods, 6, 7
True errors, 30
  assumptions about, 180, 374, 376–381
  variance–covariance matrix of, 31–32, 376, 380
True error variance, 31
  confidence interval for, 96–98, 246
True linear model, 392
True nonlinear model, 392
t-statistic. *See* Student t-statistic
Trust region. *See* Double-dogleg trust region approach
Two-dimensional simulation, versus three-dimensional simulation, 372–373
Two-parameter objective-function surfaces, 35–36, 71, 74–75
  in Exercises, 81–82
Two-tailed hypothesis test, 328

UCODE_2005 program, 3, 18–21
  capabilities of, 19–21
  convergence criteria in, 76
  correlation coefficients calculated by, 21, 59, 62t, 63t, 149t

INDEX   453

dampling parameter in, 73–74
double-dogleg trust region approach in, 68
execution times using, 346
full weight matrices and, 28
for graphs of observed versus simulated values, 105
Investigate Objective Function mode in, 36
Marquardt parameter in, 73
Monte Carlo methods and, 186
parameter limits in, 80
parameter log transformation in, 79
plotting objective function surfaces in, 36
prediction scaled sensitivities produced by, 161
prior information equations supported by, 32
runs statistic information printed by, 108
sensitivity approximation by, 47, 59
TolPar variable in, 76
using for Exercises, 25
weighting in, 32, 34, 295, 298
Unambiguous measurement errors, 300
Unbiased observations, 30–31. *See also* Bias
Uncertainty. *See* Confidence intervals; Linear uncertainty analysis; Parameter uncertainty; Prediction uncertainty; Weighting(s); Weights
Uncertainty, defined, 14
Uncertainty evaluation, 2
  utility of, 6–7
Uncertainty (UNC) Process. *See* MODFLOW-2000 program
Uncorrelated observation errors, 216, 378. *See also* Observation errors
Unintended inconsistencies, 286
Unrealistic estimated parameter values, 80, 289–290
  diagnosing cause of, 326–327
Unrealistic weightings, 302–303
Unweighted observations, versus simulated values and correlation coefficient *R*, 105–106
Unweighted residuals, plotting, 100, 363

US Geological Survey (USGS) Ground Water-Site Inventory (GWSI) data base, 357, 358
US Geological Survey (USGS) software. *See* Software; LINEAR_UNCERTAINTY program; MODEL_LINEARITY entries; MODFLOW_2000 entries; Multi-Model Analysis (MMA) computer code; Multi-Node Well (MNW) Package; OPR-PPR program; RESAN-2000 program; RESIDUAL_ANALYSIS program; UCODE_2005 entries
U.S. Geological Survey (USGS), topographic maps determining elevations from, 295–296

Value of improved information (*voii*) statistic, 167
Vapor phase hydrocarbon transport model calibration, 226
Variability
  of data, managing, 272
  highly parameterized models and, 290–291
  scale issues and, 219, 289
Variance–covariance matrix, 126, 301. *See also* Errors; Parameter variance–covariance matrix; True errors
  of a vector, 395
Vector(s). *See also* Eigenvectors
  for nonlinear regression, 383–384
  notation for full weight matrix, 28
  variance–covariance matrix of, 395
Visual inspection, identifying trends by, 317
Visualization software, 3, 355

water.usgs.gov web site, 3
WATFLOW/WTC finite-element model, 370
Weighted least-squares objective function, 26–28, 95. *See also* Objective function(s)
  with diagonal weight matrix, 27–28
  with full weight matrix, 28
  modified Gauss–Newton method and, 68

Weighted least-squares objective function (*Continued*)
  relation to maximum-likelihood objective function, 376
Weighted least-squares objective-function surfaces, 35–36, 37. *See also* Objective function surfaces
Weighted observations, versus simulated values, 105–106, 362–363
Weighted regression, 27
Weighted residuals, 35. *See also* Weighted residuals graphs and maps
  acceptable deviations from being random and normal, 111–113, 119–123
  assessing independence and normality of, 109, 249–250
  assessing randomness of, 98, 106–109, 317–320
  calculating, 35
  for concentration observations, 224
  in diagnosing the cause of poor model fit, 320
  in evaluating alternative models, 310–312
  in evaluating model fit, 93–94, 100–104, 106–108, 316–320
  in Exercises, 115–119, 247–250
  in interpreting calculated error variance, 97–98
  minimum, maximum, and average, 101
  random, normally distributed, 108–109
  with respect to independent variables, 106–108
  using, 265t
Weighted residuals graphs and maps, 99–104, 106–107, 108–109, 317, 318, 319, 362, 363, 364
  in Exercises, 116, 117, 119, 247, 248, 250
  problems with, 101–104
  versus independent variables, 106–107
  versus weighted or unweighted simulated values, 100–104
Weighted sensitivity matrix, singular value decomposition of, 132
Weighted simulated values, 100–106
Weighted true errors, 30

Weighting(s). *See also* Guideline 6; Weights
  of concentration observations, 224–225
  effect on calculated pcc, 198
  errors in, 323
  functions of, 31–32
  issues related to, 34–35
  of transient observations, 216
  unrealistic, 303
Weighting issues, in nonlinear regression, 298–305
Weighting terminology, confusion related to, 299–300
Weight matrices. *See also* Diagonal weight matrices; Full weight matrices
  determining covariances for, 298
  defining, 396–398
Weights. *See also* Guideline 6; Weighting(s)
  calculating, 39–40, 358
  calculating for observations that are sums of or differences between measurements, 297–298
  calculating using observed values, 101–103
  determining, 294–298
  determining the statistic used to calculate, 295
  difficulties in determining, 299
  effect on measures of prediction uncertainty, 302–303
  large, 301–303
  that reflect errors, 291, 294–305, 358–362
Well altitude uncertainty, 358
Well elevation measurement error, 216, 295–296
Well-posed regression. *See also* Guideline 3, 277–281
Wetland system, groundwater flow and heat transport simulation through, 288

X, importance of, 394
X, 69. *See also* Sensitivities; Sensitivity matrix
  importance of, 394
X value, effect on linear regression behavior, 51

INDEX

y. *See* Observations
y'. *See* Simulated values
$y'(b)$ vector, linearizing, 69
YCINT-2000 program, 162, 177, 184

Zonation, for parameterization, 11, 273–274, 309, 342–343

Zone arrays, 20
Zone of confidence, 342
Zone of uncertainty, 342
Zones. *See also* Capture zones; Zonation
  hydrochemical, 287
  recharge, 24, 327